THREE KEYS TO THE PAST:
THE HISTORY OF TECHNICAL
COMMUNICATION

ATTW Contemporary Studies in Technical Communication

M. Jimmie Killingsworth, Series Editor

*Published in cooperation with
the Association of Teachers of Technical Writing*

Volume 1:
Foundations for Teaching Technical Communication: Theory, Practice, and Program Design
   edited by Katherine Staples and Cezar M. Ornatowski, 1997

Volume 2:
Writing at Good Hope: A Study of Negotiated Composition in a Community of Nurses
   by Jennie Dautermann, 1997

Volume 3:
Computers and Technical Communication: Pedagogical and Programmatic Perspectives
   edited by Stuart A. Selber, 1997

Volume 4:
The Practice of Technical and Scientific Communication: Pedagogical and Programmatic Perspectives
   edited by Jean A. Lutz and C. Gilbert Storms, 1998

Volume 5: The Dynamics of Writing Review: Opportunities for Growth and Change in the Workplace
   by Susan M. Katz, 1998

Volume 6:
Essays in the Study of Scientific Discourse: Methods, Practice, and Pedagogy
   edited by John T. Battalio, 1998

Volume 7:
Three Keys to the Past: The History of Technical Communication
   edited by Teresa C. Kynell and Michael G. Moran, 1998

THREE KEYS TO THE PAST:
The History of Technical Communication

edited by
Teresa C. Kynell
Northern Michigan University
and
Michael G. Moran
University of Georgia

Volume 7 in ATTW Contemporary Studies in Technical Communication

 Ablex Publishing Corporation
Stamford, Connecticut

The following chapters were reprinted with permission from:

Chapter 1: Bazerman, C. (1991). How natural philosophers can cooperate: The literary technology of coordinated investigation in Joseph Priestley's "History and the Present State of Electricity" (1767). In C. Bazerman & J. Paradis (Eds.), *Textual dynamics of the professions: Historical and contemporary studies of writing in professional communities* (pp. 13–44). Madison, WI: University of Wisconsin Press.

Chapter 2: Zappen, J. P. (1989). Francis Bacon and the historiography of scientific rhetoric. *Rhetoric Review, 8*(1), 74–88.

Chapter 6: Whitburn, M. D., Davis, M., Higgins, S., Oates, L., & Spurgeon, K. (1978). The plain style in scientific and technical writing. *Journal of Technical Writing and Communication, 8*(4), 349–358.

Chapter 9: Connors, R. D. (1982). The rise of technical writing instruction in America. *Journal of Technical Writing and Communication, 12*(4), 329–352.

Copyright © 1999 by Ablex Publishing Corporation

All rights reserved. No part of this publication may be reproduced, stored in a retrieval system, or transmitted, in any form or by any means, electronic, mechanical, photocopying, microfilming, recording or otherwise, without permission of the publisher

Printed in the United States of America

Library of Congress Cateloging-in-Publication Data

Three keys to the past : the history of technical communication / edited by
   Teresa C. Kynell, Michael G. Moran.
   p. cm.—(ATTW contemporary studies in technical communication ; v. 7)
   Includes bibliographical references and index.
   ISBN 1-56750-393-4 (cloth)—ISBN 1-56750-394-2 (pbk.)
   1. Communication of technical information—History. I. Kynell, Teresa C.
II. Moran, Michael G. III. Series.
T10.5.T48   1999
601'.4—dc21   98-5847
                                                  CIP

Ablex Publishing Corporation
100 Prospect Street
P.O. Box 811
Stamford, Connecticut 06904-0811

To
Fiona Roberts Gibbons
and
Molly and Alison Moran

CONTENTS

Preface   ix

Introduction   1

Part I: KEY INDIVIDUALS IN THE HISTORY OF TECHNICAL COMMUNICATION

1. Landmark Essay: How Natural Philosophers Can Cooperate: The Literary Technology of Coordinated Investigation in Joseph Priestley's *History and Present State of Electricity*
   Charles Bazerman   21

2. Landmark Essay: Francis Bacon and the Historiography of Scientific Rhetoric
   James P. Zappen   49

3. Oliver Evans and His Antebellum Wrestling with Rhetorical Arrangement
   R. John Brockmann   63

4. Sada A. Harbarger's Contribution to Technical Communication in the 1920s
   Teresa Kynell   91

Part II: KEY EUROPEAN MOVEMENTS IN THE HISTORY OF TECHNICAL COMMUNICATION

5. The Emergence of Women Technical Writers in the 17th Century: Changing Voices Within a Changing Milieu
   Elizabeth Tebeaux   105

6. Landmark Essay: The Plain Style in Scientific and Technical Writing
   Merrill D. Whitburn, Marijane Davis, Sharon Higgins, Linsey Oates, and Kristene Spurgeon   123

7. Deconstructing Depression: A Historical Study of the Metaphorical Aspects of an Illness
   Henrietta Nickels Shirk   131

8. Renaissance Surveying Techniques and the 1590 Hariot-White-de Bry Map of Virginia
   Michael G. Moran   153

Part III: KEY AMERICAN MOVEMENTS IN THE HISTORY OF TECHNICAL COMMUNICATION

9. Landmark Essay: The Rise of Technical Writing Instruction in America
   Robert J. Connors   173

10. Interfacing: Multiple Visions of Computer Use in Technical Communication
   Johndan Johnson-Eilola, Stuart A. Selber, and Cynthia L. Selfe   197

11. Refining a Social Consciousness: Late 20th Century Influences, Effects, and Ongoing Struggles in Technical Communication
   Jo Allen   227

Part IV: BIBLIOGRAPHY IN THE HISTORY OF TECHNICAL COMMUNICATION

12. Studies in the History of Business and Technical Writing: A Bibliographical Essay
   William Rivers   249

Author Index   309

Subject Index   317

PREFACE

On the cover of this book you will find a reprint of a late 16th-century map depicting the Outer Banks of North Carolina. In many ways, we felt that we were "mapping" new territory as we attempted to bring together both the best of what had been written on the history of technical and scientific communication and the best of what is currently being studied and written. The pieces collected here represent a variety of approaches to the evaluation of historical shifts in technical communication, from primary research to deconstructive techniques to social and cultural analysis. No collection of this kind, of course, can do more than sample the research strategies of a field, but this eclectic sampling reflects the different means historians use in their attempts to understand the past.

We sought to collect materials on the history of technical communication that would fill in key gaps in the development of the discipline. In a way, this collection tells the story of where we've been. Thus, the pieces here are not just "key" to understanding our evolution, but also "key" in that they open doors to further exploration into significant contributions by individuals and important movements in the field. As a result, these essays show what historical evidence can reveal and how historical methods can be useful to technical communicators. This book identifies some of the underlying principles of the discipline and demonstrates the utility of historical inquiry.

Compiling a collection of any sort is always daunting, but our task was made easier by the kind assistance of colleagues and students alike. First, we would like to thank all the contributors to this collection who worked hard to produce carefully researched, polished pieces. We especially thank Charles Bazerman, Robert Connors, Merrill Whitburn, and James Zappen for granting permission for reprint of their essays. We would like to thank David Gants of the University of Georgia for sharing his expertise on computers and word processing throughout the process of assembling the essays. Special thanks go to M. Jimmie Killingsworth, ATTW Series Editor, who made himself constantly available to answer many questions that accompany such an endeavor. Without his encouragement from start to finish, this project would never have reached fruition.

We want to thank, in particular, the graduate students who put in countless hours on copy editing, formatting, and proofreading the material. Thanks go to David

Gardner and Matt Maki for their time and labor on individual pieces, and to John Murphy, who transformed the bibliographies of the classic essays into APA format. A very special thanks goes to Marcia Parkkonen, a graduate student, for contributing many hours of solid work toward the completion of this project.

While friends and family members did not contribute directly to the creation of this collection, they did contribute indirectly in ways too numerous to include here.

Mike wishes to thank his good friends Mike Simon, John Widmer, Tom Beisswenger, Jan Kimpel, and Phil Templeton, who, after five years, are still willing to listen to his ideas once a week on the golf course. And he especially thanks Alison and Molly, who always offer sound advice and warm support.

Teresa gives many thanks to her colleagues, Beth Tebeaux and Anne Youngs, who listened patiently when things were going well and when they weren't. Thanks also go to two wonderful friends, Fiona Gibbons and Donna Silta, who are always there and always willing to help. Thanks, as well, to my three girls: Kris, Suzy, and Rebecca. Special thanks, though, go to Kurt V. S. Kynell, a wonderful spouse, an understanding partner, and a superb editor.

Teresa Kynell
Michael G. Moran

Introduction

It is not surprising that, as technical communication has become an established academic area with undergraduate and graduate programs, specialized journals, and professional organizations, scholars have begun to explore the history of the subject. Within the past 10 years, several important books have appeared that fill gaps in our historical understanding of the area. Charles Bazerman's *Shaping Written Knowledge* (1988), for instance, examines the rise of the experimental scientific article from the Renaissance to the present, and Elizabeth Tebeaux's *The Emergence of a Tradition* (1997) traces the development of the how-to book from 1475 to 1640. Other scholars have contributed to our knowledge of technical communication's development as an academic discipline. David R. Russell's *Writing in the Academic Disciplines, 1870–1990* (1991) examines the growth of technical writing courses within the larger context of the writing across the curriculum movement, and Katherine H. Adams's *A History of Professional Writing Instruction in American Colleges* (1993) studies the development of advanced writing courses from the inception of higher education to the present, devoting chapter 7 to writing in agriculture, engineering, and business. Finally, Teresa Kynell has written on the development of writing in engineering schools from 1850 to 1950 in *Writing in the Milieu of Utility* (1996). Research in the area has grown to such an extent that William Rivers recently published a 50-page critical bibliography titled "Studies in the History of Business and Technical Writing" (1994), an updated version of which appears in this book.

While all these books contribute to our as-yet-incomplete history of technical communication, the aforementioned books focus on curricular shifts, emerging disciplinary patterns, and broad movements in this still-evolving discipline. The selections in this book refocus attention on four key historical areas: individuals, movements, advances, and reprinted articles which have been important in contributing to our understanding of the shaping factors in technical communication's past. The present volume contributes to this developing research area by collecting

2   INTRODUCTION

essays on the history of technical and scientific discourse to suggest the range of contemporary topics and methods. While the essays contribute to our understanding of the history of technical communication, we also hope that they will provide models for this kind of research and encourage other scholars to initiate projects in the area. As technological advances change the way we live, so will technical communicators evolve to meet these needs. We believe that studying our past more closely provides insights into not only how the discipline has evolved, but also how such historical shifts might hint at the future of technical communication. We will begin by critiquing the kinds of research done in the past, then suggesting some of the types of projects that are needed.

RESEARCH METHODS

Historical Models for the Classroom

Some of the earliest historical research on technical communication was undertaken to strengthen classroom teaching. One common assumption of such research is that technical writing students benefit from knowing about the history of the subject. By reading the writing of eminent scientists and engineers from the past, Stephen Gresham (1981) argues students will come to understand that technical communication grows out of "a solid tradition," a tradition in which student writers participate. In addition, by reading the work in this tradition, students will learn to appreciate what Gresham calls the "poetry" of scientific and technical prose. Students will therefore participate in the sense of wonder about the world that earlier scientists and engineers felt and expressed (p. 88). Since these writers often wrote "beautiful prose," students will be encouraged to emulate this beauty in their own writing (Jones, 1985, p. 116). Finally, students will learn to appreciate their technical fields better through understanding their fields' pasts. They will learn, for instance, that scientific and technical fields "are [not] rigid, monolithic, and devoted to formulaic thinking and nothing but pure objectivity" (Rutter, 1991, p. 135); instead, as the writing in them proves, technical and scientific writers of the past often expressed in their prose excitement about their discoveries.

A second assumption of this kind of research is that historical passages can form the basis of effective assignments that improve student writing. As several researchers have argued, classic pieces of technical writing can provide students with models of modes of discourse that technical writers commonly use. These modes include classification, analysis, division, process analysis, description, and definition (Gresham, 1981; Miller, 1961). Studying passages will also help students master principles other than form. As Miller argues, students will see that, while good technical writing is always clear, such writing can be interesting through its use of stylistic devices such as metaphors, similes, and analogies. Students will also learn that good technical communicators always consider their audience (Gresham,

1981). In addition, students will see how an issue, like audience analysis, came to be an important consideration in the historical move to technical communication.

We should note that all of the other methods that we discuss produce research that is valuable to teachers.

Studies of Individual Technical Communicators

A second early form of historical research is the examination of individual technical communicators, the method that R. John Brockmann designates the study of "celebrated authors or scientists" (1983, p. 155). The advantage of this approach is that it was relatively easy to identify early practitioners, study their work (or part of it), and then comment on what makes that work distinctive as technical or scientific communication. The researcher could thereby provide some depth of understanding of one writer's methods and contributions. The researcher often assumed that such writers were representative of their historical period so that understanding the work of one writer offered insight into the contributions of others. Other researchers assumed that the writing of the practitioner offered a model for emulation. At a time when no major historical studies existed to provide a paradigm to guide more specific work, the "famous writer" approach offered a methodology that generated some understanding of the tradition.

The approach, however, has potential problems. First, as Brockmann (1983) argues, it takes a 19th-century "generals-and-kings" view of history—that "history consists of the work of the famous and influential" (p. 155). While such an approach might be valuable when studying the work of an important scientist or engineer, it misleads about the nature of technical communication's history, which was often formed by the work of forgotten practitioners, technicians, and bureaucrats. Second, researchers have too often gone far afield in their search for technical communicators, sometimes claiming that writers such as Samuel Johnson and Ernest Hemingway provide valuable models of technical prose. A third problem with this approach is that it often wrenches the individual writer from any meaningful historical, social, intellectual, or cultural context. All too often such studies have implied that a single writer developed a particular strategy or technique or that he or she alone was the only important technical writer of a period. The assertion that Chaucer was the first important technical writer in English because of his *Treatise on the Astrolabe* is a case in point (see, for instance, Freeman, 1961; Lipson, 1982; and Ovitt, 1987). In "The First Technical Writer in English: A Challenge to the Hegemony of Chaucer," John Hagge (1990) argues "that claims for Chaucerian priority in early English technical writing need to be reconsidered" (p. 272) and shows that many works in medieval technical communication "either predate Chaucer or are contemporaneous with him" (p. 271). Hagge goes on to discuss numerous technical documents of the period that provide a meaningful context for understanding Chaucer's treatise. As William E. Rivers (1994) reminds us, in studies of individual writers, "we need to connect their work to other technical, scientific, or business writing done in the same period and to all the writing (and thinking) of the period" (p. 45). In this volume, we highlight the work of those writers, scholars, and teachers who, without such study, might go unrecognized for sometimes small, though significant, achievements. Indeed, some of those forgotten thinkers and writers played a key role in curricular and scholarly shifts in the discipline.

Curricular Histories

Curricular histories explore the history of technical communication as an academic discipline. The subject matter of these studies falls into three major categories: the history of textbooks, the history of courses, and the history of instructional practices. Curricular studies also provide both scholars and practitioners with some sense of the influences—both internal and external—which shape classroom content. Evaluating the interplay of teacher, text, and historical period in context can reveal insights into paradigmatic educational trends which are important historically and equally important as contemporary classroom restraints change. While these three approaches are often mixed together in individual studies, it is useful for our purposes to discuss them separately.

Studies of the history of textbooks assume that textbooks reflect a period's theory and practice. This assumption can sometimes be fallacious. Teachers do not always use textbooks the way that their authors intended them to be used. In fact, instructors often teach against the text, critiquing the text to demonstrate to students its weaknesses and inaccuracies. Textbooks often reflect not classroom but publishing practices, so the historian must approach them skeptically. However, textbooks can be useful objects of study because, while they might not reflect teaching practices, they often reflect dominant ideologies within the technical professions at a given time within a particular cultural milieu.

For example, one of the earliest technical communication texts, T. A. Rickard's *A Guide to Technical Writing* (1908), was essentially a usage guide designed primarily for practicing engineers. Frank Aydelotte's *English and Engineering* (1917), a book of essays reprinted by writers ranging from Robert Louis Stevenson to John Ruskin, was a decidedly different kind of text which emphasized broad training in the liberal arts for engineering students. By 1920, Rickard would publish the second edition of his book (very similar to the first edition in style and intent), and by 1923 Sada Harbarger would publish *English for Engineers*, a book that connected solid writing skills with the future success of the engineer. That these three books, with obvious philosophical differences and educational purposes, coexisted and were used by a variety of teachers attests to the difficulty of singling out any one text, in historical context, and arguing that it stands as the model of classroom practice.

An exemplary study is John Hagge's (1995) "Early Engineering Writing Textbooks and the Anthropological Complexity of Disciplinary Discourse," which explains the development of early engineering writing textbooks within the context of "American social and intellectual history" (p. 444). Hagge analyzes 20 early technical writing texts and identifies two major assumptions that informed their attitudes toward engineering language. The first, the need for engineering students to master the "discourse norms" (p. 456) of their profession, reflects the belief of the engineering profession at the time that these norms reflect a "specialized sub language" (p. 459) that students needed to master in order to become professionally literate members of the engineering community. To become professionally literate, students, the textbooks argued, must read widely within the professional literature of engineering. The second assumption of the texts was that engineering students also needed to be widely read in humane letters so that they became cultured gentlemen. Hagge demonstrates that this goal grew from the assumption that engineers needed not only to be able to communicate with other engineers but that they also needed to be able to communicate to a broad cross-section of the American public, ranging from politicians to laborers. Such a humane background would also place engineers on the same level as the traditional cultured professions of medicine, law, and the clergy, a goal to which the engineering profession aspired. Hagge's analysis therefore shows how these 20 texts reflect the social aspirations of the early 20th century engineering profession.

Because instructional practices vary significantly from institution to institution, one method of studying a broad curriculum involves careful investigation of the proceedings of educational societies and organizations. Such proceedings typically provide both anecdotal educational material as well as a record of curricular shifts and wide-ranging changes that often have an effect on several disciplines. The Society for the Promotion of Engineering Education is a case in point. The SPEE, which first met in 1893, addressed a variety of curricular matters, including the duration of undergraduate work, course content, and the role of the humanities in the educational life of the engineer. The society is noteworthy, write Reynolds and Seely (1993), for "a continuing commitment to the improvement of instruction at the classroom level...[and] recognition from other professional societies and from government agencies as the spokesman for engineering education" (p. 136). Scholars evaluating the move to technical communication will find the lengthy SPEE proceedings a rich source of information on instructional practice and anecdotal accounts of individual teachers' practices.

Histories of Rhetorical Strategies

Another approach to the history of technical communication is the study of particular rhetorical strategies used in technical discourse. This approach consists of two related methods. The first examines texts or groups of related texts to determine the rhetorical strategies that the authors use; the second method examines rhetorical or logical theory of a period to demonstrate how technical communication either influenced or responded to it.

The first method, the explication of texts, analyzes rhetorical strategies of a significant text or body of texts important to the technical or scientific tradition. Such an approach usually draws on standard elements of rhetoric and logic for its method of analysis. Researchers often use the three Aristotelian appeals—ethos, pathos, and logos—to evaluate basic persuasive strategies of the texts. Other researchers attempt to establish the topoi or topics of invention embedded in texts. Others examine stylistic or organizational strategies. Still others draw on logical strategies, such as Toulmin's system of logic or traditional inductive or deductive reasoning, to elucidate the text.

A good example of this kind of analysis is S. Michael Halloran and Annette Norris Bradford's "Figures of Speech in the Rhetoric of Science and Technology" (1984), which evaluates mid-20th century molecular biology using categories of style derived from classical rhetoric. Halloran and Bradford attack the traditional notion that scientific and technical discourse avoids tropes and schemes and merely presents objectively information discovered through observation. Practicing scientists, they argue, must use both kinds of figures in order to conduct research and communicate its results. Science needs tropes, especially metaphors, because it must build models to form concepts and to shape the direction of future research. Halloran and Bradford examine work on the DNA molecule that James Watson and Francis Crick proposed in 1953—the famous double helix. As Halloran and Bradford show by analyzing a body of seminal texts in molecular biology, Watson and Crick, as well as later research scientists in the field, used the genetic message metaphor both to understand the molecule's function and to communicate their findings to their audiences. The extended metaphor asserts that the molecule " 'transmits' 'information' that is 'transcribed' onto other substances in the cell and thus 'translated' onto the characteristics of a specific organism" (Halloran & Bradford, p. 184). While the metaphor adds vivacity to the prose, it also functions as a heuristic by offering a fundamental way to conceptualize the molecule's structure and genetic function. Halloran and Bradford also argue that scientific writing needs schemes in order to communicate research findings to an audience. They question the value of highly simplified prose that becomes so predictable that the reader loses interest; in place of such simplicity, they call for prose that uses schemes such as parallelism, repetition, omission, and inversion to slow readers down and draw their attention to important passages. Such techniques make the prose more comprehensible by guiding readers through the discourse.

Closely related to the explication of texts is the second method of exploring rhetorical strategy. This method examines a period's rhetorical or logical theory to show how it responded to and/or influenced the technical prose of the period or periods under consideration. Eighteenth century rhetoric, for instance, responded to the rise of scientific discourse by developing theories of expository prose that explained the function of this kind of discourse (see Howell, 1971). Once articulated, they influenced the scientific or technical prose. One classic example of this process is Joseph Priestley, who articulated the theory of analytic arrangement in his rhetoric, *A Course of Lectures on Oratory and Criticism*, while applying the strategy in his scientific prose (see Moran, 1984). The advantage of this research approach is that it demonstrates how communicators and their discourse interacted with the rhetoric of their times.

A model for this method is James Paradis's "Bacon, Linnaeus, and Lavoisier: Early Language Reform in the Sciences" (1983). Interested in the question of how early scientists created a new language that allowed a closer fit between linguistic terms and physical reality, Paradis begins with Francis Bacon's critique of rhetoric in *The Advancement of Learning*, in which he advocated the plain style, and in *Novum Organum*, in which he called for "terminological reforms" (Paradis, p. 202). Paradis argues that Bacon took a skeptical view of language, which he saw as a mixture of hastily assembled terms based on experience and the senses that lacked the exactness that the new empiricism required. Language was also contaminated with fanciful terms and meaningless abstractions from scholastic thought. The language of science, Bacon argued, could be refined only if scientists linked each term exactly to observable fact. Once scientists had developed carefully defined terms, these terms could then reveal the fundamental nature of reality and serve as the basis for meaningful abstractions from that reality. Later researchers applied Bacon's linguistic principles to various branches of science, and Paradis traces Bacon's influence on Linnaean descriptive biology and on Lavoisier's chemical nomenclature. Paradis argues, for instance, that the Linnaean lexical system was the first to standardize successfully the link between concept and name in any branch of science. He developed his system by differentiating plants according to the structure of their sex organs. Thus, the system worked because each term in it referred to one of 26 morphological features of the plant, thus allowing the biologist to place any plant in its proper genus and distinguish each plant from other members in that genus. Paradis ends by discussing Lavoisier's textbook, *Elements of Chemistry in a New Systematic Order* (1793), which presented a precise nomenclature for chemical compounds based on the common components of acids, bases, and salts. Lavoisier's new terms, such as "oxygen gas," "carbonic acid," and "metallic oxyds," replaced their less precise older terms: "dephlogisticated air," "acid of chalk," and "calces, metallic" (Paradis, p. 219).

Corporate and Government Communication Histories

Another kind of historical study is the analysis of how various corporations and government agencies have produced and used discourse to further their aims. This work has barely begun but would include a range of projects. Some projects would be synchronic in that they would study the discourse at a particular time in an entity's history; others would be diachronic in that they would study the development and change of discourse and its function over a historical period.

Kitty O. Locker's synchronic project, titled "The Earliest Correspondence of the British East India Company (1600-19)" (1985), is one of the few studies of the communication of an early corporation. While the essay is preliminary (it covers only the first 20 years of the company's existence and does not offer a full rhetorical analysis of the correspondence), it suggests some of the elements that such a project might examine. Locker, for instance, offers some history of the company to establish a context for the correspondence, and she categorizes the types of discourse that the company's factors (representatives) in the field produced. External correspondence addressed three entities: the English Crown, foreign princes with whom the company hoped to trade, and foreign companies and governments against which the company competed. The internal correspondence consisted of records (ship's journals, minutes of meetings, and remembrances or memos to the file) and correspondence or letters proper, which Locker analyzes for basic rhetorical strategies of organization, style, and tone.

Case Studies of Specific Events

Closely related to corporate and government communication histories are case studies of the ways that communication within organizations contributes to significant events, usually disasters. Such studies closely examine the rhetorical and social interactions among employees in an institution to identify problems within its communicative structure. An implicit assumption that researchers make is that writing in particular and communication in general play significant roles in the daily operation of an organization and that, if the employees within that organization could write more effectively, the disaster could have been prevented. It is important to note, however, that the writing problem is not usually a matter of prose lacking clarity or correctness. The problem is usually attributed to deeper communicative disorders within the organization, including such things as the writers' relative standings within the corporate hierarchy, the personal relationships among the writers and readers, and the differing professional backgrounds among communicants, all of which interfere with understanding. While two 20th century disasters—the Three Mile Island nuclear accident and the Shuttle Challenger explosion—have received the most attention, we believe that the method can be applied productively to events of earlier centuries, too.

Because it works out a methodology for this kind of research and because it examines both the Three Mile Island and the Shuttle Challenger disasters, Carl G. Herndl, Barbara A. Fennell, and Carolyn R. Miller's "Understanding Failures in Organizational Discourse" (1991) provides a model for the case study approach. In both disasters, the authors conclude, "[m]isunderstanding and miscommunication...were found to be contributing causes of the accident" (p. 279). As their analysis of various documents connected with the two accidents suggests, however, the miscommunication did not result from lack of clarity or control of surface features of the prose, which in all cases followed fairly well accepted conventions of correctness and clarity on the sentence level. The authors use pragmatic analysis, which examines ways that social differentiation shapes discourse, and argument analysis, which examines the argumentative shape of discourse, to analyze the memos written by employees of Babcock and Wilcox, the company that built the TMI nuclear reactor. The pragmatic analysis suggests that social concerns of individual authority and social status interfered with communication among employees about the nature of the reactor's technical problems. The argumentative analysis, based on Stephen Toulmin's notion of argumentative *fields*, suggests that engineers and operators drew their warrants from such different fields and made such different assumptions about valid evidence that the two groups did not communicate effectively. In the case of the Challenger explosion, the authors combine pragmatic and argument analysis to demonstrate that the managers and the line engineers at Martin Thiokol, the company that built the space shuttle, disagreed about whether or not to launch Challenger because they disagreed on claims and warrants based on their different positions and responsibilities. The engineers wanted the flight canceled based on their experience with the failure of O-rings during cold weather. The managers, on the other hand, based their claim that the flight should go according to schedule on their experience with past flights. Unfortunately, because the managers had the power to make final decisions, the flight was approved.

Genre Studies

Genre histories examine the rise, development, and—sometimes—fall of various types of technical and scientific discourse. Such studies, however, should be more than merely examining formal qualities of discourse. Most of these studies use Carolyn R. Miller's (1984) definition of a genre as a repeatable rhetorical strategy used to achieve goals in situations that are perceived by groups as sharing similarities. Genre studies, therefore, should not emphasize content or form of discourse alone. They should emphasize the social action that the genre is designed to achieve. Since the genre emphasizes action, it must be concerned with "situation and motive" (p. 152) because all human action has meaning within its situational context and through the attribution of motives to human participants. Because rhetorical situations repeat themselves, genres develop as stylized responses to conventional situations. For example, the letter of complaint is written when writers find themselves in the common situation of having been mistreated, either intentionally or unintentionally, in some interaction with another person or an organization. As business and technical communication textbooks show, the letter to write in response to this situation has certain conventional elements that have been shown over time to be effective in achieving the desired goal of being recompensed. These elements include maintaining good will, objectively identifying the problem at issue, requesting reasonable compensation or relief, and so forth.

We now have two book-length genre studies in technical and scientific communication. The first is Charles Bazerman's *Shaping Written Knowledge* (1988), which examines the rise of the experimental article in science and traces it to its present formulation. As Bazerman shows by examining early experimental articles in the *Philosophical Transactions of the Royal Society,* Newton's writings, the social dynamics of early scientific publication, and 20th century experimental articles in physics, the genre of the experimental article in science has become so highly conventionalized that contemporary scientists do not "think originally and creatively about how to master recalcitrant language in order to create...powerful stories" about their experimental findings (p. 59); they simply follow the dictates of the genre. The second major genre study is Elizabeth Tebeaux's *The Emergence of a Tradition* (1997). Tebeaux takes as her project the rise and development of printed how-to books during the period from Caxton's establishment of the printing press to the English Revolution. These books cover an astounding number of topics that interested mostly middle-class readers of the period. Newly literate, this class, consisting of merchants, tradesmen, and their wives, as well as other groups, enjoyed increasing wealth and demanded practical education in areas of professional and personal interest. With the rise of printing, books could be produced to meet the needs of the new market of readers who wanted access to practical knowledge. Authors wrote and printing presses disseminated volumes on personal medical care, herbals, farming, animal husbandry, gardening, household management, cooking, military science, navigation, and surveying, to cite just the major topics. Thus, the genre of the how-to book arose to meet the social need of the middle class to be empowered to conduct practical actions that improved their lives.

THE PRESENT VOLUME

Thus, previous research into the field of technical communication reflects a variety of research methods and approaches that contribute to our understanding of our as-yet-incomplete history. While the authors represented in this volume adhere to some of the established methodologies previously outlined—evaluating the contributions of individuals, examining curricular patterns, studying rhetorical strategies, analyzing specific incidents, exploring the rise of genres, and tracing broad trends—this volume contributes to the discipline a series of both new and reprinted "landmark" essays that illuminate the past and provide unique historical insight into the development of technical communication. Drawing on both primary and secondary resources, the authors of the following chapters examine not only individuals who shaped the foundation of technical communication, but also broad American and European movements that established precedents for some of our contemporary professional and pedagogical practices. We believe that these essays help fill in key gaps in our understanding of how technical communication was conceptualized, as both a professional activity and as an academic discipline. As a result, the central theme of this volume is key influences in the historical evolution of technical communication.

Rather than "straight" histories of given periods, though, the historicity of the following pieces is grounded in distinct theoretical or methodological approaches to the subject. While history provides us with a narrative account of events, it must also be more than events or people—history is often a series of occurrences and

INTRODUCTION   11

shaping factors that contribute to the slow evolution of larger patterns of significance. This volume, therefore, examines those shaping factors and suggests ways in which those factors affected the evolution of technical communication as a discipline in this country. These selections extend the literature reviewed in the first part of this introduction by addressing the disparity between historical works that are predicated on key periods and works that are predicated on key movements within periods. Following the variety of trends and movements, this work shifts easily from the 20th century to the 16th century and back again in search of significant influences that help researchers better understand the foundational issues that have shaped contemporary practice.

The four distinct sections of this volume—key individuals, key European movements, key American movements, and the bibliography of the history of technical and business communication—each contribute to our sense of the history of this discipline. This volume also affirms the historical importance of women to the discipline (see Kynell and Tebeaux), examines the role of 20th century technology to the field (see Johnson-Eilola, Selfe, and Selber), and suggests that disciplines from surveying to medical texts (see, respectively, Moran and Shirk) all illuminate the development of the field. The seeming disparity among these selections highlights another unique aspect of this book: inquiry into technical communication is virtually boundless. No longer confined to discussions of engineering, science, or business, historians of technical communication now find primary and secondary resource materials in a variety of archival environments. It is our goal in bringing together these selections to ensure that no one influence is privileged over another. In our historical diversity we isolate those consistent factors that have shaped us.

Part I: Key Individuals in the History of Technical Communication

The first section of this volume includes two reprinted landmark essays and two original essays which each examine one key individual in the history of technical communication. This section, in particular, highlights the contributions of both well-known and lesser-known figures in order to exemplify the range of players in the history of the discipline.

The first chapter, James Zappen's landmark piece "Francis Bacon and the Historiography of Scientific Rhetoric," examines three 20th century interpretations of Bacon's science and rhetoric: positivistic science and the plain style, institutionalized science and the figured style, and democratic science. In reviewing the commentators on Bacon, Zappen finds a variety of historical and historiographical interpretations significant in that they provide a range of views on Bacon in particular and scientific rhetoric in general. Zappen concludes the piece, however, by identifying his own interpretation of Bacon's rhetoric, determining that Bacon had not one but several methods and styles in his approach to science. Zappen identifies Bacon's view of democratic science as a kind of invitation to participate in the scientific method, an alternative perception of what, he concludes, is "good for both the scientific and social community." Zappen's chapter represents a specific advance in the study of technical communication's history by evaluating the range of interpretations of Bacon's scientific rhetoric and suggesting that the plain style is not just a concomitant of positivism, but a vehicle with applications in democratic science as well.

The second landmark essay we present in this section, Charles Bazerman's "How Natural Philosophers Can Cooperate: The Literary Technology of Coordinated Investigation in Joseph Priestley's *History and Present State of Electricity*" (1767), examines Priestley's attempts in the 18th century to foster cooperation among the researchers and scientists investigating the properties of electricity. Bazerman traces the investigations of Priestley into electricity, including the practical experiences of working electricians, that led to the writing of the *History and Present State of Electricity*. In evaluating this book, Bazerman examines Priestley's sense of the shared experience of scientific research and his concern for the development of common knowledge and communal participation. This landmark chapter is significant for a variety of reasons. First, the piece carefully examines Priestley's early scientific work. Second, the piece isolates those Priestleyan factors of *The Present State* that contributed to a model of scientific cooperation that contemporary scientists continue to emulate.

The third selection in Part I, John Brockmann's "Oliver Evans and His Antebellum Wrestling with Rhetorical Arrangement," analyzes rhetorically two of Evans's technical discourses, *Young Mill-Wrights and Millers Guide*, a successful manual that Brockmann describes as the most reprinted antebellum American technical guide, and *The Abortion of the Young Steam Engineers Guide*, a far less successful book that appeared in only two editions. The first book was popular in part because it carefully followed the principles of Ciceronian disposition or arrangement and therefore presented its technical information in ways that met reader expectations. The second book, on the other hand, disregarded Cicero's principles and suffered the consequences. This connecting of rhetorical arrangement with marketable technical communication provides a key to understanding one important figure in our past.

The fourth and final chapter in Part I, Teresa Kynell's "Sada A. Harbarger's Contribution to Technical Communication in the 1920s," examines an important woman in the history of American technical communication, how her contributions were particularly significant given the masculine hegemony of the scientific and engineering communities, and how her early work in teaching service courses (composition and technical communication) provided foundational models in the discipline. Advancing from instructor all the way to Chair of the English Committee within the Society for the Promotion of Engineering Education, Harbarger pioneered technical communication as a discipline in this country. Kynell's work is important because it not only isolates Harbarger's contributions, but it also examines the role of women in service-related teaching. Harbarger made the androcentric environment of technical communication more welcoming to women.

Part II: Key European Movements in the History of Technical Communication

This section of the volume includes one reprinted landmark chapter as well as three original contributions. Acknowledging some of our antecedents as a discipline in the European tradition, this section highlights those European movements that provide early insights into shaping factors and influential practices that are foundational to our profession and pedagogical activities today.

The first piece in Part II, Elizabeth Tebeaux's "The Emergence of Women Technical Writers in the Seventeenth Century: Changing Voices within a Changing Milieu," examines the first published technical books and documents by women during the English Renaissance and the 17th century. Her examination of books written by and for Renaissance women on topics such as gardening, cooking, medicine, and household management reflects not only middle-class literacy levels, but also demonstrates the ability of those writers to address their readers' needs. Her work constitutes an important contribution to this collection for two reasons. First, her chapter offers a model essay for the extensive use of primary sources; second, the chapter reveals the rich contributions that female technical writers have made to our history.

The second selection in Part II, Merrill Whitburn's reprinted landmark chapter "The Plain Style in Scientific and Technical Writing," studies the stylistic devices in use before the scientific revolution and how those devices (tropes, metaphors, similes, etc.) often obscured rather than clarified the authors' intentions. In part a reaction against such rhetorical complexity, the "plain style" became the style of scientific discourse during the 17th century. Whitburn, however, questions the effectiveness of teaching the plain style in isolation; instead, he advocates teaching students to use rhetorical devices to add a "personal touch" to student prose. The application of various figures, he argues, can be used to promote economy and clarity in technical prose.

The third piece in Part II, Henrietta Shirk's "Deconstructing Depression: A Historical Study of the Metaphorical Aspects of Illness," examines the uses of metaphor to describe depression, sometimes called melancholia, in published texts from the fifth century to the present. Shirk evaluates the historical use of metaphor as a rhetorical device, using deconstruction as her method of analysis. Shirk's work is significant not only because it analyzes medical texts, but also because it demonstrates the usefulness of deconstruction for examining professional texts in their historical contexts.

The fourth chapter in Part II, Michael G. Moran's "Renaissance Surveying Techniques and the Mapping of Raleigh's Virginia," analyzes the work of Thomas Hariot and John White who were sent by Sir Walter Raleigh to create the first maps of North America based on detailed surveys. These technically accurate maps set the standard for New World cartography. Moran discusses the instruments and techniques that Hariot and White employed, shedding light on both surveying and mapmaking, fields

14 INTRODUCTION

of endeavor that became professionalized in England during the 16th century. Moran's analysis of the Hariot-White-de Bry map in the Theodor de Bry edition of Hariot's *A Briefe and True Report of the New Found Land of Virginia* (1588/1972) demonstrates a sophisticated relationship between the written report and the map. This important essay, like Tebeaux's, demonstrates the use of primary sources to piece together the antecedents of the discipline in Europe.

Part III: Key American Movements in the History of Technical Communication

The third section of this volume includes one reprinted landmark essay and two original essays which examine key movements and important advances in the evolution of technical communication in this country. Drawing on early to late 20th century influences, the authors analyze everything from computer technology to engineering curricular shifts. This section's last essay ties together the book by evaluating and synthesizing our ongoing struggles in technical communication as we move into the 21st century.

The first chapter in Part III is Robert J. Connors's landmark essay "The Rise of Technical Writing Instruction in America," a piece widely cited by students and researchers interested in the historical connection between the American engineering curriculum and technical communication. Connors traces the history of technical writing in American colleges from 1900 to 1970, focusing on key individuals, important textbooks, and shifts in engineering pedagogical practices. He proves that, after World War II, technical communication became a viable and important part of English studies. This chapter, first published in the early 1980s, was among the first works to examine the forces that shaped technical communication courses in America and remains an invaluable resource for the historian and researcher.

The second selection in Part III, "Interfacing: Multiple Visions of Computer Use in Technical Communication," is authored by three scholars of computer applications in technical communication: Johndan Johnson-Eilola, Stuart Selber, and Cynthia Selfe. They argue that tracing the history of the computer as a series of technological inventions diminishes the role of social, cultural, economic, and political influences on computing. The authors study the relationship between computers and technical communication through meta-analysis, examining not only the ways in which computer technologies are developed and used, but also how their use intersects with the teaching and practice of technical communication. This important piece evaluates possible interfaces between technical communication and computer technologies and, perhaps more importantly, challenges our thinking about historical movements in the discipline.

The third and final piece of Part III, Jo Allen's "Refining a Social Consciousness: Late 20th Century Influences, Effects, and the Ongoing Struggle in Technical Communication," evaluates the ways that we determine the kind of history we chronicle. Since, as Allen notes, the events and circumstances that make up our history are quite often episodic and do not always follow a neat time line, her essay offers an evolutionary perspective on the social nature of technical communication. Allen evaluates how ideas have changed the forces behind the changes, the factors that shaped the changes, and the consequences of those changes. She examines the maturation of our discipline and, in so doing, provides a valuable final chapter to this section.

Part IV: Bibliography in the History of Technical Communication

The book ends with a completely updated and revised version of William E. Rivers's (1994) "Studies in the History of Business and Technical Writing: A Bibliographic Essay." In many ways the most important selection in this volume, Rivers's chapter provides students and scholars with a bibliographic guide to the wealth of research on the history of technical communication in this country and in Europe. The chapter functions as the capstone selection of this volume; its length, thoroughness, and completeness testify to the increasingly important role of history in our discipline.

The chapters selected for this volume represent some of the best work that has been done and is being done in the history of technical communication. Our history as a discipline, however, is still being written as technological changes and rapid advancements in the workplace call us to reconsider our choices. We hope that this volume will not only reveal patterns and movements in our past, but will also help inform the decisions that we will make in the future.

REFERENCES

Adams, K. H. (1993). *A history of professional writing instruction in American colleges: Years of acceptance, growth, and doubt.* Dallas, TX: Southern Methodist University Press.
Aydelotte, F. (1917). *English and engineering.* New York: McGraw-Hill.
Bacon, F. (1965). *The advancement of learning* (C. W. Kitchin, Ed.). New York: Dutton.
Bacon, F. (1975). *The new organon and related writings* (F. H. Anderson, Ed.). Indianapolis, IN: Bobbs-Merrill.
Bazerman, C. (1988). *Shaping written knowledge: The genre and activity of the experimental article in science.* Madison, WI: Wisconsin University Press.
Brockmann, R. J. (1983). Bibliography of articles on the history of technical writing. *Journal of Technical Writing and Communication, 13,* 155–65.
Brockmann, R. J. (1988). Does Clio have a place in technical writing? Considering patents in a history of technical communication. *Journal of Technical Writing and Communication, 18,* 297–304.
Evans, O. (1795). *The young mill-wrights and millers guide.* Wallingford, PA: The Oliver Evans Press.
Evans, O. (1805). *The abortion of the young steam engineers guide.* Wallingford, PA: The Oliver Evans Press.

Freeman, W. A. (1961). Geoffrey Chaucer, technical writer. *Society of Technical Writers and Publisher's Review, 8*(4), 14–15.

Gresham, S. (1981). From Aristotle to Einstein: Scientific literature and the teaching of technical writing. In D. Stevenson (Ed.), *Courses, components, and exercises in technical communication* (pp. 87–93). Urbana, IL: National Council of Teachers of English.

Hagge, J. (1990). The first technical writer in English: A challenge to the hegemony of Chaucer. *Journal of Technical Writing and Communication, 20,* 269–289.

Hagge, J. (1995). Early engineering writing textbooks and the anthropological complexity of disciplinary discourse. *Written Communication, 12,* 439–491.

Halloran, S. M., & Bradford, A. N. (1984). Figures of speech in the rhetoric of science and technology. In R. J. Connors, L. S. Ede, & A. Lunsford (Eds.), *Essays on classical rhetoric and modern discourse* (pp. 179–192). Carbondale, IL: Southern Illinois University Press.

Harbarger, S. A. (1923). *English for engineers.* New York: McGraw-Hill.

Harriot, T. (1972). *A briefe and true report of the new found land of Virginia: The complete 1590 Theodore de Bry edition.* New York: Dover. (Original work published 1588)

Herndl, C. G., Fennell, B. A., & Miller, C. R. (1991). Understanding failures in organizational discourse: The accident at Three Mile Island and the Shuttle Challenger disaster. In C. Bazerman & J. Paradis (Eds.), *Textual dynamics of the professions: Historical and contemporary studies of writing in professional communities* (pp. 279–305). Madison, WI: University of Wisconsin Press.

Howell, W. S. (1971). *Eighteenth century British logic and rhetoric.* Princeton, NJ: Princeton University Press.

Jones, D. R. (1985). A rhetorical approach for teaching the literature of scientific and technical writing. *The Technical Writing Teacher, 12,* 115–125.

Kynell, T. (1996). *Writing in the milieu of utility: The move to technical communication in American engineering programs, 1850-1950.* Norwood, NJ: Ablex.

Lavoisier, A. (1793). *Elements of chemistry in a new systematic order* (R. Kerr, Trans.). London: Creech.

Lipson, C. S. (1982). Descriptions and instructions in Medieval times: Lessons to be learned from Geoffrey Chaucer's scientific instruction manual. *Journal of Technical Writing and Communication, 12,* 243-256.

Locker, K. O. (1985). The earliest correspondence of the British East India Company (1600–19). In G. H. Douglas & H. W. Hilderbrand (Eds.), *Studies in the history of business communication* (pp. 69–86). Urbana, IL: Association for Business Communication.

Miller, C. R. (1984). Genre as social action. *Quarterly Journal of Speech, 70,* 151–167.

Miller, W. J. (1961). What can the technical writer of the past teach the technical writer of today? *IRE Transactions on Engineering Writing and Speech, 4,* 69–76. Rpt. in D. Cunningham & H. Estrin (Eds.), *The teaching of technical writing* (pp. 198–216). Urbana, IL: National Council of Teachers of English.

Moran, M. G. (1984). Joseph Priestley, William Duncan and analytic arrangement in 18th century scientific discourse. *Journal of Technical Writing and Communication, 14,* 207–215.

Ovitt, G., Jr. (1987). History, technical style, and Chaucer's Treatise on the astrolabe. In M. Amsler (Ed.), *Creativity and the imagination: Case studies from the Classical Age to the Twentieth Century* (pp. 34–58). Newark, DE: University of Delaware Press.

Paradis, J. (1983). Bacon, Linnaeus, and Lavoisier: Early language reform in the sciences. In P. V. Anderson, R. J. Brockmann, & C. R. Miller (Eds.), *New essays in technical and scientific communication: Research, theory, practice* (pp. 200-224). Farmingdale, NY: Baywood.

Priestley, J. (1767). *History and present state of electricity*. London: J. Dodsley.

Reynolds, T. S., & Seely, B. E. (1993). Striving for balance: A hundred years of the American society for engineering education. *Journal of Engineering Education, 82*, 136–151.

Rickard, T. A. (1908). A *guide to technical writing*. San Francisco: Mining and Scientific Press.

Rivers, W. E. (1994). Studies in the history of business and technical writing: A bibliographical essay. *Journal of Business and Technical Writing, 8*, 6–57.

Russell, D. R. (1991). *Writing in the academic disciplines, 1870-1990: A curricular history*. Carbondale, IL: Southern Illinois University Press.

Rutter, R. (1991). History, rhetoric, and humanism: Toward a more comprehensive definition of technical communication. *Journal of Technical Writing and Communication, 21*, 133–153.

Tebeaux, E. (1997). *The emergence of a tradition: Technical writing in the English Renaissance, 1475–1640*. Amityville, NY: Baywood.

part I
Key Individuals in the History of Technical Communication

1

Landmark Essay: How Natural Philosophers Can Cooperate: The Literary Technology of Coordinated Investigation in Joseph Priestley's *History and Present State of Electricity* (1767)*

Charles Bazerman
University of California at Santa Barbara

> Cheerfulness and social intercourse do, both of them, admirably suit, and promote the true spirit of philosophy. (Priestley, 1775, vol. 2: p. 164)

Recent studies of the rhetoric of science have emphasized the competitive struggle played out through scientific texts. Scientific publications are seen as persuasive briefs for claims seeking communal validation as knowledge (Latour & Woolgar, 1979; Knorr-Cetina, 1981). Moreover, individual texts have been seen as part of a negotiation process among competing interests that may result in statements of

---

* This essay was supported by released time granted by the Dean of Liberal Arts and Sciences, Baruch College. I also thank Rachel Laudan, John McEvoy, Michael G. Moran, Greg Myers, Simon Schaffer, and Harriet Zuckerman for their comments and criticisms.

knowledge different than those proposed in the initiating texts (Collins, 1985; Myers, 1985; Latour, 1987). During these struggles authors draw on many extra textual resources (social, economic, intellectual, and empirical) which are deployed in the text (Collins & Pinch, 1982; Rudwick, 1985; Callon, Law, & Rip, 1986). Only after communal acceptance do these claims take on the appearance of irrefutable truths stated with objective authority transcending the urging of an author (Latour, 1987).

Genres of scientific writing can be seen as recurrently successful rhetorical solutions to the persuasive problem of advancing claims within an empirical research community. Communally persuasive forms of representing empirical experience and structuring compelling arguments upon that experience have resulted in claims appearing to be proven knowledge, except to those who know of the local struggles. The standardization of textual form has helped to regularize and focus the struggle of scientific writing, even while it has served to hide that struggle (Bazerman, 1988).

Nonetheless, an older tradition has considered scientific activity as more than competitive play. Scientific communication has most often been conceived of as part of a cooperative endeavor. The charter myth of this tradition is Sir Francis Bacon's description of Salomon's House in *The New Atlantis* (1627). Here, Bacon describes a cooperative bureaucracy of 36 field researchers, reviewers of the literature, experimenters, experimental designers, theorists, and applied technologists. Bacon anticipated no particular communication difficulty in this cooperative project, beyond the general linguistic problem of the four idols. Later in the 17th century, this cooperative, bureaucratic model inspired a number of organizational decisions of the French Royal Academy and the British Royal Society. However, personal interests and disagreements soon tore at the fabric of such an untroubled plan, and a communication system which facilitated and structured disagreement took shape over the next century (Bazerman, 1988).

Despite the systemic competitiveness of modern science, when we remove ourselves from the daily hand-to-hand combat of scientific argumentation, we can perceive large patterns of cooperation and the communal construction of a shared knowledge. This knowledge is not dictated by a single text or monumental figure (whether God, Aristotle, or Newton), but is advanced (sometimes slowly, rapidly, or spasmodically) through the joint endeavors of large numbers of people. Not only are little details filled in and puzzles worked out within static paradigms, but major novel findings appear and are absorbed, theories are modified and replaced, and knowledge moves in startling and unanticipated directions. To be persuaded of the overall cooperative pattern of scientific work, one need only contemplate the remarkable changes currently being wrought and absorbed by diverse researchers in "hot" areas such as superconductivity, fundamental forces, viral biochemistry, and neural physiology. Indeed, many modern commentators of science make cooperation an essential component of scientific activity and communication (Ziman, 1968; Merton, 1973; Garvey, 1979).

THE PUZZLE OF COOPERATION

Noticing that cooperation seems to occur, however, does not let us know how it happens, nor why the cooperation should seem to be as enduring and fundamental as it appears to be in science. Persuasion and cooperation, as we know from political and other familiar everyday realms, are uncertain and fragile phenomena. Beliefs seem to change rapidly, alliances fall apart, and cooperation often needs to be cemented by laws, money, and coercion. If even the degree of cooperation we manage in everyday affairs remains beyond our full comprehension, how can we begin to account for the much more remarkable cooperation evident in scientific work, a cooperation which seems to span religions, philosophies, national boundaries, and centuries? However, until we have as concrete, detailed accounts of the microprocesses by which cooperation and coordination occur as we do of competitive processes, cooperation and coordination may only appear to be value-laden suppositions rather than actual social activities. This chapter, accordingly, offers a microanalysis of the cooperative mechanisms of one 18th-century text that was self-consciously constructed to foster cooperation and that foreshadows a number of features of modern scientific papers. This analysis reveals the many levels on which coordination needs to be achieved through language and the tension which needs to be maintained between cooperation and competition, codification and originality, if the communal endeavor of science is to move forward.

Certainly early science did not seem to achieve the cooperative complexity and coordination of contemporary science, despite Bacon's high hopes. Rather than building on one another's theories, authors were as likely to attempt to supplant each other's claims. Authors rarely constructed claims that explicitly integrated a wide range of the claims of others. Even the Baconian hopes for an appeal to the facts did not lead to philosophic harmony, as facts themselves became a matter of dispute. Within this atmosphere, local cooperation was only created by the dominance of strong individuals who set the national theoretical terms and research agendas, supported by institutionalized power; Newtonianism and Cartesianism, although occasionally communicating in individual ad hoc circumstances, more often fired salvos across the English Channel (Bazerman, 1988).

While science remained small, with relatively few results to coordinate and few compelling challenges to the hegemony of the brilliant works of early giants, such ad hoc cooperation as existed through unsystematic familiarity with each others' works from travel, correspondence, and publications was perhaps adequate to carry the communal work of science forward. The emergence of societies and journals helped create regular forums for communication among scientists and organize the communication practices (Bazerman, 1988); however, as natural philosophic findings proliferated in the 18th century, cooperation had to be explicitly achieved within the substance of the communications. Textual mechanisms needed to be developed to coordinate the work and emerging perceptions of researchers who were widely dispersed temporally, geographically, and theoretically.

Joseph Priestley's 1767 book *The History and Present State of Electricity* explicitly takes up the challenge of fostering cooperation among the growing number of electricians and drawing new participants into this emerging research community. Besides expressing concern for the benefits of joint work, the book employs many textual mechanisms that integrate past, present, and future work in the field. Through a comprehensive review of the literature, Priestley establishes the corpus of communal experience and organizes it around problems and principles that define an evolving state of knowledge and research agenda. A list of generalizations emerging from that communal history provides a common knowledge base for continuing work; a discussion of the major theories sorts out the conceptual meaning of research; a list of open issues suggests directions for research; and a historical review of the development of apparatus and practical suggestions for construction provide a common material basis for generating phenomena to be investigated. Besides trying to establish coherence and focus within a research front emerging from a shared understanding of past work, Priestley is concerned to draw new researchers into the communal project, so he provides practical suggestions for carrying out experiments, as well as a series of amusing experiments to attract and train neophytes. Finally, he provides narratives of his own work to demystify the process of investigation and to provide exemplars of work that might be carried on with only humble means. With our current, limited knowledge of the development of textual features of scientific writing, we cannot unequivocally credit Priestley with invention of the textual devices he employs nor can we trace a direct line of evolution to current cooperative literary practices.[1] Yet Priestley's thoroughgoing interest in fostering coordinated work of an extensive community offers a striking starting point for examining the complexity of cooperative textual machinery that has developed to coordinate the voluminous and undeniably competitive work of contemporary science.

PRIESTLEY AND 18TH-CENTURY ELECTRICITY

Electricity by the mid-18th century was a proliferating area and presented much that could use coordination. The modern study of electricity is usually dated from William Gilbert's *On the Magnet* (1600), which includes a chapter on the attractive power of rubbed amber, known since classical times. Gilbert noted a number of other substances that showed a similar property. During the 17th century, a few items were added to the list of electricals, various theories were presented to account for the phenomenon, and electrical repulsion was noticed for the first time. At the beginning of the 18th century, however, the invention of the electrostatic generator made possible the discovery and investigation of such phenomena as luminosity, sparks, shocks, conduction, induction, and the difference between two varieties of electricity. The improvements of these machines and the 1745 invention of the Leyden jar (the modern condenser) permitted experiments with charges of increasingly great power; both medicinal and lethal effects were noted. By 1750 Benjamin Franklin had presented evidence of the equivalence of lightning and electricity, setting off a series of investigations into atmospheric electrical phenomena. Electricity was literally exploding across the mid-18th century natural philosophic scene.[2]

Although Joseph Priestley (1733–1804) had an interest in natural philosophy during his own education and early career as dissenting minister and schoolmaster, he did not actively pursue scientific studies until the mid-1760s when Matthew Turner, his colleague at Warrington Academy, offered a course of lectures in chemistry (Schofield, 1966). Priestley was to achieve his greatest fame in this area through the discovery of oxygen in 1774. Nonetheless, electricity, not chemistry, provided the subject of Priestley's first investigations and publications.

We do not know exactly when Priestley began to work on electricity, but by late 1765 on a trip to London he arranged an introduction to Franklin and several other prominent electricians to gain their support (see the letter from John Seddon to John Canton in Schofield, 1966, p. 14). Franklin encouraged him in his plan to write a "history of discoveries in electricity," and helped arrange for the requisite books (Priestley, 1970). John Canton, William Watson, and Richard Price, as well as Franklin, remained his correspondents, mentors, and benefactors over the next year as he wrote *The History and Present State of Electricity*.

The first and longer half of the lengthy book (432 of 736 quarto pages in the first edition) is a detailed history of all investigations and discoveries in electricity from the time of the ancients to his day. In its synoptic command, attention to empirical details in the literature, and its open-ended attitude, it can be seen as one of the earliest versions of the modern genre of review of the literature.

The second half of the work, not indicated in any of Priestley's early plans, consists of seven additional parts: a list of then-known general properties of electricity; a discussion of the history of electrical theories, including a detailed comparison of two major theories; some general considerations on the current state of electrical research, and a series of queries to direct future work; descriptions and directions for constructing electrical machines; a set of procedural advice (or practical maxims) for those wishing to carry out electrical experiments; directions for carrying out entertaining demonstration experiments; and a description of his own new experiments on the subject. As the first half may be designated the history, this latter half may be said to be the "present state of electricity." Much of this material is presented in no other previous work on electricity. Although today we might find the various kinds of materials presented in this latter half in a variety of places, ranging from children's activity books to advanced textbooks, equipment manuals, and research journals, we are not likely to find them all under the same cover.

Doing Natural Philosophy by Doing History:
Priestley's Philosophic Framework

Although this odd mixture of things may appear to be a neophyte's grab bag—talking about everything he sees with little sense of design—such lack of design is unlikely, for each part of the book is introduced by several pages of explicit description and rationale for the literary procedures that follow. Moreover, at Warrington Academy, Priestley had regularly delivered a series of lectures on oratory and criticism (eventually published in 1777) as well as a course of lectures on the theory of language (printed privately in 1762). He was a self-conscious user of language, and his procedures in *The History and Present State of Electricity* are consistent with his teachings on rhetoric (see Moran, 1984). Furthermore, his rhetorical practices are a self-conscious attempt to realize his millenarian vision of human progress. Particularly relevant here is his understanding of the role of historical discourse in increasing human wisdom, for it is a history of electricity that he tells and it is as participants in a historical process that he addresses his readers.[3]

As instructor at Warrington Academy since 1761, Priestley had been delivering a series of lectures on history (later published in 1788). In these lectures he argues that the study of history "strengthens the sentiments of virtue" by showing us the characters of the many kinds of humans and "improves the understanding" by extending our experience. In particular, study of the history of natural philosophy presents edifying portraits "of genius in such men as Aristotle, Archimedes, and Sir Isaac Newton, [which] give us high ideas of the dignity of human nature, and the capacity of the human mind" (Priestley, 1788, p. 120).

Moreover, the history of natural philosophy increases our individual empirical experience by attaching us to a community of experience. Priestley declares, "the most exalted understanding is nothing more than the power of drawing conclusions, and forming maxims of conduct, from known facts and experiment, of which necessary materials of knowledge the mind is wholly barren" (1788, p. 108). Understanding is based on experience to form the proper associations.[4] But each individual is limited, so only through history can we come to share in the experience of others. For "the improvement of human kind and human conduct, and to give mankind clear and comprehensive views of their interest, together with the means of promoting it," Priestley felt "the experience of some ages should be collected and compared, that distant events should be brought together" (p. 108). Natural philosophy gives order to the accumulated human experience, so that we may then choose wisely about our lives and improve the human condition.

Priestley himself seemed to have a strong synoptic grasp of history revealed in his invention of the historical time line. Using the bar graph for the first time to represent historical duration (Funkhouser, 1937), Priestley published in 1765 an extremely popular book titled *A Chart of Biography,* which printed over 15 editions by 1820, and in 1769, an equally popular *A Chart of History,* which also had at least 15 editions by 1816 (Fulton, 1937). By this now-common technique he was able to give graphic shape to the sweep of history. This sense of the sweep of history is essential to the vision of *The History of Electricity.* Additionally, Priestley reveals an open-ended attitude toward the historical process by leaving a blank space at the far end of *A Chart of History* for the reader to fill in the developments of the last decades of the 18th century. Again, this open-ended sense of the historical process imbues the electricity book. He does not pretend that electrical knowledge is complete and history ends with his account. Rather, Priestley views electricity as an evolving practice and investigation, caught at a present moment or state of development and leading into an unknown future.

Priestley's concern for progressive historical improvement of life is founded in his millennial theological positions concerning the perfectibility of man led by a benevolent deity (see Laboucheix, 1977; Heibert, 1980; McEvoy, 1979, 1984) and consonant with his radical politics, including support of both the American and French Revolutions (see Fruchtman, 1983; Kramnick, 1986; Crosland, 1987; Priestley, 1970).[5] In his writings on education, Priestley turns this concern into a practical program of training young men for a life of worldly activity to replace the purely clerical education common at his time (*An Essay on a Course of Liberal Education for Civil and Active Life* [1765]). Natural philosophy has an obvious place within such a theology, politics, educational plan, and progressive view of history. Natural philosophy reveals the benevolence and wisdom of God's plan and offers humans a way to participate actively in the fulfillment of that plan.[6]

Within these theological, historical, educational, and rhetorical contexts, Priestley's aim for his account of electricity becomes clear: to further the communal work of electrical investigation. The preface to the first edition of *The History and Present State of Electricity* keeps returning to the theme of how this history and others like it may advance the progress of science; for example, "once the entire progress, and present state of every science shall be fully and fairly exhibited, I doubt not but we shall see a new and capital aera commence in the history of all sciences" (vol. 1: p. xviii).[7] Later he states even more directly, "To quicken the speed of philosophers in pursuing this progress, and at the same time, in some measure, to facilitate it, is the intention of this treatise" (vol. 2: pp. 53–54). Moreover, in a simplified version published the year after (1768), he reveals his plan comprehensively: "My principal design was to promote discoveries in Science, by exhibiting a distinct view of the progress that had been made hitherto, and suggesting the best hints that I could for continuing and accelerating that progress" (p. v).

A HISTORY OF NATURAL PHILOSOPHY: PART 1

The first step in this project of furthering the communal work is to gather together the accumulated experience of electricity by natural philosophers. As Priestley comments in the preface to the first edition, "At present, philosophical discoveries are so many, and the accounts of them are so dispersed, that it is not in the power of any man to come to the knowledge of all that has been done, as a foundation for his own inquiries. And this circumstance appears to me to have very much retarded the progress of discoveries" (vol. 1: p. vii). Although this comment may be familiar in the 20th century, it represents an attitude not generally reflected in natural philosophic texts before this time. Generally references to the work of others was perfunctory, if present at all, and little attempt was made to make systematic sense of the previous literature. Often the writers seem unfamiliar with relevant published work. Franklin, in the distant colonies, presents an extreme example; he began his work with only the aid of a popular summary of contemporary work published in the *Gentleman's Magazine*. Even after he became familiar with a wider range of work, his publications rarely mentioned any historical work and gave only passing mention to the work of his contemporaries.

The most extensive discussion of the electrical literature Priestley had seen before writing his history was the four-page bibliographic appendix to Desaguliers' 48-page pamphlet in 1742, called *A Dissertation Concerning Electricity*, which elaborates and gives citations for items mentioned in the main text. A more extensive German summary of the literature and annotated bibliography by Daniel Gralath (1747–1757) did not come into Priestley's hands until after the first edition had been published; Priestley used Gralath's work to revise the second edition.

Priestley took the task of gathering the accumulated experience of prior electricians very seriously. He insisted on reading, wherever possible, the original texts of all his predecessors. Much of Priestley's correspondence in this period concerns his attempt to obtain rare volumes (Schofield, 1966), and he apparently incurred large expenses in this regard (see Crosland, 1983). In the bibliography of his book he also requests his readers to send him volumes he has not yet seen.

To establish the immediate connection between the sources and his retelling, he footnotes with specific page references each text quoted, summarized, or discussed. About half of the pages of the historical section have at least one reference note and some have as many as five or six. Except for secondary format differences, Priestley follows modern footnote practice. He also includes a bibliography of all items on electricity of which he had heard (63 items in the first edition and 75 in the second and ensuing editions) and also notes the volumes that he had consulted (30 for the first edition and 43 for later editions). Also, a detailed index identifies where each author is discussed.

In giving such care to identify sources, Priestley emphasizes that he is writing a history of natural philosophy embodied in publications rather than a Baconian natural history of the phenomena themselves. He comments, for example, on Franklin's *Experiments and Observations on Electricity*: "Nothing was ever written upon the subject of electricity which was more generally read and admired in all parts of Europe than these letters. There is hardly any European language into which they have not been translated" (1775, vol. 1: p. 192). The history of electricity is in part the history of the appearance and circulation of texts which carry accounts of experiences. The experiences are not separable from the people who encounter them nor from the texts in which accounts are transmitted.

Historical Consciousness within Progressive Knowledge

This historical awareness of the evolving human accounting for natural phenomena allows him to treat earlier findings within historically appropriate knowledge, while still using later developments to comment on, evaluate, or interpret the findings.[8] Typically, we see Priestley's historical awareness of the current state of knowledge in his discussion of Boyle's work: "We should now be surprised that any person should not have concluded *a priori,* that if an electric body attracted other bodies, it must, in return, be attracted by them, action and reaction being universally equal to one another. But it must be considered, that this axiom was not so well understood in Mr. Boyle's time, nor till it was afterwards explained in its full latitude by Sir Isaac Newton" (1775, vol. 1: p. 8).

Priestley even includes material that was by his time considered in error, so as to make the account of communal experience complete. Although sometimes he labels the discredited results as delusions, elsewhere he presents them with no comment, and in other places as productive challenges. Priestley's account includes so many cases of at first implausible results later accepted as common knowledge that he is chary to exclude any result. Discredited theories are respected for their appropriateness to the state of knowledge in their times and their heuristic value for new discoveries.

Where results evoked controversy and troubled attempts at replication, Priestley gives accounts of the processes by which the community came to pass judgment, as when J. A. Nollet travels to Italy to investigate claims about the medicinal effects of an electrical device and becomes "convinced that the accounts of cures had been much exaggerated" (1775, vol. 1: p. 187). Priestley then recounts other unsuccessful attempted replications, including some performed "in the presence of a great number of witnesses, many of them prejudiced in favour of the pretended discoveries; but they were all forced to be convinced of their futility, by the evidence of facts" (vol. 1: p. 187). He comments, "After the publication of these accounts properly attested, every unprejudiced person was satisfied, that the pretended discoveries from Italy and Leipsick, which had raised the expectation of all electricians in Europe, had no foundation in fact" (vol. 1: p. 188). Priestley describes judgment as being passed by the accumulated experience, which is recorded and circulated in a sequence of documents.

Specific Accounts and General Claims

In the attempt to represent fairly the experience and thinking of previous electricians, Priestley offers lengthy accounts, staying self-consciously close to the original presentations. Rarely is any publication given less than a full paragraph's discussion and often several pages are devoted to describing crucial findings. In both the preface and in passing Priestley shows self-conscious awareness of the responsibilities and difficulties of accurate summary and he often quotes at length, sometimes for more than a page at a time. In the preface he comments "that I might not mis-represent any writer, I have generally given the reader his own words, or the plainest translation I could make of them" (1775, vol.1: p. x). And throughout Priestley explains and justifies any liberties he takes with the text or the chronology.

Priestley's discussion of each electrician is built on specific empirical experiences or experiments which that individual was the first to notice or verify. These are recounted in sufficient detail for the particular event to be pictured, and in a number of cases to be replicated. Further, Priestley seems to have replicated many of the experiments he recounts. He explicitly notes the few cases when experiments present practical obstacles for replication, such as the need for unusual, costly, or sensitive apparatus.

A description of one of Francis Hauksbee's experiments is typically particular yet concise, relying as it does on the familiarity of typical apparatus and general procedures.

> Having tied threads round a wire hoop and brought it near to an excited globe or cylinder, he observed, that the threads kept a constant direction towards the center of the globe, or towards some point on the axis of the cylinder, in every position of the hoop; that this effect would continue for about four minutes after the whirling of the globe ceased, and that this effect was the same whether the wire was held above or under the glass; or whether the glass was placed with its axis parallel, or perpendicular to the horizon.
>
> He observed, that the threads pointing towards the center of the globe were attracted and repelled by a finger presented to them; that if the finger, or any other body, was brought very near the threads, they would be attracted; but if they were brought to the distance of about an inch, they would be repelled, the reason of which difference he would not seem to understand. (Priestley, 1775, vol. 1: p. 10)

Specific observations are the core of the account, but they are introduced and punctuated by a presentation of experimental procedures and followed by a brief discussion. By such accounts of experiments Priestley makes available a vicarious experience of essentially all the significantly novel experiments performed by all electricians to that point in time and opens up the possibility of actual repetition of the experiments.

These accounts of experiences are not, however, presented as isolated events. Priestley organizes the experiments around general principles of electrical behavior. The Hauksbee experiment described above is preceded by statements classifying the experiment at two levels of generalization, with Priestley's italics emphasizing the significant general concepts: "I shall first relate the experiments [Hauksbee] made concerning *electrical attraction and repulsion*.... The most curious of his experiments concerning electrical attraction and repulsion are those which shew the direction in which these powers are exerted" (1775, vol. 1: pp. 19–20).

Moreover, Priestley presents a series of experiments as coherent sequential investigations into particular phenomena, so that one experience seems to lead to the next according to the dictates of rational investigation. The Hauksbee experiments quoted above are immediately followed by further experiments to explore attraction and repulsion phenomena. This sense of a coherent program is extended beyond the work of individual researchers to be used as an organizing device for the work of many researchers. Priestley imposes a rational shape to communal work, which by this gesture becomes a communal research program. Such generalizing coherence, naturally enough, appears at the head of chapters and sections, such as at the beginning of section 9: "Electricians, after observing the great quantity of electrical matter with which clouds are charged during a thunderstorm, began to attend to the lesser quantities of it which might be contained in the common state of the atmosphere and the more usual effects of this great and general agent in nature" (1775, vol. 1: p. 421).

At times, the coherence of a communal program is identified through a fundamental problem being investigated, rather than through the phenomena discovered, as at the beginning of section 2: "One of the principal desiderata in the science of electricity is to ascertain wherein consists the distinction between those bodies which are conductors, and those which are non-conductors of the electrical fluid" (Priestley, 1775, vol. 1: p. 241). As we shall see in considering the later section on desiderata, the concept of research questions is to Priestley an important device for organizing current work and helps project the discipline into the future. And even in this historical part, such open questions can be used to make sense of and evaluate work already accomplished. The opening of section 2 quoted just above continues:

> All that has been done relating to this question, till the present time, amounts to nothing more than observations, how near these two classes of bodies approach one another; and before the period of which I am now treating, these generalizations were few, general and superficial. But I shall now present my reader with several very curious and accurate experiments which, though they do not give us entire satisfaction with respect to the great desideratum above mentioned; yet throw some light on the subject. (vol. 1: p. 241)

Overall, the book presents a progressive historical account of increasing knowledge, organized around the accretion of general principles that give order to the accumulated experience. This textual structure does require some chronological adjustments and conceptual relabelings. Priestley admits his imposition of an after-the-fact logic on events, such as commenting that his accounts of Hauksbee's experiments are "related not exactly in the order in which he published them, but according to their connection. This method I have chosen, as best adapted to give the most distinct view of the whole" (1775, vol. 1: p.19). Moreover, as mentioned previously, Priestley makes connections between earlier observations and later-developed general principles. Although such anachronistic use of generalizations may offend modern historiography, it does give order to prior empirical experience and establish broad empirical grounding to current generalizations. By creating an account of all prior experiences using current generalizations, yet remaining close to the original experimental particulars, Priestley demonstrates the general force of his generalizations. In a late chapter, Priestley even goes back to examine ancient Roman accounts of phenomena that only since the time of Franklin could have been considered electrical. This procedure, later articulated by Pierre Duhem as "saving the phenomenon," ensures that the history of experience is not ignored when new concepts are developed. This is also the historical standpoint of contemporary reviews of the literature that use current concepts and research questions to make sense of the previous work in the field.[9]

Despite the organizing power Priestley finds in his contemporary concepts, he does not discard those experiences that do not fit under any concept or contradict current categories. To make room for anomalous and aconceptual material, Priestley vigorously uses the category of *miscellaneous* both at the end of chapters and as full chapters, as in the "Miscellaneous Discoveries of Dr. Franklin and his Friends in America During the Same Period" and the final chapter of the historical section, "Miscellaneous Experiments and Discoveries Made Within This Period."

Because Priestley believes in the power of anomalies to reveal new truths, he carefully notes them. In introducing his discussion of tourmalin he remarks, "This period of my history furnishes an entirely new subject of electrical inquiries; which, if properly pursued, may throw great light upon the most general properties of electricity. This is the *Tourmalin:* though, it must be acknowledged, the experiments which have hitherto been made upon this fossil stand like exceptions to all that was before known of the subject" (1775, vol. 1: p. 367).

Codification and Access to Ordered Experience

In the collection, representation, and codification of all recorded experience of electrical phenomena, Priestley has made accessible and given order to the communal empirical experience. By rescuing from obscurity early and unread work and showing that work consistent with later work and contemporary concepts, he draws a wider range of participants and experience into the cooperative effort of coming to terms with nature. Furthermore, in making the previous work available, intelligible, and experienceable (vicariously or in actual practice) to his readers, Priestley enriches each person's experience and provides a common base of experience and knowledge for all new participants in the field. All electricians will now know and have contact with essentially the same range of experience, with whatever local additions they might have access to or create themselves. With the history of the field available and codified, and all participants knowing the same thing, work may then proceed more rapidly, efficiently, and cooperatively. Throughout the history Priestley had noted as admonitory examples just those instances where lack of access, ignorance of previous work, or lack of shared assumptions led to duplication of effort or unnecessary conflict.

The shared history Priestley presents has not reached conceptual or empirical closure. He indicates the open questions, the anomalies, and the incompletely understood phenomena. In the preface he promises to provide updates on future research (as is provided in the second edition and as a separately published pamphlet). Last-minute prepublication addenda were also included in both first and second editions. The third edition had only limited revisions, for Priestley promised to write a *Continuation to the History*—a promise never fulfilled. Even more significantly, the latter half of the book points to an open-ended future by establishing the shared basis for continued work and offering practical guidance for further experiments. Priestley presents the extensive history of the first half as only a necessary prologue to the ongoing practice of knowledge creation.

GENERAL PROPOSITIONS AND OBSERVABLE KNOWLEDGE: PART 2

The product of history, as Priestley tells it, is emergent principles which order the accumulated experiences. In part 2 of the book, Priestley abstracts these generalizations in a seven-page "series of propositions comprising all the general properties of electricity." These propositions describe the observable effects of electricity, rather than present ontological statements about the nature of electricity itself.

The propositions are largely cast in terms of generalized experimental events: for example, "It is the property of all kinds of electrics, that when they are rubbed by bodies differing from themselves (in roughness or smoothness chiefly) to attract light bodies of all kinds which are presented to them" (Priestley, 1775, vol. 1: p. 4). Accordingly, many statements begin with "if" clauses to indicate the generalized experimental conditions that may be experienced by all observers. "If an electric shock, or strong spark pass through or over the belly of a muscle, it forces it to contract as in a convulsion" (vol. 1: p. 9). Even the occasional existential statement is elaborated in generalized operational experimental terms: "Electricity and lightning are, in all respects, the same thing. Every effect of lightning may be imitated by electricity, and every experiment in electricity may be made with lightning, brought down from the clouds, by means of insulated pointed rods of metal" (vol. 2: p. 10).

The succinctness and generality of these claims is to Priestley a sign of the advance of knowledge: "For the more we know of any science, the greater number of particular propositions we are able to resolve into general ones" (1775, vol. 2: p. 2). Since these propositions are not a priori projections, but rather inductive generalizations, they compose an order created out of accumulated experience. The ability to find encompassing propositions of increasing generality indicates understanding of more powerful and fundamental principles of phenomena.

Nonetheless, the propositions presented by Priestley, although generally following a sequential expository order, are not tightly organized around a single account of the nature of electricity, although such an account might have led to even greater succinctness, as Priestley notes (1775, vol. 2: p. 2). They are largely disjunct statements about separately observable phenomena, with only a few logical connectives. Priestley is very careful to distinguish these general propositions, which may be separately observable by all electricians from any coherent account of what electricity might be, for in his time that was only a matter of theoretical speculation about unobservable matters.

By establishing a succinct codification of what is currently known, generally agreed to, and observable, Priestley clarifies the extent of shared knowledge. This brief, yet comprehensive, list allows for coordination of continuing work, recognition of novelties and anomalies in new observations, rapid socialization of neophyte electricians into the current state of knowledge, and easy reference. The list of propositions thus serves the functions of both the modern handbook and the modern textbook. Furthermore, by separating those statements which are generally agreed to be empirical truths from uncertain theories, Priestley allows for a differentiation of discussion in ensuing work. He does not propose a unitary system of knowledge, as Newton does where general theory appears inseparable from representation of empirical experiences, so that the theorizing is made invisible and denied (see Bazerman, 1988). Rather Priestley establishes agreement on the level at which all can agree and focuses debate on less certain matters. Thus he not only codifies the existing knowledge, but codifies the levels and manners of discussion. He provides literary technology first for coordination of areas of agreement by slowing down the communal ascent up the ladder of generalization (as Bacon cautions the individual researcher to do), and second for domestication of conflict by limiting the arena of disagreement.

HISTORICIZING THEORY: PART 3

Having historicized experience and discovery in the early parts of his work, Priestley historicizes theory in the third part. Theories, as Priestley presents them, historically precede knowledge. Theories, which he uses interchangeably with hypotheses, help frame experiments and lead to newly observed phenomena, but they themselves are not substantiated knowledge. When the hypothesized phenomenon is made observable through experiment, it passes out of the realm of theory into the realm of operational knowledge. As Priestley states in his introductory comments to part 3:

> Hypotheses...lead persons to try a variety of experiments, in order to ascertain them. In these experiments, new facts generally arise. These new facts serve to correct the hypothesis which gave occasion to them. The theory, thus corrected, serves to discover more new facts, which, as before, bring the theory still nearer to the truth. In this progressive state, or method of approximation, things continue; till, by degrees, we may hope that we shall have discovered all the facts, and have formed a perfect theory of them. By this perfect theory, I mean a system of propositions, accurately defining all the circumstances of every appearance, the separate effect of each circumstances, and the manner of its operation. (1775, vol. 2: pp. 15–16)

At the end of investigation, then, theory changes from a conjecture about causes to an empirically based operational account.

Theories, then, are useful but uncertain and historically bounded accounts. They are heuristic. Discussion of theories leads to difficulties only because investigators present their hypotheses as general truths and become too attached to them. Thus they do not allow the replacement or modification of theory in relation to new findings, nor do they admit new hypotheses that would serve as heuristic for new discoveries. Priestley found this attachment to speculative theories particularly rife within electricity because, "As the agent is invisible, every philosopher is at liberty to make it whatever he pleases, and to ascribe to such properties and powers as are most convenient for his purpose" (1775, vol. 2: p. 16).

In his own account of electrical theory, Priestley adopts several literary methods to identify the limited and transient utility of theories and to decrease his own and the reader's attachment to any particular theory. He describes the historical state of knowledge out of which each theory arises and which it is meant to account for, identifies the new findings that the theory led to, and finally presents the empirical results the theory could not adequately account for. Unlike the timeless presentation of general propositions, theories are given a specific time and place. Moreover, Priestley casts theories aside after they have played their role in the generation of empirical truths and have been made obsolete by further discoveries.

Priestley adopts this historical attitude even to theories viable in his own time, including his favored account: Franklin's theory of positive and negative electricity. Priestley discusses how Franklin's theory provides a satisfactory account of a number of phenomena, especially that of the Leyden jar—the phenomenon which the theory was first developed to explain. But he also discusses a number of phenomena for which the theory remains inadequate, such as the influence of points and the electrification of clouds. Furthermore, Priestley points out that the theory is in a state of flux, being subject to modification by a number of electricians. On the other hand, he finds that the ability of this theory to incorporate findings and ideas from previous theory very much in its favor, as it does not abandon collective experience. Finally, although Priestley ends this chapter with a panegyric to Franklin, the quality he praises most is Franklin's diffidence about his own theory and his just "sense of the nature, use and importance of hypotheses" (1775, vol. 2: p. 39), which attributes more importance to the facts produced than the general accounts.

The most significant feature of Priestley's presentation of theories is that the chapter on his favored theory is followed by an almost equally long chapter on a contending theory to which he also attributes great utility. Priestley comments, "I shall, notwithstanding the preference I have given to Dr. Franklin's theory, endeavour to represent [the theory of two electrical fluids] to as much advantage as possible, and even to do it more justice than has yet been done to it, even by Mr. Symmer himself" (1775, vol. 2: p. 41). After an extensive summary of the theory, Priestley points out certain phenomena for which this theory appears useful and plausible, in some instances providing less-tortured accounts than Franklin's single-fluid theory. Like Franklin's theory, Robert Symmer's theory offers no inconsistency, but lacks insight into certain phenomena. Priestley then modifies the theory to answer the chief objection made to it and comes to the conclusion that the theory is consistent with all available evidence. Priestley cites another electrician, Cigna, granting Franklin's theory the upper hand because of its overall greater simplicity, but no final strong judgments are made. The section ends with Priestley inviting readers to communicate "any other theory, not obviously contradicted by facts" (vol. 2: p. 52).

Thus, even while maintaining a favored theory, Priestley manages to distance himself from it and develop a dispassionate method for discussing and evaluating competing theories. Not only does his tone and literary method allow for modification and theory change, it separates the advocacy of theory from the discovery of facts, even while recognizing the dialectical connection between fact and theory. Priestley's mode of discussion diffuses the argumentative gap that results from theory differences, allows cooperative research on the level of general propositions, and offers an orderly procedure for discussing and evaluating theories. He offers a means for communal theory development and modification short of total replacement. Finally, by reducing the status of theory, Priestley reduces the stakes in theory wars.

"A GREAT DEAL STILL REMAINS TO BE DONE": PART 4

To Priestley, the codification (or gathering together and conceptual organization) of prior work only served to highlight the incompleteness of our communal electrical knowledge. In the first three parts—the history, the general propositions, and the theory—he is at pains to point out what issues are left open, what is unknown, and what is puzzling. One of the great dangers Priestley finds in the individual's adherence to a single theory is that the individual may feel that electricity has been solved and therefore find little motivation to extend researches. Such is his accusation against Nollet (1775, vol. 2: p. 25). Systematic codification, to the contrary, identifies specific areas needing investigation and unsolved research problems (or desiderata, as Priestley calls them).

To make these incompletenesses even more visible, and therefore to guide future work, Priestley gathers them together in the middle chapters of section 4, in the form of "queries and hints," following on the exhortation to continuing research of the opening chapter. These queries and hints are presented as lists of questions. Questions, even while they invite unknown answers, constrain the form of the answers. A series of questions can set an agenda for communal work and provide a framework for comparing competing answers.

Priestley's questions are set under various headings corresponding to phenomena identified and elaborated in the previous work. Under each heading, the opening questions tend to be the more fundamental questions, which are then elaborated more specifically in the following questions. For example, the "Queries and Hints Concerning Excitation" begins with a fundamental question of structure, moves to elaborating phenomena, and then specific experiments:

> What is the difference, in the eternal structure of electrics, that makes some of them excitable by friction, and others by heating and cooling?
> What have friction, heating, cooling, and the separation after close contact in common to them all? How do any of them contribute to excitation? And in what manner is one, or the other electricity produced by rubbers and electrics of different surfaces?
> Is not Mr. Aepinus's experiments of pressing two flat pieces of glass together, when one of them contracts a positive and the other a negative electricity, similar to the experiments of Mr. Wilcke concerning...?

By explicitly mentioning recent and ongoing work in relation to open questions, Priestley identifies common problems that can draw researchers into a common endeavor and sets an example of public discussion of unresolved work. Rather than leaving the workshop of the various researchers closed until the individuals are ready to present a claim substantiated by a public demonstration of a successful experiment, Priestley invites the entire community to share in open discussion, drawing hints from each other. Furthermore, he invites others to present their open questions, for

> Many persons can throw out hints, who have not leisure, or a proper apparatus for pursuing them: others have leisure, and a proper apparatus for making experiments, but are content with amusing themselves and their friends in diversifying the old appearances, for want of hints and views for finding new ones. By this means, therefore, every man might make the best use of his abilities for the common good. Some might strike out lights, and others pursue them; and philosophers might not only enjoy the pleasure of reflecting upon their own discoveries; but also upon the share they had contributed to the discoveries of others. (1775, vol. 2: pp. 60–61)

Publicly shared questions help coordinate the work of many hands and help further the incomplete ideas of many minds. As that work goes forward, questions are answered, discarded, or changed. Questions, like theories, are historically bounded. Priestley comments that his questions may likely soon appear "idle, frivolous, or extravegant" (1775, vol. 2: p. 58), and he makes a number of emendations in the second and third editions, eliminating some questions and adding new ones. Priestley clearly has the idea of a moving research front composed of evolving problems, and he wishes to establish a textual means of communally articulating the current state of questions.[10]

To Priestley, the advancement of electrical knowledge also depended on coordination with other forms of knowledge that bear on the investigations. The closing section of part 4 discusses the other branches of knowledge which electricians ought to become familiar with: in particular, the studies of chemistry, light and color, atmosphere, anatomy, mathematics, mechanics, and perspective. His inclusion of mechanics and perspective recognizes that research is an empirical and social practice. The philosopher who wishes to explore new phenomena must be able to construct new philosophical machines, and the philosopher who wishes to share his findings by publication must be able to draw precisely.

MECHANICAL COORDINATION: PART 5

Since Priestley recognized that the advance of electrical knowledge was dependent on the advance of machines, he felt the need to engage electricians in mechanical construction. In the historical first part of the book, he often pointed out the importance of machinery to specific discoveries. In the fifth part, Priestley summarizes the historical progress of machinery to codify and coordinate the state of the art. Machines provide the originators with access to new phenomena, but even more they provide the entire community of electricians access, for reproduction of machines allows reproduction of phenomena. Shared machinery makes possible shared experiences and cooperative investigation of the phenomena generated and displayed by the apparatus. Moreover, to maximize communal development, the best machinery and principles of construction should be made available. Finally, to coordinate work, you must coordinate apparatus, so that investigators are making claims concerning the same things.

Explanations of the principles of the machines and specific guidelines for the construction of the most effective machines are, therefore, important to a coordinated experimental science. Priestley devotes about 20 pages to the description of guidelines for the construction of electrostatic generators, metallic points, batteries of Leyden Phials, and electrometers. Just as there are open questions about electrical phenomena, there are open questions about the optimum designs: for example, "It has not yet been determined by electricians what kind of glass is the most fittest for electrical purposes" (1775, vol. 2: p. 91). Priestley then devotes another dozen pages to discussing the advantages and disadvantages of particular machines developed and used by various investigators. The mechanical descriptions are supported by illustrations.

As indicated by an advertisement accompanying the *Familiar History* of 1768, Priestley himself engaged in the construction and distribution of electrical machines. (For further discussion of the importance of mechanical practice in Priestley's work, see Schaffer, 1984.)

THE INCREASE OF EMPIRICAL EXPERIENCE: PARTS 6 AND 7

Access to machines does not guarantee successful, copious, and progressive results. People must use those machines and use them correctly. Experience has been the traditional teacher of successful experimental practice, but in parts 6 and 7, Priestley aims to expedite successful experience for neophytes, so that they will produce more results more rapidly.

Part 6 is devoted to "Practical Maxims for the Use of Electricians," describing the craft knowledge that Priestley has gained through his own experience. He hopes to make the path of young electricians less arduous, for "it is in the interest of science in general, that everything be made as easy and inviting as possible to beginners. It is this circumstance only that can increase the number of electricians, and it is from the increase of this number that we may most reasonably expect improvements in the science" (1775, vol. 2: p. 119). What follow are 15 pages of homely craft advice and warnings, such as "A little bees wax drawn over the surface of a tube will greatly increase its power" (vol. 2: p. 120). And "let no person imagine that, because he can handle the wires of a large battery without feeling any thing, that therefore he may safely touch the outside coating with one hand, while the other is upon them. I have more than once received shocks that I should not like to receive again" (vol. 2: p. 32). Such advice currently is conveyed in laboratory manuals and other training documents. But as Collins forcefully argues, much craft knowledge remains inarticulate and certainly unpublished, so it is learned, if at all, directly over the laboratory bench.

In this same spirit of introducing neophytes into experimentation and ultimately increasing the communal experience, Priestley offers in part 7 directions for performing "The Most Entertaining Experiments Performed by Electricity." He opens this part with an enthusiastic account of the delights and wonders of simple experiments: "What can seem more miraculous than to find, that a common glass phial or jar, should, after a little preparation (which, however, leaves no visible effect, whereby it could be distinguished from other phials or jars) be capable of giving a person such a violent sensation, as nothing else in nature can give and this shock attended with an explosion like thunder, and a flash like that of lightning?" (1775, vol. 2: p. 135). Moreover, Priestley encourages would-be experimenters by suggesting that these effects, although entertaining, are far from trivial: "So imperfectly are these strange appearances understood, that philosophers themselves cannot be too well acquainted with them.... It is possible that, in the most common appearances, some circumstance or other, which had not been attended to, may strike them; and that from thence may be reflected upon many other electrical appearances" (vol. 2: p. 136). Thirty pages of detailed instructions follow on how to carry out experiments involving explosions, shocks, flashes of light, ringing bells, dancing dolls, puppets with hair standing on end, and gunpowder. They are all described with much enthusiasm for their amusing qualities. Such descriptions now are reserved for children's introductions into experimental sciences of the "Scientific Tricks You Can Do" genre, with much of the same motivation: recruitment of the young into science. Priestley well understood that no communal research program will prosper without the personnel to carry it forward.

EXTENDING KNOWLEDGE: PART 8

In the last part of the book, Priestley presents his own experiments, developed out of his replications of experiments in the literature. In a sense he presents himself as the ideal reader of his account of the history and present state of electricity, using the text as a foundation for new activity.[11] Moreover, he presents his new work as an example of the dynamics of knowledge, that the history of natural philosophy is open-ended and that new work proceeds out of old. This message of continuing, coordinated work outweighs the specific findings that Priestley presents; he comments that his method of presentation is "less calculated to do an author honour as a philosopher; it will, probably, contribute more to make other persons philosophers, which is a thing of much more consequence to the public" (1775, vol. 2: p. 165).

Priestley's new experiments are arranged according to topics first raised in the historical chapters, continually referring to the literature and private correspondence with other electricians. He opens sequences of experiments with comments, such as he found certain of Beccaria's experiments not quite adequate (1775, vol. 2: p. 232), or that he had certain doubts about an aspect of Franklin's theory (vol. 2: p. 184), or that there was a matter of dispute among electricians (vol. 2: p. 201). Similarly Priestley indicates obtaining materials (e.g., vol. 2: p. 187) and ideas for experiments from other electricians (e.g., vol. 2: p. 187). In addition, he reports corresponding with them about procedures (vol. 2: p. 179). Finally, he compares his results with those of others (e.g., vol. 2: pp. 273–276). In such repeated cross-reference to the work of others, Priestley anticipates modern practice rather than following the practice of his contemporaries.

Yet, although Priestley embeds his accounts in the literature and indicates his constant interaction with the rest of the community, he does not present his work as necessarily occupying a fixed or stable place in the literature, as is often the case in modern articles, where one uses the literature as a matrix around the new claim to assert its centrality, meaning, and solidity (see Dudley-Evans, 1986; Swales & Najjar, 1987). Priestley presents his work as part of an ongoing and unsettled process. The starting place in the literature does not fix where the investigation ends nor is the opening hypothesis (derived from the literature or by analogy with other reported phenomena) necessarily the final one confirmed or denied. He presents his research as a mode of dialectical discovery which sometimes ends in a strong claim, but more often does not. Sequences of experiments are left with unresolved questions or invitations to others to continue the work.

Priestley called this discovery path form of presentation an analytical argument and believed it was more deeply persuasive than the prooflike argument which supports a claim asserted at the opening of the essay, which he called the synthetic argument. In analysis you could carry your reasonable audience down the same path of experience and reasoning that led you to your conclusions, rather than coercing belief through a constraining set of arguments. Intellectual coordination would come from a more complete sharing of the research experience (1775, vol. 2; see also Priestley, 1777/1965; Lawson, 1954; Moran, 1984).

Additionally, Priestley felt that this analytical method would aid others in their investigations by indicating all the false leads and mistakes. Others could avoid these mistakes, find meanings in details that did not seem so significant to the original investigator, and take advantage of the full range of thinking. Moreover, such naturalistic accounts would demystify the research process for neophytes so that they would be less intimidated to take up their own researches.[12] Priestley, in fact, criticizes Newton for hiding his thinking process, with the result that we hold the man too greatly in awe to follow in his footsteps:

> If a man ascend to the top of a building by the help of a common ladder, but cut away most of the steps after he has done with them, leaving only every ninth or tenth step; the view of the ladder, in the condition in which he has been pleased to exhibit it, gives us a prodigious, but an unjust idea of the man who could have made use of it. But if he had intended that any body should follow him, he should have left the ladder as he constructed it, or perhaps as he found it, for it might have been a mere accident that threw it in his way. (1775, vol 2: pp. 67–68)

Priestley's ladders, as presented in over 200 pages, are many, but few get far off the ground. He presents the different lines of investigation sometimes as having strict logical direction, with each hypothesis or problem suggesting a new experiment and the results suggesting a new problem or hypothesis. Yet he rarely concludes in any statement of great certainty or generality. At other times the sequencing of experiments seems weaker as accidental observations set the sequence in motion with no strong direction. Priestley ends such sequences just by varying the experiment to see if he can find any leads. In this vein, he comments at the beginning of the sequence of experiments on circular spots:

> In the courses of experiments with which I shall present the reader in this and the following two sections, I can pretend to no sort of merit. I was unavoidably led to them in the use of a very great force of electricity. The first appearance was, in all the cases, perfectly accidental, and engaged me to pursue the train; and the results are so far from favouring any particular theory or hypothesis of my own, that I cannot perfectly reconcile many of the various phenomena of any hypothesis. (1775, vol 2: p. 260)

This sharing of wide-ranging, but imperfectly accounted for, experience leads to a diffuse presentation, where few forests emerge from many trees. Priestley here does not even have the ordering potential of retrospective categories, as he did in the historical account. As I have noted elsewhere (Bazerman, 1988, chap. 3), this kind of discovery account gained popularity in *Philosophical Transactions* at just about this time (possibly led by Priestley's enthusiasm for it), but it did not last until the turn of that century. Although it allows for sharing of open-ended work and recognition that the truths of natural phenomena cannot be directly and self-evidently read from individual experiments, it seems to have been too equivocal in its message for the codification necessary for the coordination of continuing work. A century earlier Newton had also used a discovery narrative in his "New Theory of Light and Colours," but he had found it insufficiently persuasive to forestall serious controversy. In order to compel assent Newton developed the prooflike form of argument which Priestley criticized him for (Bazerman, chap. 4).

Modern scientific articles have found a solution for codifying the research which combines compelling assertion in the fashion of Newton with a recognition of the evolving character of the literature and the communal investigation in the fashion of Priestley. By asserting claims within a constructed matrix of the literature, modern articles attempt a kind of rolling codification. They typically pretend their claims are already accepted and integrated into a literature reconstructed in the article introduction (see Berkenkotter, Huckin, & Ackerman, 1991). The body of the argument then appears to act as an irrefutable inductive proof, although the value and meaning of both the literature and the current investigation may be far from settled within the knowledge-validating research community. In this manner, each new article takes part in the sorting out of knowledge claims even as they are proposed.[13]

CONCLUSION: THE DILEMMA OF LARGE COOPERATIVE ENDEAVORS

In forging a way of talking about science that would coordinate the work and experience of many, while not holding them accountable to any a priori or individually conceived theory, Priestley was attempting to create a broad-based science open to all who wished to respect and extend the common experience. This project was founded on his radical theological, philosophical, and political beliefs. He believed democratic participation and open-ended negotiation of phenomena would lead to discovering the true accounts of nature, encompassing the experience of all humankind. Priestley understood that such an endeavor must be coordinated on many levels, from experimental findings to machine construction to research problems. But he desired that codification emerge only from the shared wisdom, experience, and responsible negotiation of humanity, excluding none of the verifiable variety of life. Priestley's organized philosophy and sociology of science relied on his developing an appropriate rhetoric of science that would facilitate cooperation and coordination of current communal empirical investigation, while respecting the experience embodied in the history of science.

Priestley was partly successful in creating rhetorical means to assert codification, yet still keeping the door open to the full range of experience. To avoid the cultural amnesia of codification of history, Priestley stays close to the literature, which he attempts to reproduce with some historical sensitivity and copious detail, even where it does not fit into his contemporary categories. Priestley treats codifications of present activities of theorizing, experimenting, and machine construction as useful but temporary accounts, to be rewritten as events progress. Codifying the future, however, is trickier; it can shut down the open-ended processes of experience and discovery by enforcing a closed system of bureaucratic definition of what can and should be done. Yet not codifying the future strongly enough leads to an uncoordinated proliferation of actions of little meaning—an open end that unties the threads of the past drawn together in the present moment. This is a rhetorical problem that Priestley had not yet solved, although many of his rhetorical techniques for dealing with past, present, and future have been used in scientific writing since his time.

Perhaps the most important consequence of the rhetorical dynamics put in motion by Priestley is a form of discourse directed at an inward-facing community concerned with shared research problems and developing a communal experience. He creates textual means for researchers to look toward each other to create a common knowledge. In support of this prototype discipline, Priestley also offers textual means of recruiting and socializing new participants into the communal project.

What Priestley perhaps undervalues, however, are individual assertion and competition within the coordinated communal activity. In eschewing individual glory in the name of the communal advance, in letting all into his unsettled workshop, and in refusing to pretend to certainty in the face of historical flux, Priestley has inadequately allowed for the hypothesizing force of science that has allowed individuals to assert bold leaps of knowledge and then to await to see if the world lives up to their educated intuitions. Priestley creates a machinery for benign cooperation, but that machine also has seemed to need the drive of agonistic struggle, to help force claims up the ladder of generalization and power. Despite Priestley's amiable sociability, science has maintained an important role for aggressive assertion of theories, embattled competition, and Nobel prizes. In fact, these have become part of science's sociability.

Nonetheless, the coordinating mechanisms of the kind Priestley advances are precisely what make the agonism more than a war of all against all. These mechanisms have given order to the accumulating corporate experience and have provided common assumptions, comparable terms, and similar empirical procedures with which to advance the shared work. Priestleyan codification has created an evolving and contingent—but at any moment predominantly stable and communally recognizable—playing field, upon which focused and fruitful struggle can take place.[14]

NOTES

[1] Rachel Laudan's current investigation of early histories of science may reveal more about the textual tradition out of which Priestley was writing (private communication).

[2] J. L. Heilborn's (1979) standard modern history of electricity to 1800 affords fewer than 60 pages to 17th-century developments, but devotes about 180 pages to the two thirds of the 18th century that preceded Priestley's publication.

³ Unexplored in this essay is how Priestley's vision of history and the historical progress of knowledge fits in with developing attitudes of enlightenment toward history and the accumulated wisdom of the human race. Encyclopedism in its direct French and modified British forms are of course relevant here. Also unexplored are the roots of Priestley's ideas of cooperative communities, which may have their foundations in radical Puritanism of the 17th century.

⁴ Priestley's psychology is explicitly Hartleyian associationist in both *A Course of Lectures on Oratory and Criticism* and *The History and Present State of Electricity*. See also the introduction to the modern edition of the former by Vincent Bevilacqua and Richard Murphy. In light of Priestley's view of empirical natural philosophy, it is important to emphasize the role Priestley sees for experiences in forming associations. Associations are for him not just arbitrary connections among mental representations. Progress (and thereby fulfillment of the divine plan) comes for Priestley from the incorporation of empirical experience of the world into the set of mental associations and the readjustment of those associations so as to be harmonious with and useful for the ordering of the experience. Increasing empirical experience becomes, for him, a moral duty. This is a curious theological variant on Fleck's observation that modern science is characterized by the active pursuit of passive constraints.

⁵ Laboucheix reconciles Priestley's radical politics and progressivism with his theological and physical determinism by examining Priestley's dynamic view of materialism, necessity, and decision making, which creates the opportunity for human intelligence to understand and abstract the laws of nature, so that humans may accommodate themselves and live in harmony with those principles that determine their existence.

⁶ Priestley, by associating natural philosophy with a life of action to be pursued by males in fulfillment of a divine plan, and then by framing the study of natural philosophy within a corresponding male educational system, furthers the gender coding of human action in Western culture. There are consequences here for genderization of rhetoric as well as the more general genderization of society, but both these issues must remain beyond the scope of this essay. I would, however, point out that the cooperative technology fostered by Priestley differs significantly from the forms of cooperation involving acquiescence and subordination often stereotypically gender coded as female. Priestley also notes certain characterological correlates of the philosophic activity he promotes for men: "Nor is the cultivation of piety useful to us only as men, it is even more useful to us as philosophers: and as true philosophy tends to promote piety, so a generous and manly piety is reciprocally, subservient to the purposes of philosophy" (Priestley, 1775, vol. 1: p. xxiv).

⁷ Page references throughout will be to the third edition of 1775 (in two octavo volumes), which is available in a modern reproduction (New York: Johnson Reprint, 1966), instead of the first edition of 1767 (a single-quarto volume), available only in the original. The texts of the two editions (and the intermediate second, 1769) are in most details the same, except for updated information, presented largely through whole paragraph insertions describing more recent work and a few deletions of research questions which have been superseded. The later fourth (1775) and fifth editions (1794) follow the third in all respects. A French translation in three volumes (1771) follows the first edition, and a single-volume German translation follows the second.

⁸ Hoecker examines in greater depth the tension between Priestley's historical sensitivity and his progressive vision of divinely inspired historical development.

⁹ However, Priestley's history does differ from modern reviews of the literature in its comprehensiveness of coverage, historical extensiveness, and detail of reportage. In part this

may be because modern findings usually occur within highly codified systems of knowledge, practice, and questions. Thus new findings usually come presorted into categories, as elaborated in introductory review sections; only novel, unexpected, or anomalous work stands out and calls for attention. Otherwise most findings simply confirm or elaborate the already codified system. Reviews of literature necessarily focus on those unusual details that raise questions, and are selective about the many reports that only add "more gory details." Only revolutionary new claims need to go back to examine the entire file of gory details to reinterpret them consistent with the novel concepts and new questions, and even the reinterpretation may be carried on through translation of large groups of material under general headings. Priestley, on the other hand, was creating the codification which made sense of the extensive history. He was first putting the material into conceptual categories, although the concepts had been emerging through the entire period he examines. On the modern review article, see Myers, 1991, chap. 2.

[10] In listing questions Priestley varies a practice used a century earlier in the early *Philosophical Transactions* and then early in the century by Newton in the *Opticks*. In the early *Philosophical Transactions*, however, these questions were aimed at gaining specific data from world travelers who could report back on life forms, geologic and astronomical phenomena, and cultural knowledge from the far corners of the earth. The list of questions for travelers were not set as research problems so much as specific informational requests. The respondents were asked to cooperate in providing useful information, but they were not invited to participate in a research front. Newton, on the other hand, used his queries as ways of asserting his beliefs on topics about which he thought he had certain answers but about which he did not have compelling arguments. To Newton there was no research front, only settled issues, imperfectly proven. The questions are often in the coercive form of "Is it not true that..." The form of cooperation he sought (and often obtained among the Newtonians) was acquiescence to his suggestions. Priestley here, however, phrases his questions as genuinely open invitations to cooperative investigation.

[11] The process of writing the book was indeed his apprenticeship into the community of electricians, introducing him to the literature, machines, experiments, and active investigators. For a discussion of the relation between reading, activity, and writing in the formation of working scientists' knowledge and plans, see Bazerman, 1988, chap. 8.

[12] In this impulse Priestley anticipates Medawar by two centuries.

[13] Huckin has noticed in some fields an increased emphasis on the news value of articles, at the expense of the empirical argument. This carries the sometimes useful fiction of rolling codification one step further. If the pretended codification leads the actual evaluation by too great a distance, however, consequences may go beyond problems in examining the claims of each article to a disorganization of the communal knowledge which allows coordination of work.

[14] In Fujimura's (1987) terms, by creating means to allow alignment of disciplinary and individual work, Priestley has made possible the identification of doable problems within modern science, against which the individual may assess his or her own resources.

REFERENCES

Bacon, F. (1627). *The new Atlantis*. London.
Bazerman, C. (1988). *Shaping written knowledge: The genre and activity of the experimental article in science*. Madison, WI: University of Wisconsin Press.

Berkenkotter, C., Huckin, T., & Ackerman, J. (1991). Context and socially constructed texts: The initiation of a graduate student into a writing research community. In C. Bazerman & J. Paradis (Eds.), *Textual dynamics of the professions: Historical and contemporary studies in writing in professional communities* (pp. 191–215). Madison, WI: University of Wisconsin Press.

Callon, M., Law, J., Rip, A. (1986). *Mapping the dynamics of science and technology.* London: Macmillan.

Collins, H. (1985). *Changing order: Replication and induction in scientific practice.* Beverly Hills, CA: Sage.

Collins, H., & Pinch, T. (1982). *Frames of meaning: The social construction of extraordinary science.* London: Routledge and Kegan Paul.

Crosland, M. (1983). A practical perspective on Joseph Priestley as a pneumatic chemist. *British Journal for the History of Science, 16,* 223–238.

Crosland, M. (1987). The image of science as a threat: Burke versus Priestley and the "Philosophic Revolution." *British Journal for the History of Science, 20,* 277–307.

Desaguliers, J. T. (1742). *A dissertation concerning electricity.* London.

Dudley-Evans, T. (1986). Genre analysis: An investigation of the introduction and discussion sections of MSc dissertations. In M. Coulthard (Ed.), *Talking about text* (pp. 128–145). Birmingham, England: English Language Research.

Franklin, B. (1754). *Experiments and observations on electricity.* London.

Fruchtman, J. (1983). *The apocalyptic politics of Richard Price and Joseph Priestley: A study in late eighteenth-century English republican millennialism.* Philadelphia: American Philosophical Society.

Fujimura, J. H. (1987). Constructing "do-able" problems in cancer research: Articulating alignment. *Social Studies of Science, 17,* 257–294.

Fulton, J. F. (1937). *Works of Joseph Priestley, 1733–1804: Preliminary short title list.* New Haven, CT: Yale University School of Medicine.

Funkhouser, H. G. (1937). Historical development of the graphical representation of statistical data. *Osiris, 3,* 269–404.

Garvey, W. D. (1979). *Communication: The essence of science.* Oxford, England: Pergamon Press.

Gilbert, W. (1893). *De magnete* (P. F. Mottelay, Trans.). New York: J. Wiley. (Original work published 1600)

Gralath, D. (1747–1757). "Geschicte der electicitaet" including "electrische bibliothek." *Versuche und Abhandlung, 1,* 175–304; *2,* 355–460; *3,* 492–556.

Heibert, I. A. (1980). The integration of revealed religion and scientific materialism. In *Joseph Priestley scientist, theologian, and metaphysician.* Lewisburg, PA: Bucknell University Press.

Heilborn, J. L. (1979). *Electricity in the 17th and 18th centuries.* Berkeley, CA: University of California Press.

Hoecker, J. J. (1979). Joseph Priestley as a historian and the idea of progress. *Price-Priestley Newsletter, 3,* 29–40.

Huckin, T. (1987). *Surprise value in scientific discourse.* Paper presented at the Conference on College Composition and Communication, Atlanta, GA.

Knorr-Cetina, K. D. (1981). *The manufacture of knowledge: An essay on the constructivist and contextual nature of science.* Oxford, England: Pergamon Press.

Kramnick, I. (1986). Eighteenth-century science and radical social theory: the case of Joseph Priestley's scientific liberalism. *Journal of British Studies, 2,* 1–30.

Laboucheix, H. (1977). Chemistry, materialism, and theology in the work of Joseph Priestley. *Price-Priestley Newsletter, 1,* 31–48.

Latour, B. (1987). *Science in action.* Milton Keynes, UK: Open University Press.

Latour, B., & Woolgar, S. (1979). *Laboratory life: The social construction of scientific facts.* Beverly Hills, CA: Sage.

Lawson, C. (1954). Joseph Priestley and the process of cultural evolution. *Science Education, 38,* 267–276.

McEvoy, J. G. (1979). Electricity, knowledge, and the nature of progress in Priestley's thought. *British Journal for the History of Science, 12,* 1–30.

McEvoy, J. G. (1984). Joseph Priestley, scientist, philosopher, and divine. *Proceedings of the American Philosophical Society, 128,* 193–199.

Merton, R. K. (1973). *The sociology of science.* Chicago: University of Chicago Press.

Moran, M. G. (1984). Joseph Priestley, William Duncan and analytic arrangement in 18th century discourse. *Journal of Technical Writing and Communication, 14,* 207–215.

Myers, G. (1985). Texts as knowledge claims: The social construction of two biology articles. *Social Studies of Science, 15,* 595–630.

Myers, G. (1991). Stories and styles in two molecular biology review articles. In C. Bazerman & J. Paradis (Eds.), *Textual dynamics of the professions: Historical and contemporary studies in writing in professional communities* (pp. 45–75). Madison, WI: University of Wisconsin Press.

Newton, I. (1704). *Opticks.* London.

Philosophical Transactions of the Royal Society of London. (1665). London.

Priestley, J. (1762). *A course of lectures on the theory of language and universal grammar.* Warrington, UK.

Priestley, J. (1765). *A chart of biography.* Warrington, UK.

Priestley, J. (1765). *An essay on a course of liberal education for civil and active life.* London.

Priestley, J. (1767–1794). *The history and present state of electricity* (Vols. 1–2). London.

Priestley, J. (1768). *Familiar introduction to electricity.* London.

Priestley, J. (1769). *A chart of history.* London.

Priestley, J. (1788). *Lectures on history and general policy.* Birmingham, England.

Priestley, J. (1962). *Selections from his writings* (I. V. Brown, Ed.). University Park, PA: Pennsylvania State University Press.

Priestley, J. (1965). *A course of lectures on oratory and criticism* (V. M. Bevilacqua & R. Murphy, Eds.). Carbondale, IL: University of Southern Illinois Press. (Original work published 1777)

Priestley, J. (1970). *Autobiography.* Teaneck, NJ: Farleigh Dickinson University Press.

Rudwick, M. (1985). *The great devonian controversy.* Chicago: University of Chicago Press.

Schaffer, S. (1984). Priestley's questions: An historiographic survey. *History of Science, 22,* 151–183.

Schofield, R. E. (1966). *A scientific autobiography of Joseph Priestley: Selected scientific correspondence.* Cambridge, MA: MIT Press.

Swales, J., & Najjar, H. (1987). The writing of research article introductions. *Written Communication, 4,* 175–191.

Ziman, J. (1968). *Public knowledge.* Cambridge, England: Cambridge University Press.

2

Landmark Essay: Francis Bacon and the Historiography of Scientific Rhetoric

James P. Zappen
Rensselaer Polytechnic Institute

Throughout the 20th century, historians of science and scientific rhetoric have turned regularly to Francis Bacon and the 17th century in search of models or precedents for our own time. These historians have upheld several different, even incompatible and conflicting, views of science and scientific rhetoric and have found in Bacon a precedent for each of these views: positivistic science and the plain style; institutionalized science and a more complex and highly figured style; and, most recently, democratic science, which, I shall argue, also has a complement in the plain style.

These differences in point of view are both historical and historiographic. On the one hand, these differences reflect the richness of Bacon's texts, which helps to explain why differences in interpretation are possible. On the other hand, they also reflect the perceptions of individual historians, which help to explain why different interpretations appear at different times; why one interpretation appears rather than another; and why they appear at all. Insofar as these differences are historiographic, they reflect individual historians' perceptions of what science and scientific rhetoric might be, or ought to be. Thus, they are differences of ideology, the "particular arrangement of economic, social, and political conditions" of both the subject of history and the historian (Berlin, qtd. in Octalog, 1988, p. 11). They are differences in historians' perceptions of "the good of the community" (Murphy, qtd. in Octalog, pp. 5–6), its nature and definition, whether it is a scientific or a social community, whether it is good to serve the one or the other or both.

From a perspective both historical and historiographic, the historians of science and scientific rhetoric to whom I have alluded are important not only because they provide a range of possible interpretations of Bacon and the 17th century (and so purport to be descriptive), but also because they articulate different perceptions of how science and scientific rhetoric might serve the good of the community, whether scientific or social (and so seek to be normative as well). In this paper, I review three 20th-century interpretations of Bacon's science and his rhetoric: positivistic science and the plain style; institutionalized science and its more highly figured style; and democratic science. I then present my own interpretation of Bacon's rhetoric as a plain style suitable for general participation in a particular kind of democratic science and show how this rhetoric was adopted by the Puritan reformers of the 1640s and 1650s. I conclude that each of these interpretations, including my own, reflects a different ideology, a different perception of the good of the scientific and the social community, an alternative vision of what science and scientific rhetoric might and ought to be.

POSITIVISTIC SCIENCE AND THE PLAIN STYLE

Historians of science and scientific rhetoric from the 1920s through the 1950s found in Bacon a precedent to both positivistic science and the plain style, and they have upheld his positivistic view of science. Burtt notes the pervasive influence of Bacon's positivistic science—his "conception of science as an exalted co-operative enterprise, his empirical stress on the necessity and cogency of sensible experiments, his distrust of hypothesis and general analysis of inductive procedure"—in the mid-17th century (1954, pp. 125–126). Willey claims that this reading of Bacon was stereotypical throughout the 18th and 19th centuries and that even early 20th-century readers still feel the "rightness" of Bacon's "continually dwelling among things" (1953, pp. 32–34). Haydn similarly cites Bacon as the most famous of the 17th-century empiricists committed to "the conviction that the world of empirical reality contains 'things as they are'" (1950, pp. 253–255).[1]

Historians of scientific rhetoric during this same period assumed this positivistic view of science and affirmed the suitability of Bacon's rhetoric, in particular his theory of the plain style, for this kind of science. These historians found more or less a direct relationship between Bacon and later attempts by members of the Royal Society of London to establish the plain style as the standard in scientific prose. Croll identifies Bacon with "the growth of scientific and positive rationalism" in the 16th and 17th centuries and with the anti-Ciceronian (Attic or Stoic) prose style, which set in opposition to the Ciceronian ornate style the Aristotelian ideals of clearness, brevity, and appropriateness ("Attic Prose," 1969, pp. 68–72, 82–85; "Muret," 1969, pp. 110–112). Croll argues that Bacon's theory of style was not a direct precursor of the Royal Society's practice of the plain style, but was related to it through their common origin in Aristotle's *Rhetoric* ("Attic Prose," pp. 75–77). Jones identifies Bacon with the method of observation and experiment characteristic of positivistic science and with the "naked" or plain style that was the "stylistic standard" for this kind of science (1951a, pp. 76–78; 1951b, pp. 143–145). Unlike Croll, he distinguishes the anti-Ciceronian style from the plain style and argues that Bacon rejected the former in favor of the latter and so may have directly influenced the development of the plain style (1951a, pp. 104–110). Williamson distinguishes a variety of anti-Ciceronian (Senecan) prose styles and, like Croll, associates Bacon most closely with the Stoic style, which "compromised least with rhetoric" (1951, pp. 150–151, 170–185). He argues that Bacon anticipates most of the Royal Society's requirements for a plain style suitable for experimental science (pp. 178, 289–294).

INSTITUTIONALIZED SCIENCE AND THE FIGURED STYLE

Historians of science and scientific rhetoric from the 1960s to the present have rejected this view of Bacon as a precedent to positivistic science and the plain style and have found in him instead a precedent to institutionalized science and its more complex and highly figured style. Historians of science in the 1960s turned attention from Bacon's role in the development of positivistic science and emphasized instead his role in the development of scientific communities—modern scientific institutions, their disciplines or fields of inquiry, represented by academic departments and professional associations. Kuhn, probably the most influential figure during this period, rejects Baconian "histories" as facts devoid of theory that have little to do with the development of the "paradigms" that govern scientific research (1970, p. 16). He argues that "the community structure of science," its "shared commitments" and "shared examples" rather than its facts, constitutes these paradigms (pp. 176, 186–187). Other historians have emphasized Bacon's positive role in the development of scientific communities or institutions. Boas credits Bacon and the 17th century with the development of both the method of observation and experiment and the notion of corporate science, both "organisation of method, and organisation of men" (1966, pp. 238–239, 250–254). Purver presses this view further when she argues that Bacon did not only advocate induction and experiment, but rather a union of the empirical and rational faculties within one person and cooperation between persons—"a corporate approach" to science, "living sciences" through successive generations, funded by "some public person or large institution" (1967, pp. 34–35, 51–53). In this corporate approach, she finds "the real significance of Bacon's plan," the magnitude of which "cannot be too much emphasized" (p. 52).[2]

Historians of scientific rhetoric in the 1970s and 1980s have similarly emphasized the communal or institutional structure of modern science and have found in Bacon a precedent to a more complex and highly figured style that they believe is more suitable for this kind of science. Even Jones had noted that Bacon's theory of style did not match his practice (1951a, pp. 76–78). And numerous historians subsequently have observed that Bacon's theory of style is actually complex, that it embraces several methods of presentation and several styles, not just the plain style.[3] These historians emphasize the complexity of Bacon's theory of style and the suitability of each of several methods or styles for different audiences and purposes.

These and other historians explain and defend this more highly figured style on grounds that this style is suitable for communal or institutionalized science as it is or as it ought to be. Stephens argues that Bacon explored several styles in an attempt "to seduce the intellectual community into the movement on behalf of the advancement of learning" (1975, pp. 1–3, 14–15). Lipson rejects the notion that Bacon is the "father of the plain English prose style" and, like Stephens, argues that he developed a theory of style that encompasses several methods or styles designed for use in "specialized, knowledge-seeking fields" (1985, pp. 143–144). Lipson suggests that this theory may provide a "ballast" for dealing with scientists who still cling to a positivistic view of science (pp. 152–153). Halloran and Bradford observe that Bacon's style, in practice, was highly figured and argue that this style is still in use in certain fields of science and is potentially useful "in solving the rhetorical problems of a given scientific or technical field" (1984, pp. 180–181, 191–192).

DEMOCRATIC SCIENCE AS AN IDEOLOGY

Historians of several ilks, including historians and philosophers of science and technology, feminist critics of science, and contemporary Bacon scholars, have challenged this view of institutionalized science and, like the more traditional historians, have sought a precedent in Bacon and have found in him an alternative ideology of science that they believe is more democratic and humanitarian than institutionalized science. These historians have sought to extend access to science and technology to social groups traditionally excluded and to foster greater responsiveness by science to social needs. In this way, they have sought to reform science from within, thus creating a closer identification of the scientific with the social community. Specifically, these historians have questioned the attitudes usually attributed to Bacon and commonly associated with institutionalized science: its power, its domination over nature, and its control by a scientific and bureaucratic elite. Instead, they have found in Bacon attitudes and beliefs that they associate with their own democratic view of science: an emphasis upon the utility of science for human life, respect for nature, and a belief in democratic participation in science and its applications.

Numerous historians and philosophers of science and technology have traced to Bacon the origins of the ideology of institutionalized science and have called into question the attitudes associated with this ideology. In his summary of this work, Winner argues that institutionalized science owes to Bacon its goals of "the delivery of power" through "the domination of nature" and its conception of a ruling class of "scientists, technicians, technologists, industrial managers, bureaucrats, and the like" (1977, pp. 24–25, 115–117, 148–149). Other historians and philosophers of science and technology have challenged this view of Bacon and have found in him instead a precedent to a democratic and humanitarian view of science. Pacey, for example, acknowledges Bacon's goals of power and domination over nature, but points out that Bacon also emphasizes the utility of technology for human life: "compassion and discipline in the use of knowledge," Christian charity construed as "practical action for relieving the suffering of individuals," and works of compassion directed to "the benefit and use of life" (1983, pp. 87–88, 114–115, 172–173). Pacey claims that contemporary experts who control technology ignore the "user sphere," and he recommends a form of "interactive innovation" designed to achieve "democratic" rather than "expert" control of technology (pp. 50–51, 142, 157).

Feminist critics of science and contemporary Bacon scholars express a similar point of view. Keller notes that it was Bacon "who first and most vividly articulated the equation between scientific knowledge and power, who identified the aims of science as the control and domination of nature" (1985, p. 33). But Keller also argues that Bacon's view of science finds expression in "the sexual dialectic implicit in his metaphors," a dialectic that balances power with the humane use of power, domination over nature with service and obedience to nature: "Science is to be aggressive yet responsive, powerful yet benign, masterful yet subservient, shrewd yet innocent" (pp. 34, 37). Harding challenges the traditional view of control of science by a scientific and bureaucratic elite and finds in Bacon and in the Puritan reform movement of the 1640s and 1650s a precedent for a more democratic and egalitarian science (1986, pp. 219–222). Quoting from Van den Daele's summary of Webster's research on the Puritan reform movement, Harding argues that "Bacon's vision for the advancement of learning was entirely consistent 'with the democratic, participatory impulse of the Puritan era'" (p. 220). Like Keller, Harding acknowledges the presence of metaphors in science, but she cautions that metaphors, however useful in scientific explanation, appear in time to be literally true and so tend to fix in the popular imagination stereotypes such as Newton's mechanical conception of nature or androcentric views of science (pp. 237–239).

So also Weinberger argues that the "problem of technology" is usually supposed to be the possession of "the very means that grant us power" and "our impious mastery over nature," but that the real "problem of technology" is political: the right that all share equally to partake of "the promise of technology," a right complicated not by scarcity but by abundance (1985, pp. 17–18, 330–332). Whitney finds in Bacon a concern for the "technological application" and "utility" of science; a vitalist conception of nature as "natura naturans," "nature-engendering nature"; and a program of "radically nonauthoritarian, socialist political engagement" that is necessarily pluralistic (1986, pp. 122–125). He suggests that such a program calls for "a vision of rational, universalized discourse" such as Habermas' theory of communication.

DEMOCRATIC SCIENCE IN BACON

Consistent with this alternative ideology of science, my interpretation of Bacon's rhetoric suggests that the plain style might be construed not as a vehicle for either positivistic or institutionalized science, but rather as a vehicle suitable for general participation in democratic science and its applications as Bacon envisioned it. Specifically, my interpretation suggests that Bacon's scientific method was not one but several methods, his theory of style likewise embracing several methods and several styles, the plain style reserved for that part of his method designed for general participation in science.

Bacon's scientific method was not one but several methods, or several stages in a single method. As Bacon describes it in *The Great Instauration* and *Novum organum*, the scientific method has several parts or stages, each directed to a different faculty, sense, memory, or reason. In *The Great Instauration*, the scientific method is a component in Bacon's plan for the advancement of learning in his time and has two parts among the six in this plan (Anderson, 1971). The six parts are "The Divisions of the Sciences"; "The New Organon; or Directions concerning the Interpretation of Nature"; "The Phenomena of the Universe; or a Natural and Experimental History for the foundation of Philosophy"; "The Ladder of the Intellect"; "The Forerunners; or Anticipations of the New Philosophy"; and "The New Philosophy; or Active Science" (vol. 8: pp. 38–54). The two parts of the scientific method are "The New Organon," Bacon's new inductive and experimental method, set forth in *Novum organum*; and "The Phenomena of the Universe," the natural and experimental histories, the facts upon which the method will operate, set forth in *Parasceve*, an introduction to the histories, and in numerous histories either planned or completed. The only other part that Bacon completed during his lifetime is "The Divisions of the Sciences," a classification and summary of knowledge in his time, which appears in *The Advancement of Learning* and its Latin version, *De augmentis scientiarum*. The last three parts were to present results, albeit tentative, of the operation of the scientific method. "The Ladder of the Intellect" and "The New Philosophy" were to present results based upon induction and so necessarily awaited the completion of the natural and experimental histories. "The Forerunners" was to be based upon Bacon's own experiments and so was to be necessarily provisional.

In *Novum organun*, the scientific method has three stages, which are directed to the faculties of sense, memory, or reason. These three stages work together to join the empirical or sensible and the rational faculties (Purver, 1967; Wallace, 1943). The three stages are a "Natural and Experimental History," "Tables and Arrangements of Instances," and "Induction," the first and third stages corresponding to the second and third parts of Bacon's plan for the advancement of learning, though in reverse order. As Bacon explains them, the three stages are three "ministrations": a ministration to the sense, which is the experimental and natural history; a ministration to the memory, the tables and arrangements of instances; and a ministration to the reason (or understanding), the inductive method.[4] The ministration to the memory joins the empirical faculty (sense) to the rational faculty (reason) by ordering the experimental and natural histories in tables and arrangements so that the inductive method can interpret them. In practice, the tables and arrangements of instances do not appear independently but do appear in Bacon's description of the inductive method in the second book of *Novum organum.*

IMAGINATIVE AND PLAIN STYLES IN BACON

Bacon's theory of style provided several methods of presentation and at least two styles, an imaginative style and the plain style, suitable for addressing different purposes and audiences and different faculties. In this context, the plain style was not a general prescription applicable to all scientific communication but a style peculiarly suitable for the writing of natural and experimental histories, the part of Bacon's scientific method directed to the sense, and the only part of the method designed to permit general participation in science and its applications with an attitude of respect for nature and hope in the ultimate utility of the method.

As Bacon explains in *Advancement* and *De augmentis,* his theory of style encompasses several pairs of methods, the first and most important of which are the magistral and initiative methods, the exoteric and acroamatic methods, and aphorisms and "Methods" (vol. 6: pp. 288–296; vol. 9: pp. 121–130). These methods are arranged in pairs because they address different purposes and audiences, the magistral and exoteric methods and "Methods" being designed for the use of knowledge and "the crowd of learners"; the initiative and acroamatic methods and aphorisms for the progression of knowledge and "the sons...of science" (vol. 6: pp. 288–292; vol. 9: 122–125). These methods are suitable for use in the various parts of Bacon's plan for the advancement of learning, including the parts that constitute the scientific method (Jardine, 1974; Lipson, 1985; Stephens, 1975; Williamson, 1951). The initiative method is designed to present knowledge "in the same method wherein it was invented" and is suitable for *Novum organum* and perhaps for the natural and experimental histories (vol. 6: pp. 288–290). The method of aphorisms presents only "the pith and heart of sciences" and is also suitable for *Novum organum* and for a few of the histories (vol. 6: pp. 291–292). The magistral (teaching) method and "Methods" ("the arrangement and connexion and joining of the parts") are suitable for *Advancement* and *De augmentis;* the acroamatic (enigmatical) method for *The New Atlantis* and numerous other works; and so on (vol. 6: pp. 288–292).

Bacon's theory of style also provided at least two styles, suitable for addressing different faculties and therefore suitable as well for use in different parts of his scientific method. This theory distinguished an imaginative style suitable for addressing the reason and a plain style suitable for addressing the sense. Some historians of Bacon have noted contradictions in his theory of style or inconsistencies between his theory and his practice (Halloran & Bradford, 1984; Lipson, 1985). But Bacon's theory of style has a basis in his distinctions among the faculties (Stephens, 1975; Wallace, 1943). Bacon distinguishes as many as six faculties of the mind—reason, understanding, imagination, memory, appetite, and will. He includes imagination rather than sense among these faculties of the mind and explains that imagination serves as an intermediary between the sense and the reason. In this way, he justifies his distinction between an imaginative style, which addresses the reason through the imagination, and a plain style, which reports directly to the sense. (He does not justify or explain the tables or arrangements of instances, which address the memory.)

The imaginative style has justification in Bacon's definition of rhetoric and is a suitable style for that stage in his scientific method directed to the reason, the inductive method. This style encompasses both Bacon's highly figured style, which presents figures to the imagination in order to move the reason to action; and an unfigured (but not plain) style, the aphorism, which demands that the imagination create its own figures.[5] As Bacon explains in *Advancement* and *De augmentis,* the imagination is an agent or messenger between sense and reason, "for Sense sendeth over to Imagination before Reason have judged," and in this role it turns to rhetoric, whose "duty and office...is *to apply Reason to Imagination* for the better moving of the will" and "to fill the imagination to second reason, and not to oppress it" (original emphasis; vol. 6: pp. 258–260, 297–298; vol. 9: pp.60–62, 131–132).[6] The imaginative style, in its highly figured form, employs the "force of eloquence and persuasion" and the "ornament of words" to move the reason to action (vol 6: p. 299; vol. 9: pp.132–133). In its unfigured form, the aphorism, the imaginative style does not provide figures or other ornaments of style but demands that the imagination create its own figures (Stephens, 1975). In this form, the imaginative style cuts off the "discourse of illustration," "recitals of examples," "discourse of connexion and order," and "descriptions of practice" so that the style will not—as "Methods," its opposite, does—satisfy the reason (understanding) and so prevent further inquiry (vol. 6: pp. 291–292). The imaginative style in its unfigured form, the aphorism, is the most suitable style for that stage in Bacon's scientific method directed, through the imagination, to the reason.

The plain style has justification in Bacon's general remarks about style and is a suitable style for that stage in Bacon's scientific method directed to the sense, the writing of natural and experimental histories. The plain style has its justification in Bacon's general remarks about style in the first book of *Advancement*, some of the more frequently quoted lines in all of his work, which condemn the "affectionate study of eloquence and copie of speech" in favor of "weight of matter, worth of subject, soundness of argument, life of invention, or depth of judgment" (vol. 6: pp. 117–120). Bacon may or may not have meant these remarks as a general prescription (and proscription) for all scientific communication, but he applied them specifically to the writing of natural and experimental history and not to other kinds of scientific communication. In *Parasceve,* the plain style has a justification similar to that in *Advancement* but applies specifically to the histories. In this context, the plain style must omit all "ornaments of speech, similitudes, treasury of eloquence, and such like emptinesses" in favor of "things...set down briefly and concisely, so that they may be nothing less than words" (vol. 8: p. 359). The histories must present these things to the sense but not to the reason (understanding). They must ensure "that nothing will remain unprovided whereby the sense can be equipped for the information of the understanding," but they must not direct themselves to it, "for the world is not to be narrowed till it will go into the understanding...but the understanding...expanded and opened till it can take in the image of the world, as it is in fact" (vol. 8: pp. 361–363). As the style that is closest to the things of the world, the plain style is the most suitable style for that stage in Bacon's scientific method directed to the sense and not to the reason. The plain style, thus construed, is a suitable vehicle for positivistic science, but it is equally suitable as a vehicle for democratic science, for it is the style reserved for the writing of natural and experimental histories, the only part of Bacon's scientific method designed to permit general participation in democratic science with attitudes of respect for nature and hope of utility for human life. Democratic science, in Bacon's view, is an invitation to almost everyone, or at least every man, to participate in the task of gathering the things of the sense that are the foundation of the scientific method. In this view, democratic science is both enabled and constrained by the limitations of the scientific method, but it nonetheless demands that its participants accept Bacon's attitudes of respect for nature and hope in the utility of the method for human life, attitudes that in turn hold the promise of mastery of the plain style most suitable for this stage of the method. As Bacon explains in *Parasceve*, his invitation to almost every man to participate in science is enabled, even required, and at the same time constrained by the limitations of the scientific method. It is required because the task that Bacon proposes is "a thing of very great size, and cannot be executed without great labour and expense" (vol. 8: pp. 353–355). It is also constrained because this same task is open to the sense and therefore "to almost every man's industry," but not open to the reason ("the work itself of the intellect"), a work that Bacon reserves for himself both because he feels able to master it and because he is unwilling to spend his time doing something that others can do as well (vol.8: pp. 353–355). This invitation, in short, allows participants to engage in the task of the sense but not of the reason or the understanding.

Nonetheless, democratic science in Bacon's view demands that its participants accept his attitudes of respect for nature and hope in the utility of his scientific method and promises that if they do so they will achieve a mastery of the plain style most suitable for their task. Democratic science in this view demands that its participants have respect for nature "as faithful secretaries" who "do but enter and set down the laws themselves of nature and nothing else" and promises that by so doing they will in their writing be "content with brevity, and almost compelled to it by the condition of things" (vol. 8: p. 370). It also assumes that they have hope in the utility of the scientific method, but demands that they not consult "the pleasure of the reader" or even the "utility which may be derived immediately from their narrations" and again promises that if they will remember these proscriptions they will "find out for themselves the method in which the history should be composed" (vol.

8: p. 358). Although participation in democratic science in Bacon's view is constrained by the nature of its task, it nonetheless demands attitudes of respect for nature and hope of utility for human life and upholds these attitudes as essential to this task and to the achievement of the plain style most suitable for it.

DEMOCRATIC SCIENCE AND THE PLAIN STYLE IN THE PURITAN REFORM MOVEMENT

Bacon's view of the scientific method and his theory of style, thus construed, was adopted by the Puritan reformers of the 1640s and 1650s, who also fostered and broadened his view of democratic science. The Puritan reformers—Samuel Hartlib, John Dury, and others—sought to promote the public good by extending educational opportunities and fostering a utilitarian approach to science (Webster, 1970, 1975). These reformers upheld Bacon's view of the scientific method and his theory of style and sought to expand participation in that method by extending educational opportunities, especially in science, to all social classes. As Dury explains in "The Reformed School," the reformers' view of the scientific method, like Bacon's, is based in the faculties of sense, memory, and reason, to which they add imagination, and then tradition: "Sense is the first, because it conveighs unto our Imagination the shapes and images of all things, which memory doth keep in store, that Reason may make use thereof: nor can any Tradition be entertained with profit, but that, whereof the Imagination hath received from Sense the originall representations" (1970, p. 153). Their view of the scientific method also governs their educational practice, for its three parts correspond to the three periods in the educational process: "Sense and Imagination, with the beginnings of Memory," belong to the first period; "Imagination and Memory, with the beginnings of Reasoning," to the second; and "all the Acts of Reasoning," to the third (p. 159).

The reformers' view of language, again like Bacon's, includes several styles, each suitable for a different part of the scientific method and a different period in the educational process. In the first period, a child is taught "to name that which is obvious to his Sense by its proper name in his Mother–Tongue"; in the second, to memorize "the Rudiments and necessary Rules of Grammaticall constructions" in Latin, Greek, and Hebrew; and in the third, "to Analyse Authors, and observe their Art of Reason and Utterance to perswade; and then how to order their own thoughts and expression, to search out Truths and to declare the same, Historically, Philosophically, Oratorically, Poetically" (Dury, 1970, pp. 161–162, 164). These views of the scientific method and of language the Puritan reformers sought to bring to all social classes, including the poorer classes (but not to women), by extending educational opportunities to these classes.[7] Their view of democratic science thus was larger than Bacon's, for they sought to extend to almost every man participation in every stage of the scientific method: sense, memory, and reason.

CONCLUSION

Bacon's scientific method and his theory of the plain style, insofar as they provide a vehicle for democratic science, were not influential beyond the Restoration and doubtless are not feasible as a model for science in the late 20th century. Nonetheless, this scientific method and this theory of style provide a historical precedent to an alternative ideology of science that seeks to reform science from within and so to effect a closer identification of the scientific with the social community. The proponents of democratic science to whom I have alluded do not propose general participation in the writing of natural histories of the kind Bacon envisioned. They do seek to reform science in several other ways. Pacey (1983) seeks active involvement in the design of technologies by users outside bureaucratic institutions, especially in local or third-world cultures. Feminist critics of science seek to purge science of its traditional racism, classism, and androcentrism. Keller seeks thereby to reclaim science from within "as a human instead of a masculine project" (1985, pp. 177–178). Harding seeks to ensure an interaction between science and social life and so to reclaim "the moral and political meanings" inherent within the scientific enterprise (1986, pp. 238–239). Contemporary Bacon scholars maintain the right that all share equally to partake of "the promise of technology" (Weinberger, 1985, p. 331) and note the "unequal distribution of goods, knowledge and power" in "today's economically integrated global village" (Whitney, 1986, pp.124–125).

My interpretation of Bacon's scientific method and his theory of style emphasizes the need for a rhetoric suitable for these or other kinds of democratic participation in science. This interpretation need not imply that a plain style suitable for general participation in the writing of natural history is necessarily suitable for other kinds of participation as well. The advocates of democratic science suggest other possibilities, such as Pacey's interactive innovation, Keller's and Harding's analyses of the use of metaphors in science, or Whitney's rational, universalized discourse. Pacey's proposal for interactive innovation addresses the problem of the design of technologies in local or third-world cultures, but includes a set of queries on expert and user views that is applicable in any culture and is compatible with recent trends in user-oriented computer documentation, for example. Keller's analysis of the use of metaphors in science shows how metaphorical language can promote democratic as well as institutionalized science. Harding's caution about this use of metaphors points to a need for critical analysis of metaphorical language with a view toward not only the use of metaphors in scientific explanation, but also the negative stereotypes that these metaphors create and sustain. Finally, Whitney's allusion to Habermas' theory of communication suggests how a "theory of communicative competence" might foster the growth of democratic science by identifying "a priori normative standards" for rational social discourse (Burleson & Kline, 1979, p. 424).

These possibilities are not exhaustive or conclusive, of course, but merely provide some suggestions on what a democratic science and a scientific rhetoric might be. Such a science and a scientific rhetoric constitute an alternative to the traditional ideologies of positivistic science and the plain style and institutionalized science and its more highly figured style. They represent an alternative ideology of science and scientific rhetoric, an alternative perception of what is good for both the scientific and the social community. Thus, they represent not only another historical interpretation of Bacon's science and his rhetoric, but also an alternative vision of what science and scientific rhetoric might and ought to be.

NOTES

[1] Bredvold similarly ascribes to Bacon a positivistic view of science as "merely collection and classification of specimens," but he argues, on Whitehead's authority, that science was in the 17th century and remains in the 20th century primarily mathematical (1956, p. 50).

[2] Ornstein provides a point of contrast. In an early (1928) book ostensibly concerned with the development of scientific communities, she mentions Bacon only briefly and solely in his capacity "as the promoter of experimental science" (pp. 39–40).

[3] See Howell, 1961, pp. 369–371; Jardine, 1974, pp.169–248; Lipson, 1985, pp. 144–149; Stephens, 1975, pp. 98–171; Vickers, 1968, pp. 30–201; Wallace, 1943, pp. 16–24; and Williamson, 1951, pp. 153–158, 176–183. Vickers provides another point of contrast. In his later (1968) study of Bacon's style, he rejects Croll's and Williamson's characterization of Bacon's style as anti-Ciceronian and describes his syntax as "pure English Ciceronian" (pp. 116–117).

[4] Stephens points out that Bacon sometimes distinguishes between the reason and the understanding, reason being the equivalent of understanding in its discursive role (1975, pp. 57–58). In his description of the scientific method in *Novum organum* and *Parasceve,* Bacon uses the terms *reason* and *understanding* and also *intellect* and *mind* interchangeably. In his description of the methods of presentation and in his definition of rhetoric in *Advancement* and *De augmentis,* he most often (but not always) uses the term *reason*. I use the term *reason* and include parenthetical references to indicate where Bacon's usage differs.

[5] Stephens distinguishes these two styles and labels them the aphorism, or the philosophical style, and the acroamatic method (1975, pp. 98–99).

[6] Cogan argues that Bacon's definition of rhetoric applies to the faculties within an individual (1981, pp. 221–223). But it applies equally well to the faculties of different individuals or groups in science, whether institutionalized or democratic.

[7] The Puritan reformers advocated extending educational opportunities to women but only such as would "fit them...to become modest, discreet, and industrious house-keepers" ("Some Proposalls towards the Advancement of Learning" [Webster, 1970, p. 190]).

REFERENCES

Anderson, F. H. (1971). *The philosophy of Francis Bacon.* New York: Octagon-Farrar.
Bacon, F. (1869). *Advancement of learning.* In J. Spedding, R. L. Ellis, & D. D. Heath (Eds.), *The works of Francis Bacon* (vol. 6, pp. 77–412). New York: Hurd.

Bacon, F. (1869). *The great instauration*. In J. Spedding, R. L. Ellis, & D. D. Heath (Eds.), *The works of Francis Bacon* (vol. 8, pp. 13–55). New York: Hurd.

Bacon, F. (1869). *New Atlantis*. In J. Spedding, R. L. Ellis, and D. D. Heath (Eds.), *The works of Francis Bacon* (vol. 5, pp. 355–413). New York: Hurd.

Bacon, F. (1869). *The new organon (Novum organum)*. In J. Spedding, R. L. Ellis, & D. D. Heath (Eds.), *The works of Francis Bacon* (vol. 8, pp. 57–350). New York: Hurd.

Bacon, F. (1869). *Of the dignity and advancement of learning (De augmentis scientiarum)*. In J. Spedding, R. L. Ellis, & D. D. Heath (Eds.), *The works of Francis Bacon* (vol. 8, p. 383; vol. 9, p. 357). New York: Hurd.

Bacon, F. (1869). *Preparative towards a natural and experimental history (Parasceve)*. In J. Spedding, R. L. Ellis, & D. D. Heath (Eds.), *The works of Francis Bacon* (vol. 8, pp. 351–371. New York: Hurd.

Boas, M. (1966). *The scientific Renaissance 1450–1630*. New York: Harper Torchbooks Science Library-Harper.

Bredvold, L. I. (1956). *The intellectual milieu of John Dryden: Studies in some aspects of seventeenth-century thought*. Ann Arbor, MI: Ann Arbor Paperbacks-University of Michigan Press.

Burleson, B. R., & Kline, S. L. (1979). Habermas' theory of communication: A critical explication. *Quarterly Journal of Speech, 65*, 412–428.

Burtt, E. A. (1954). *The metaphysical foundations of modern physical science* (Rev. ed.). Garden City, NJ: Anchor-Doubleday.

Cogan, M. (1981). Rhetoric and action in Francis Bacon. *Philosophy and Rhetoric, 14*, 212–233.

Croll, M. W. (1969). "Attic prose" in the seventeenth century. In J. M. Patrick, R. O. Evans, & J. M. Wallace (Eds.), *"Attic" and baroque prose style: The anti-Ciceronian movement* (pp. 51–101). Princeton, NJ: Princeton University Press.

Croll, M. W. (1969). Muret and the history of "Attic prose." In J. M. Patrick, R. O. Evans, & J. M. Wallace (Eds.), *"Attic" and baroque prose style: The anti-Ciceronian movement* (pp. 107–162). Princeton, NJ: Princeton University Press.

Dury, J. (1970). The reformed school. In C. Webster (Ed.), *Samuel Hartlib and the advancement of learning*, pp. 139–165. London: Cambridge University Press.

Halloran, S. M., & Bradford, A. N. (1984). Figures of speech in the rhetoric of science and technology. In R. J. Connors, L. S. Ede, & A. A. Lunsford (Eds.), *Essays on classical rhetoric and modern discourse* (pp. 179–192, 284–286). Carbondale, IL: Southern Illinois University Press.

Harding, S. (1986). *The science question in feminism*. Ithaca, NY: Cornell University Press.

Haydn, H. (1950). *The counter-Renaissance*. New York: Harbinger-Harcourt.

Howell, W. S. (1961). *Logic and rhetoric in England, 1500–1700*. New York: Russell.

Jardine, L. (1974). *Francis Bacon: Discovery and the art of discourse*. London: Cambridge University Press.

Jones, R. F. (1951a). Science and English prose style in the third quarter of the seventeenth century. In *The seventeenth century: Studies in the history of English thought and literature from Bacon to Pope* (pp. 75–110). Stanford, CA: Stanford University Press.

Jones, R. F. (1951b). Science and language in England of the mid-seventeenth century. In *The seventeenth century: Studies in the history of English thought and literature from Bacon to Pope* (pp. 143–160). Stanford, CA: Stanford University Press.

Keller, E. F. (1985). *Reflections on gender and science*. New Haven, CT: Yale University Press.

Kuhn, T. S. (1970). *The structure of scientific revolutions*. Chicago: University of Chicago Press.
Lipson, C. S. (1985). Francis Bacon and *plain* scientific prose: A reexamination. *Journal of Technical Writing and Communication, 15*, 143–155.
Octalog. (1988). The politics of historiography. *Rhetoric Review, 7*, 5–49.
Ornstein, M. (1928). *The rôle of scientific societies in the seventeenth century*. Chicago: University of Chicago Press.
Pacey, A. (1983). *The culture of technology*. Cambridge, MA: MIT Press.
Purver, M. (1967). *The Royal Society: Concept and creation*. London: Routledge.
Some proposalls towards the advancement of learning. (1970). In C. Webster (Ed.), *Samuel Hartlib and the advancement of learning*, pp. 165–192. London: Cambridge University Press.
Stephens, J. (1975). *Francis Bacon and the style of science*. Chicago: University of Chicago Press.
Van den Daele, W. (1977). The Social construction of science: Institutionalization and definition of positive science in the latter half of the seventeenth century. In E. Mendelsohn, P. Weingart, & R. Whitley (Eds.), *The social production of scientific knowledge* (pp. 27–54). Dordecht, Holland: Reidel.
Vickers, B. (1968). *Francis Bacon and Renaissance prose*. London: Cambridge University Press.
Wallace, K. R. (1943). *Francis Bacon on communication and rhetoric: Or, the art of applying reason to imagination for the better moving of the will*. Chapel Hill, NC: University of North Carolina Press.
Webster, C. (1970). Introduction. In C. Webster (Ed.), *Samuel Hartlib and the advancement of learning* (pp. 1–72). London: Cambridge University Press.
Webster, C. (1975). *The great instauration: Science, medicine and reform 1626–1660*. London: Duckworth.
Weinberger, J. (1985). *Science, faith, and politics: Francis Bacon and the Utopian roots of the modern age: A commentary on Bacon's* Advancement of learning. Ithaca, NY: Cornell University Press.
Whitney, C. (1986). *Francis Bacon and modernity*. New Haven, CT: Yale University Press.
Willey, B. (1953). *The seventeenth century background: Studies in the thought of the age in relation to poetry and religion*. Garden City, NJ: Anchor-Doubleday.
Williamson, G. (1951). *The Senecan amble: Prose form from Bacon to Collier*. Chicago: Phoenix Books-University of Chicago Press.
Winner, L. (1977). *Autonomous technology: Technics-out-of-control as a theme in political thought*. Cambridge, MA: MIT Press.

3

Oliver Evans and His Antebellum Wrestling with Rhetorical Arrangement

R. John Brockmann
University of Delaware

Rhetoric is the strategic disposition of ideas to influence the decisions of others, and authors continually make choices to create the most strategic disposition for their message. Some of technical communication's most interesting research has examined how communicators choose between rhetorical alternatives, and some of the most important lessons for new technical communicators can be drawn from how others made their decisions in the past (Odell, Gotswami, Herrington, & Quick, 1983). An interesting series of rhetorical decisions of the past can be observed in the work done by Oliver Evans (1755–1819), during the creation of his two books: One, his 1795 *The Young Mill-wright and Millers Guide*, went through 15 editions in English and was the most reprinted technical book written by an American prior to the Civil War; while his second, *The Abortion of the Young Steam Engineers Guide* (1805), went through only two English editions.[1] The first book gave readers the ability to recreate Evans's newly patented, automated assembly line that could be used for the milling of flour by waterpower, while the second gave readers the ability to recreate Evans's newly patented, high-pressure steam engine that could dramatically increase the power of low-pressure steam engines normally employed at the time.

In my earlier examination of Oliver Evans's work, "Oliver Evans and the Weave of Text and Graphics for Antebellum Millwrights," I pointed out many similarities in Evans's physical circumstances when he composed each book (Brockmann, 1998, chap. 1). Each incurred large expenses for copperplate engravings, and thus each put

Evans's family on the verge of abject poverty. In *Oliver Evans to His Counsel Who Are Engaged in Defense of his Patent Rights* (1817), a work that functioned very much like a memoir, Evans recounts concerning the creation of *The Guide* in the 1970s:

> I was reduced to such abject poverty that my wife sold the tow clothe which she had spun with her own hands for clothing for her children, to get bread for them, my head was covered with many gray hairs and I required spectacles. (p. 16)

Later, after working on *The Abortion*, Evans also recounted in his memoir:

> I was left in poverty at the age of 50, with a large family of children and an amiable wife to support, for I had expended my last dollar in putting my 'Columbian' steam engine in operation, and in publishing the 'Steam Engineer's Guide'...(1817, p. 18)

Not only were the physical circumstances surrounding the composition of the books similar, but so were Evans's goals for in both works he combined altruistic as well as entrepreneurial goals. On the one hand, he claimed to write his books to overcome the situation he himself had encountered as an apprentice *mechanic* (the 18th and 19th century word for *engineer*); there were no texts from which to learn. Evans was remembered as saying to a colleague:

> much...on the difficulties inventive mechanics labored under for want of published records of what had preceded them, and for works of reference to help the beginner. In speaking of his own experience, he said that everything he had undertaken he had been obliged to start at the very foundation; often going over ground that others had exhausted and abandoned; leaving no record. (Ferguson, 1965, p. 38)

In a similar way, when he advertised for *The Abortion*, Evans had the following printed in Philadelphia's *Aurora* newspaper:

> ...an investigation of the principles, construction and powers of different Steam Engines...he [had] laid down a new and true theory, and developed principles of nature that may lead to highly important discoveries and improvements. (Bathe & Bathe, 1972, p. 107)

The appearance of such scientific and theoretical disinterestedness was a key rhetorical goal for Evans, because in achieving such a tone he could escape the charge of being a *projector* (the late 18th and early 19th century word for *huckster*).[2] Evans himself felt it was important to avoid looking like a "projector" because, when he wrote a letter requesting a loan from a sponsor, he cautioned: "I will observe to you that I having the name of a Projector will probably find difficulty in obtaining money from most men" (Evans, 1794).

Yet for all his disinterestedness both real and feigned, Evans did indeed want to make money by using his books as vehicles for introducing and selling his patented designs for mills and steam engines. Thomas Jones, who edited *The Guide* after Evans's death, added a preface in 1826, where he noted that "the main object of Evans's writing this work was to introduce his invention to public notice..." Jones's observation echoed what Evans himself wrote concerning *The Guide*[3]:

> I commenced the work, intending only to explain my own improvements, but being advised to add something respecting building mills, as well as on the manufacture of flour, to induce millwrights and millers to purchase and read it, I was led into the investigation of the principles of water acting on wheels to drive mills, and of the construction of mills. (1824, pp. 16–17)

This entrepreneurial motivation is also what Hunter observed about *The Abortion*:

> Evans made The Abortion the vehicle of an appeal to potential customers, naming the industrial operations to which his engine could be applied to great advantage... (1985, p. 39)

Furthermore, this entrepreneurial motivation can be observed quite clearly in key illustrations in the two books that Evans also printed on the back of the licenses he offered for sale for either the mill or the engine. Thus, Evans wrote each book under very difficult similar circumstances and with very similar goals of both disinterested technological communication and profit. The difference between the two books arose because of Evans's radically altered expectations which, in the early book, motivated him to be deeply engaged with his audience while, in the later book, motivated him to be disengaged.

In the early 1790s, when Evans was writing *The Guide*, he anticipated both a favorable reception that he felt his invention would receive, and the money he thought it could generate through licensing. The auguries for a favorable reception were certainly good; after all, celebrated figures such as Washington, Jefferson, Randolph, Morris, and Burr had already signed on as subscribers to underwrite the publication of *The Guide*. Additionally, Matthew Carey, editor of the *American Museum or Universal Magazine*, sang of Evans's eventual success when he wrote:

> We are happy to learn that mr. Evans is now likely to reap the fruits of his labour; upwards of 100 mills have adopted his improvements, and they are daily increasing in public estimation. (1792, pp. 225–226)

However, by 1805, Evans had found the auguries false. He was informed in January that the patent rights for his automated mill would not be extended by Congress, and thus his ability to capitalize on his licensing possibilities was to be prematurely curtailed. Furthermore, in place of Carey's huzzahs, Evans not only found his high-pressure steam designs under attack,[4] but himself publicly accused of the theft of another's engine design.

The difference these altered expectations played in Evans's attitude toward his books can be seen quite dramatically by comparing Evans's words in the conclusions of the two books. In the conclusion of the original edition of *The Guide*, Evans wrote:

> Sensible of the expense, time, labor, and thought that this (though small) work has cost me, and hoping that it may be well received by, and prove serviceable to my country-I wait to see its fate; and feel joy at being ready to say FINIS. (1795, p. 9)

Yet, in the postscript to the second book, he wrote:

> Having read my work in sheets, as it came from the press, I am highly sensible of its deficiencies. It is truly an "Abortion of the Young Steam Engineer's Guide"... (p. 134)

Moreover, Evans's contemporary critiques seem to have concurred with his own assessment. Of *The Guide*, the editor of the *American Monthly Review* concluded his review in September 1795 by observing:

> Upon the whole, the author appears to have executed his plan with fidelity and great perspicuity; and though the critic may find a number of verbal or grammatical inaccuracies in the work, yet we are persuaded it will be esteemed, both by the theoretical and practical mechanic, a valuable acquisition. (p. 5)

However, Tredgold's 1827 history of the development of the steam engine offered the following words to evaluate Evans's second book, and the words Tredgold chose for his review could not have been more piercing, considering Evans's intended balance of disinterestedness and entrepreneurial pitch in his books. Tredgold wrote:

> The 'Abortion' is a curious work; it betrays that strange mixture of absurd speculation and indistinct perception of truth, which distinguishes the generality of enthusiastic projectors, and is valuable only to those who can select by means of previous knowledge or experience. (p. 43)

Today, we can gain particular insight into the relative effectiveness of Evans's writing efforts in these two works if we carefully observe how Evans fulfilled or failed to fulfill the methods of rhetorical organization that were then fashionable, the "disposition" of classical Greek and Roman rhetoric.[5]

THE YOUNG MILL-WRIGHTS AND MILLERS GUIDE

When Thomas Jones revised and edited *The Guide* for its fifth edition in 1826, he added the following in a new preface:

> Mr. Evans made no pretensions to literature; he considered himself, as he really was, a plain, practical man; and the main object of his writing his work was to introduce his invention to public notice... (p. iv)

Jones may have thought Evans had no pretensions to literature, appearing as plain and as practical as fellow Philadelphian Benjamin Franklin appeared in his own *Autobiography* (1789). However, such appearances did not mean that neither Philadelphian wrote without careful strategies derived from their knowledge of classical rhetoric.

Dorothy Bathe and Grenville Bathe list six books that Evans's 1819 will bequeathed to his children, and one was Dobson's American version of the third edition of the *Encyclopaedia Britannica-The Encyclopedia; or A Dictionary of Arts, Sciences, and Miscellaneous Literature*. Evans and Dobson crossed paths many times: Dobson was listed in 1795 as one of those who would collect subscription monies for the publication of *The Guide* (Bathe & Bathe, 1972), and Dobson was the one who was supposed to have introduced Ellicott to Evans on the eve of the publication of *The Guide*:

> I called on Thomas Dobson, printer of the Encyclopaedia; and asked him if he would accept of a small treatise on mill-wrighting: he said Oliver Evans had been there a few days before; and proposed such a work, which I thought would save me the trouble. (1795, p. vii)

Later, Evans included Ellicott's work verbatim as Part V of *The Guide*. Moreover, Evans explicitly quotes at least seven times in *The Abortion* from the 17th volume of Dobson's 18-volume *Encyclopaedia*, and boldly reprints over ten pages of its accounts of three key steam engine inventors. Evans reprinted these pages because he found the *Encyclopaedia* to be "a rare and expensive philosophical work" and he worried that "some may not have an opportunity of reading *the Encyclopaedia*."

It is necessary to understand the role Dobson and the *Encyclopaedia* played for Evans throughout his life because it is in the pages of the *Encyclopaedia* that today we can catch a glimpse of what Evans understood about classical rhetoric or "oratory," as it was printed in 1795 in volume 13 of the *Encyclopaedia*. In two columns, for over 100 oversized pages, Dobson restated the canons of Ciceronian rhetoric—invention, arrangement (or *disposition* as the *Encyclopaedia* puts it), style, memory, and delivery.[6] Ciceronian disposition, as laid out in Dobson's *Encyclopaedia*, includes six divisions:

1. An Introduction, whose purpose is to make the readers receptive to the author's message;
2. a Narration or a statement of background facts, which lays the foundation of understanding for the central message;

3. a Partition, which quickly previews the parts of the confirmation to come;
4. the Confirmation, which is the central statement of the author's argument for the point or subject;
5. the Confutation, where the author seeks to discredit any opposing views in a kind of preemptive strike;
6. and the Conclusion, which sums up the argument as a whole, moves the audience to action, and often amplifies the conclusions to a more important plane of significance.

Table 3.1 summarizes how the parts of *The Guide* fit into Ciceronian disposition, and what their positioning suggests about their various functions in *The Guide*. Each part of the disposition is then explored in depth.

Introduction of *The Guide*

The purpose of an introduction in Ciceronian rhetoric is to "prepare the minds of the hearers for a suitable reception of the remaining parts" (Ward, 1759, vol. 1: p. 179). Evans's Preface achieves this purpose in two ways. On the one hand, it draws readers into a text by suggesting that "the subject is great, necessary, or for the interest of those to whom the discourse is addressed." (Dobson, 1795, vol. 13: p. 394) Accordingly, Evans begins the Preface with the following statement:

> The reason why a book of this kind although so much wanted did not sooner appear, may be-because they who have been versed in science and literature, have not had practice and experience in the arts; and they who have had practice and experimental knowledge, have not had time to acquire science and theory, those necessary qualifications for completing the system and which are not to be found in any one man. (1795, n.p.)

TABLE 3.1.
Parts of *The Guide* with Their Functions in Ciceronian Disposition

Title	Comprising Articles	Rhetorical Function in Ciceronian Disposition
Preface		Introduction
Part I: Mechanics and Hydraulics	pp. 1–69	Narration
Part II: Young Mill-wright's Guide	pp. 70–86	Narration
Part III: Containing Evans's Patented Improvements on the Art of Manufacturing Grain into Meal and Flour	Introduction	Partition
	pp. 88–103	Confirmation
Part IV: Young Miller's Guide	Contents	Partition
	pp. 104–118	Confirmation
Part V: The Practical Mill-wright	pp. 1–40	Confutation
Appendix: Rules for Discovering New Improvements		Conclusion

More importantly, however, Evans's "Preface" gains the "good opinion of his hearers" by creating a humble persona. According to the *Encyclopaedia*'s section on oratory, such humility or modesty is an excellent way to make the audience receptive:

> It is certain, that what is modestly spoken is generally better received that what carries in it an air of boldness and confidence. Most persons, though ignorant of a thing, do not care to be thought so; and would have some deference paid to their understanding. But he who delivers himself in an arrogant and assuming way seems to upbraid his hearers with ignorance, while he does not leave them to judge for themselves, but dictates to them, and as it were, demands their assent to what he says: which is certainly a very improper method to win upon them. (Dobson, 1795, vol. 13: p. 389)

The *Encyclopaedia*'s advocacy of modesty finds an interesting echo in Franklin's *Autobiography,* where Franklin discusses the advantage of appearing hesitant to assert himself in argumentative situations:

> ...retaining only the Habit of expressing my self in Terms of modest Diffidence, never using when I advance any thing that may possible be disputed, the Words **Certainly, undoubtedly**, or any others that give the Air of Positiveness to an Opinion; but rather say, **I conceive**, or **I apprehend** a Thing to be so or so, **It appears to me**, or **I should think it so or so for such & such Reasons**, or **I imagine** it to be so, or **it is so if I am not mistaken**.-This Habit I believe has been of great Advantage to me, when I have had occasion to inculcate my Opinions & persuade Men into Measure that I have been from time to time engag'd in promoting. (1789, p. 18)

With such an air of modesty that tries to not seem bold or confident, Evans wrote the following in the Preface to *The Guide*:

> Wherefore it is not safe to conclude that this work is without error-but that it contains many, both theoretical, practical and grammatical; is the most natural, safe, and rational supposition.... Therefore I request the reader, who may prove any part to be erroneous, can point out its defects, propose amendments, or additions; to inform me thereof by letter; that I may be enabled to correct, enrich, and enlarge it, in case it bears another edition, and I will gratefully receive their communications: For if what is known on these subjects by the different ingenious practitioners in America could be collected in one work, it would be precious indeed, and a sufficient guide to save thousands of pounds from being uselessly expended. For a work of this kind will never be perfected by the abilities and labors of one man.[7] (1795, n.p.)

Evans certainly achieves the requisite modesty and humility in his introduction, both by not seeming bold or confident, and by establishing himself on the same level as his "ingenious" practitioner readers; even offering to publish their information.

Background Facts of *The Guide—The Narration*

Next, Evans explained the "first principles" on which he bases his work, and these he put into his Narration, Part I and Part II. The *Encyclopaedia* suggested the importance of the narration by noting:

> For the foundation of his reasoning afterwards is laid in the narration, from whence he takes his arguments for the conformation. And therefore it is a matter of no small importance that this part be well managed, since the success of the whole discourse so much depends upon it. (Dobson, 1795, vol. 13: p. 395)

Moreover, the *Encyclopaedia* noted that the orator should establish the probability of the information:

> Things appear probable when the causes assigned for them appear natural; the manner in which they are described is easy to be conceived; the consequences are such as might be expected; the character of the persons are justly represented; and the whole account is well stated, consistent with itself, and agreeable to the general opinion. (Dobson, 1795, vol. 13: pp. 395–396)

For Evans to show his own novel ideas were "agreeable to the general opinion," he composed Part I and Part II, an exposition on mechanics and hydraulics derived from such British Royal Society luminaries as Smeaton, Waring, Martin, and Ferguson:

- John Smeaton's *Experimental Enquiry Concerning the Natural Powers of Wind and Water to turn Mills and other machines* (1794)
- William Waring's "Observations on the Theory of Water Mills" (1793)
- Benjamin Martin and his *Philosophia Britannica* (1771)
- James Ferguson's *Lectures on Select Subjects* (1772)

Evans not only drew his requisite "general opinions" from luminaries, but, by including them in Part I and Part II in his text, Evans also bolstered his own unknown inventions by using their established names and widely respected books (Gross, 1990).[8]

The Partition of *The Guide*

Once the orator had attracted his readers in his introduction and given them the requisite background of facts to establish the "general opinion" and the probability of his ideas, Dobson's *Encyclopaedia* suggested an orator use a partition in which

> the orator acquaints his hearers with the several parts of his discourse upon which he designs to treat. Thus Cicero states his plea in his defence of Muraena: "I perceive the

accusation consists of three parts: the first respects the conduct of his life; the second his dignity and the third contains a charge of bribery. (1795, vol. 13: p. 398)

In a single page between the narration in Part I and Part II, and the confirmation in Part III and Part IV, Evans has a very short, concise overview of what was to come—very much in the manner of Cicero:

INTRODUCTION
The improvements consist of the invention, and various applications, of the following machines, viz
1. The Elevator
2. The Conveyor
3. The Hopper-boy
4. The Drill
5. The Descender

The *Encyclopaedia* also strongly suggests that the partition be short:

the orator does not show what he is then speaking of, but only what he designs to discourse upon. (Dobson, 1795, vol. 13: p. 398)

Not only is this what Evans does in quite an expeditious manner to introduce Part III, but his list of the contents of Part IV on a single page achieves the same end.

Central Statement of *The Guide*

Part III and Part IV constitute the *conformation*, or the central statement of Evans's message. Of this part of the disposition, the *Encyclopaedia* observed:

The orator having acquainted his hearers in the proposition *[partition]* with the subject on which he designs to discourse usually proceeds either to prove or illustrate what he has there laid down. For some discourses require nothing more than an enlargement or illustration to set them in a proper light, and recommend them to the hearers... (Dobson, 1795, vol. 13: p. 399)

Consequently, it is apropos that Evans chose this part of the disposition to discuss his milling inventions in detail and to use it later on his license to represent what he allowed others to use.

Based upon the "advice" of others, Evans added "Part IV, The Young Miller's Guide" to his confirmatio. Part IV described the output of all his machinery and expands the reach of *The Guide* so as to induce millwrights, the builders of mills—as well as millers and the users of the mills—to purchase, read, and buy the licenses.

Furthermore, Part IV deals with what Evans described to Young in a 1791 letter as "the manner of laying out millstones with observations and reasons."

Confutation of *The Guide*

"Part V, The Practical Mill-wright," functioning as the confutation, has customarily been described as the result of a fortunate accident (Bathe & Bathe, 1972, p. 48; Ferguson, 1980, p. 32). Earlier, we saw how Thomas Ellicott, the author of this section, came to meet Evans:

> I have often thought, that if I could spare time I would write a small treatise on millwrighting myself, (thinking it would be of much use to young millwrights), but fearing I was not equal to the task, I was ready to give it up, but on further consideration, I called on Thomas Dobson, printer of the Encyclopedia; and asked him if he would accept a small treatise on millwrighting: he said Oliver Evans had been there a few days before, and proposed such a work, which I thought could save me the trouble. But some time afterwards, the said Evans, applied to me, requesting my assistance in his undertaking; this I was the more willing to do, having built several mills with his additional improvements; and draughted several others... (1795, p. vii)

As apt as Ellicott's story appears, it is too coincidental. Of all the millwrights to appear on Evans's doorstep at just the right time, what a coincidence that the millwright happened to be a member of the first family to sign a testimonial supporting Evans's inventions in exchange for a free license (1795). Further coincidences accumulate when one realizes that Ellicott repeats, with varied words and pictures, much of the material that Evans covered in Part I, Part II and Part III, as well as the all-important Plate VIII of Part IV that appears on Evans's licenses for his mill designs. Such repetition has been dismissed as confused:

> There is overlapping and duplication in the discussion and table of the collaborators, which at times may have been a source of confusion to the reader. (Hunter, 1979, p. 94)

But Part V, by its very presence, countered the number-one argument from those who had dismissed Evans's earlier attempts to sell his design licenses: None of Evans's neighboring millers had adopted them (1817):

> At first they *[Evans's agents]* found no one willing to accept them on those terms *[gratis]*, because they had never heard nor known of any such machines being in use; a few had heard something about mine, but could not be convinced of their usefulness; would ask if the Brandywine millers *[Evans's neighbors]* had adopted them, and an answer in the negative was to them convincing proof that the improvement was not worthy their notice. (p. 12)

Thus Part V, in repeating much of Evans's material, verified the material because it came from someone other than Evans! In fact, in the introduction to Part V ("To the

OLIVER EVANS AND RHETORICAL ARRANGEMENT 73

Reader"), Ellicott not only presents the partition to his material, which he does so in the last short paragraph of six pages, but for the last three pages Ellicott waxes poetic about Evans's inventions:

> These improvements are a curiosity worth the notice of the philosopher and statesman, to see with what harmony the whole machinery works in all their different operations. (1795, p. x)

Thus, Ellicott's repetition subtly counteracted the argument against adopting Evans's invention because *The Guide*'s readers received verification of the efficacy of Evans's designs from a second, local source. Having Ellicott say what he had said gave Evans's detractors the exact information they had missed in exactly the way the wanted to hear of it; for as the *Encyclopaedia* pointed out concerning the confutation:

> in confutation, what the adversary has advanced ought carefully to be considered, and in what manner he as expressed himself. (Dobson, 1795, vol. 13: p. 403)

Conclusion of *The Guide*

The ten-page Appendix that concludes *The Guide* is entitled "Rules for discovering new Improvements-exemplified in improving the Art of thrashing and cleaning Grain, hulling Rice, warming Rooms, and venting Smoke by Chimneys, &c." It seems odd that Evans would present these varied inventions in a book on automated flour manufacturing. However, Evans probably included this material because these ten pages function as the conclusion of his book. According to classical rhetoricians, one of the functions of a disposition's conclusion is to amplify the consequences of a work (Corbett, 1990).[9] The *Encyclopaedia* spoke of amplification in these words:

> Now by amplification is meant, not barely a method of enlarging upon a thing; but to represent it in the fullest and most comprehensive view, as that it may in the liveliest manner strike the mind, and influence the passions. (Dobson, vol. 13: 1795, p. 409)

These ten pages give Evans an opportunity to amplify the consequences of his own particular flour mill invention by extending his particular work to the general understanding of "Rules for discovering new Improvements." In amplifying his conclusion, Evans was probably attempting to raise his book beyond the level of a mere projector's piece for marketing patent licenses to a more general discussion of the eternal truths of invention and discovery. This may have been his intention because Evans himself commented on the book's eternal truths of invention and discovery in his 1817 memoir:

> ...the Millwrights and Miller's Guide, containing some eternal truths, true theories, which will, like Euclid's Elements, stand the test of time... (pp. 16–17)

Review of *The Guide's* Disposition

Evans crafted the disposition for the presentation of his flour manufacturing invention by having each part work persuasively:

- In his Preface, his introduction, Evans worked to make the reader receptive to his ideas by creating a humble and modest persona;
- in Part I and Part II, his narration, Evans gave the reader a foundation of scientific facts and assembled a list of names and books that would surely impress the reader and establish Evans's inventions within the "general opinion";
- in the introductions to Part III and Part IV, his partition, Evans briefly previewed the central statements of his conformation to come;
- in the most of the rest of Part III and Part IV, his conformation, Evans presented the central statement of the book in his presentation of his inventions;
- in Part V, his confutation, Evans countered his critics' major argument in ways best understood by them by having Evans's own ideas and inventions expounded by another, Thomas Ellicott;
- and in the Appendix, his conclusion, Evans sought to rise far above the role of projector when he amplified the consequences of his work to the level of eternal truths that will stand the test of time.

In *The Guide*, Evans is deeply engaged with his audience, their past criticisms of his approaches, and their sense of who are the authorities of "general opinion." Thus from first to last, Evans develops a persuasive architecture to frame his ideas. Evans continues this deep engagement with the audience and this subtle crafting of persuasive structures in proposals and pamphlets written right up to the eve of *The Abortion*.

Advertisements, Turnpike Proposals, and *The Principles* (1803–1805): Evans's Rhetorical Dexterity Is No Fluke

The first published notice of Evans's steam engine appeared in Philadelphia's *Aurora* newspaper in February, 1803:

> THE STEAM ENGINE, which has been very justly considered among the first and most useful of inventions, is receiving further improvements, Mr. Oliver Evans of this city, has lately constructed one upon new principles, by which the power of the agent is greatly increased. The Engine, in its construction, is so simplified, that it can be easily managed by ordinary mechanics. The Cylinder is only about six inches diameter. The engine although, in an imperfect state, ground in 12 hours, the other day, no less than one hundred bushels of hard plaster. This immensely powerful agent can, on the new principles, happily applied by Mr. Evans, serve with profit, all the various purposes of machinery; particularly Breweries and Distilleries. It may be made so as to pump the water, grind the malt, and boil the water that is necessary on

the largest scale—As the construction is so much simplified, and the expenditure of fuel so lessened, these valuable and important improvements commence a new Era in the history of the STEAM ENGINE. Very few men have, indeed in the United States, made so many useful inventions and discoveries in mechanics as Mr. Oliver Evans, we therefore hope he will receive that reward which is merit claims, from a discerning and liberal public. (qtd. in Bathe & Bathe, 1972, pp. 79–80)

Steam engines based upon the Newcomen and Watts design, the standard engine design of the time, primarily used steam during its condensation phase when it would shrink in mass and develop a vacuum that could *pull* a piston. These engines were termed low pressure or atmospheric steam engines, and for their size and weight could only develop low levels of power. Moreover, they required a vast amount of fuel and cold water to produce this low level of power. Evans's breakthroughs were first to apply the steam directly to *push* a piston out, and, second, to build up high pressure in the boiler to create more power. History shows it is Evans's design that laid the groundwork for most of the steam engines of the 19th century, including those used by most of the steamboats on America's western rivers. However, Evans's design was not the standard one, and he was not a disinterested inventor.

The *Aurora* piece appears to be a laudatory editorial as suggested by its use of the third person pronouns to address Evans. However, this short piece appeared some 20 times in the *Aurora* through the rest of the year, strongly suggesting that it was a disguised advertisement, and that Evans, the projector, paid for it himself (Bathe & Bathe, 1972). The tone is positive, offering many applications, and certainly frames Evans in a most favorable light, while he maintains the guise of modesty; after all, he himself is not saying that "Very few men have, indeed in the United States, made so many useful inventions and discoveries in mechanics as Mr. Oliver Evans...."

In September 1804, Evans began to apply his engine designs by suggesting to the directors of the Lancaster Turnpike Company that they replace horse-drawn wagons with steam-powered wagons,[10] and, by writing in November to the Philadelphia Board of Health, requesting a grant of money to build a steam-powered scow that could dredge the city's waterfront (Bathe & Bathe, 1972). The Lancaster Turnpike Company turned down his plan, but the Philadelphia Board of Health authorized the building of the scow on December 10, agreeing to pay Evans for its construction as he proceeded and guaranteeing him a commission not to exceed $500. Two weeks later on December 24, buoyed by his patent, and by the first paying application of his engine, Evans wrote to Jefferson to request his aid in passing an extension of his automated flour mill patent through Congress.

Patent protection had operated for 14 years by 1804. However, with the time needed to publicize the patent to the dispersed population of the United States, as well as to enforce the protection in the courts, Evans had not received much profit from his licensing endeavors. Later in the year, Evans was encouraged that his and all patentees' protection would be extended for another seven years because Congress had just passed a bill doing that for copyrights, and the Congressional committee appointed to investigate the matter for patents had reported favorably on extending patent protection. Moreover, the extension of his flour mill patent was crucial to Evans: He had to expend his own monies in building the dredging scow for Philadelphia, and he was going to have to pay for the plates, illustrations, and printing of the steam engine book. As he wrote in the preface to *The Abortion*, he felt he needed $9,000 for all his upcoming expenses and that his patent extension would yield some $10,000.[11]

To aid the passage of his request through Congress, Evans published a small pamphlet for Congress in late 1804 entitled *The Principles of Steam Engines*. In this pamphlet, Evans exhibited all the rhetorical finesse of *The Guide*'s disposition and all the engagement of his audience with wit and charm. Furthermore, it is interesting to compare *The Principles* to the proposal to the Lancaster Turnpike Road Company—a quite different audience from Congress with a decidedly personal financial interest in the matter. By observing the two, one can see quite clearly that Evans's rhetorical finesse was not a fluke event in the composition of *The Guide*, nor was it a skill forgotten in the decade between the publication of the two books.

As with *The Guide*'s introduction, Evans again attempted in the beginning of *The Principles* to create a humble and modest persona:

> The SUBSCRIBER with diffidence presumes to lay before the honorable Senators and Representatives in Congress, individually, (hoping it may be well received) the following short description of the principles. (1805, p. 98)

In like manner, in the turnpike proposal, Evans wrote: "Permit me to lay before your respectable body the following statement" (p. 52).

As with *The Guide*'s narration, Evans again attempted in the first four paragraphs of *The Principles* to layout the "general opinion" of established engine designs (single stroke designs in paragraph one, double stroke designs in paragraph two, and atmospheric engine designs in paragraph three), as well as including references to such well-known scientists at the time as James Watts. Taking the time to establish the "general opinion" on steam engine design gave Evans an opportunity to favorably frame his own novel design, as well as establish his own scientific passive-non-projector-role. Consider, for example, the tone in paragraph two:

> The celebrated James Watts, improved on this engine by making his steam of power equal to the weight of the atmosphere, and letting it in at the top of the cylinder, to supply the place of the atmosphere to push down the piston, while the steam was condensed below, and also at the bottom while condensation was going on above, making a double stroke; and to avoid the loss occasioned by the jet cooling the cylinder, he led the steam off from each end of the cylinder into a separate vessel into which he let the jet of cold water, to condense the steam; and he found by these means he could make a more

perfect vacuum, and computed the power of his engine at between 11 to 13 lbs. to the inch, and the expense of fuel greatly lessened. (1805, p. 98)

In contrast to such a scientific disinterestedness, Evans's turnpike proposal letter, rather than establishing the "general opinion" in terms of scientific principles and theories, established how the turnpike was then operating with horses in terms of dollars and cents (p. 53):

It requires 5 horse waggons, of 5 horses each, to transport 100 barrels the same distance in 3 days. The expense I estimate as follows:

	Dolls
5 Horse waggons at the cheapest rate, 100 dolls. each	500
25 horses at 100 dolls. each	2500
Gears for 25 horses at 7.75 dolls.	193.75
5 waggon covers at 7 dolls.	35
30 bags for feed at 1 doll.	30
5 jack screws at 6 dolls.	30
5 whips at 75 cents	3.75
5 feed troughs at 2 dolls.	10
5 grease cans at 33 cents	1.65
Total	3304.15

Thus, both *The Principles* and the turnpike proposal effectively use the narration in nearly identical ways to *The Guide*. Evans deftly alters the content and the style to fit his varied audiences.

Evans then proceeds onto the longest part of *The Principles* (6 of the 12 paragraphs), which like the longest part of *The Guide* is his conformation wherein he describes in detail his own invention of the steam engine concept and three improvements. In like manner, Evans's turnpike proposal moves directly into totaling up the profits and expenses of the steam wagons in the dollars and cents fashion as he had with his narration: "First, cost in favor of the steam waggon, exclusive of drivers," and "Profits of the steam waggon per journey." This section of his turnpike proposal takes up about one third, the longest part, of the letter.

Finally, in the last two paragraphs of *The Principles*, in his confutation and conclusion, Evans moves to the reason behind this seemingly disinterested description of novel new principles to the members of Congress. Moreover, just as he had described in "Rules for Discovering New Improvements" in his conclusion to *The Guide*, and, in doing so, sought a way to rise above his projector status to a discoverer of eternal truths, so too, in *The Principles*, Evans seeks to amplify his argument to rise to a disinterested scientific role:

> Viewing the subject on the broad scale of public good, will it not be to the interest of this country, that the successful inventor shall benefit by the pursuit of his useful discoveries?...As often as a similar case occurs [the Copyright Law], the benevolent design of Congress in passing the patent law, will be totally frustrated. (qtd. in Bathe & Bathe, 1972, p. 302)

In like manner, in his turnpike proposal, Evans again amplifies his argument in a way appropriate to his businessman audience in terms of dollars and cents:

> I have no doubt but that my engines will propel boats against the current of the Mississippi, and waggons on turnpike roads with great profit. I now call upon those whose interest it is to carry this invention into effect. All which I respectfully submit to your consideration. (1805, p. 55)

Thus, in late 1804, on the eve of publishing *The Abortion* in 1805, Evans was engaged with his audiences and projects their benevolence in sure hope of persuading them to continue his patent protection or, alternately, to purchase steam waggons. For the Congressman, Evans illustrates his engagement with them by producing a short, seemingly disinterested pamphlet that follows all the architectural characteristics of classical rhetoric's disposition just as he had when similarly engaged with his audience in *The Guide*. For the turnpike directors, on the other hand, Evans produces a short sales proposal nearly devoid of complete sentences and looking very much like the debits and credits of a business balance sheet. And, in the proposal, Evans was not above including a number of profit-making pitches: "Upon the whole it appears that no competition could exist between the two. The steam waggons would take all the business on the turnpike roads" (p. 55). The focus in *The Principles* is clearly on the description of his own inventions and applications, with both the narration and confutation maintaining secondary roles, just as the focus in the turnpike proposal is on the profits to be made from the steam wagons.

The rhetoric of *The Principles* captures the spirit of patents because Evans hoped to project a disinterested purveyor of scientific information who, in exchange for novel inventions, is provided a time of protection and monopoly on the information. On the other hand, in the spirit of dollars and cents and profit in the turnpike proposal, Evans created a sharp contrast in tone, language, and involvement.

By looking at both *The Principles* and the turnpike proposal, one can see that Evans had the ability to be a deft rhetorician in late 1804, using the classical structure of disposition to produce two quite entirely different pieces, with different purposes, for two entirely different audiences. Congress must have received Evans's pamphlet in the manner he sought to convey it because one of their members, Samuel Lathan, a senator from New York, published Evans's pamphlet in its entirety in their new journal, *The Medical Repository and Review of American Publications*, in spring 1805 (vol. 8: pp. 317–318).

THE ABORTION OF THE YOUNG STEAM ENGINEERS GUIDE

In mid-January 1805, Evans sought to enlist subscribers who would underwrite the cost of publishing the steam engine book, just as he had done with *The Guide*. At this point in time, the title Evans publicized was to be an echo of his earlier book, *The Young Steam Engineers Guide* (1805, p. iv.). Yet by the end of January, two events sent him reeling and caused him to abruptly change the title of his steam engine book. First, in late January, Congress turned down his request for a patent extension for his automated flour mill. Their refusal immediately put Evans in a financial straitjacket for both building the scow, working out the applications of his engine, and publishing the new book. At the same time, Colonel John Stevens in New York obtained a copy of Evans's *Principles* (Turnbull, 1928). What Stevens read and what he knew of Evans's work caused him to feel that Evans had stolen his own high-pressure engine designs and was purveying faulty information. Thus, Stevens launched a public attack on Evans by publishing his letter in the very same issue of the *Medical Repository and Review of American Publications* that carried a reprint of *The Principles* (vol. 8: 317, 321–327).

Feeling rejected by Congress, in a financial straitjacket, and attacked for theft and falsehood, Evans resigned from the field: "His plans have thus proved abortive, all his fair prospects blasted, and he must suppress a strong propensity for making new and useful inventions and improvements" (1805, pp. iv–v). Within a month's time, Evans went from the prospect of a $10,000 profit on his patent extension to what he described as abject poverty: "I was left in poverty at the age of 50, with a large family of children and an amiable wife to support, for I had expended my last dollar in putting my 'Columbian' steam engine in operation, and in publishing the 'Steam Engineer's Guide'..." (1817, p. 18).

The result of these dramatic turn of events was that Evans's sense of the book had changed:

> Having read my work in sheets, as it came from the press, I am highly sensible of its deficiencies. It is truly an "Abortion of the Young Steam Engineer's Guide"...

His sense of the book changed perhaps in one way, as the title seems to indicate, because he felt the book was untimely born:

> The time which should have been occupied in examining the different authors who treat of the principles could not be spared. To compare their experiments, results and observations would have swelled the volume to a large size; and the drawings and engravings, necessary to have done justice to the subject, would have been very expensive. (1805, p. 134)

But more importantly, his sense of the book changed because Evans disengaged from his audience as he himself writes in the Postscript to *The Abortion* (p. 139):

I now conclude, and renounce all further pursuit of inventions and discoveries, at least until it shall appear clearly to be in my interest; lamenting that it should so often prove unprofitable and even ruinous as it has been to many.

Evans's secession from his rhetorical situation can be observed quite clearly in three ways. First, the materials written by Evans in 1804 and described above—The Principles addressed in a disinterested scientific manner to Congress and his turnpike proposal addressed in dollars and cents and profit to businessmen—are both reprinted in *The Abortion* without any apparent effort to reconcile their tones or design nor to adjust them to an audience of young steam engineers. Second, the careful balance of scientific disinterestedness and entrepreneurial pitch crafted by Evans in *The Guide* and *The Principles* is bluntly discarded in the advertisements, of which he includes:

- a three-page advertisement for the steam engine design, which appears before the table of contents;
- an appeal for compensation for his patented plans for employing his steam engine for distilling in Article 14;
- a five-page explanation and advertisement for a screw mill invention, which appears in Article 17;
- Article 18, which not only advertises non-steam engine patents, a straw-cutter, and flour press, but the patents are not even Oliver's, but are his brother's.[12]

Moreover, any hope to appear disinterested is upset by an appendix to *The Abortion* that includes vitriolic letters written between Evans and Stevens.

However, third, Evans's secession from his rhetorical situation is no where more apparent than in the choices he made for *The Abortion*'s disposition; a disposition which varied quite dramatically from his former close adherence to Cicero's rhetoric's architectural dictates. Table 3.2 summarizes how the parts of *The Abortion* fit into Ciceronian disposition, and what their positioning suggests about their various functions in *The Abortion*. Each part of the disposition is then explored in depth.

Introduction of *The Abortion*

In *The Abortion*'s introduction, Evans again sought to draw his readers in by using the introduction "corrective" to show that the subject of high-pressure steam engines had been misunderstood:

> As the title of this work arises from a peculiar circumstance, which may be deemed somewhat singular, it may therefore be gratifying to the reader to give an explanation. (1805, p. iii)

But where his persona appeared humble and modest in *The Guide*, his turnpike proposal, and *The Principles*, Evans's persona in *The Abortion* is accusatory, complaining, and arrogant. Speaking of himself, Evans wrote:

His plans have thus proved abortive, all his fair prospects are blasted, and he must suppress a strong propensity for making new and useful inventions and improvements; although, as he believes, they might soon have been worth the labour of one hundred thousand men. (p. iv)

TABLE 3.2
Five Parts of *The Abortion* with their Roles in Classical Disposition

Title	Comprising Articles	Rhetorical Function in Ciceronian Disposition
Preface	Preface	Introduction
Articles I–XI	Of Steam…to Scales of Heat and A Table of Strength of Metals	Narration
	—missing—	Partition
Articles XII–XVIII	Description of a steam engine on the new principle & Explanation of the screw mill invented by the author Useful inventions by different persons	Conformation
Appendix	Stevens/Evans/Mitchell debating letters	Confutation
Conclusion	Rules for Discovering New Improvements	Conclusion

It is gratifying to have others believe one's work "worth the labour of one hundred thousand men," but not very humble or modest to claim it for oneself. Surely, here is an introduction that will not make readers open and receptive, as *The Encyclopaedia* observed:

> If the orator set out with an air of arrogance and ostentation, the self-love and pride of the hearers will be presently awakened, and will follow him with a very suspicious eye throughout his progress. (Dobson, 1795, vol. 13: p. 389)

Perhaps it was an introduction like this that led Thomas Tredgold, Evans's British contemporary, to suggest:

> The 'Abortion' is a curious work; it betrays that strange mixture of absurd speculation and indistinct perception of truth, which distinguishes the generality of enthusiastic projectors, and is valuable only to those who can select by means of previous knowledge or experience. (1827, p. 43)

Background Facts of *The Abortion*

As in *The Guide*, Evans gives a *narration* by explaining the basic principles concerning steam, heat, and distillation. He again bolstered and framed the presentation of his own inventions for some 48 pages in Articles 1 to 11 by referring to such well-known scientists as Dalton,[13] Leslie,[14] and Black,[15] and reprinting material directly from the *Encyclopedia*. In this section of his disposition, Evans seems on familiar ground working with tables, footnotes, and lengthy quotes. This section is very similar in structure and approach to his earlier works, and one could conclude that this part of *The Abortion* was completed before the turn of events in late January, 1805. In fact, this section fulfills quite nicely the early advertisement Evans gave to prospective subscribers:

> an investigation of the principles, construction and powers of different Steam Engines...he [had] laid down a new and true theory, and developed principles of nature that may lead to highly important discoveries and improvements. (Bathe & Bathe, 1972, p. 107)

The Partition and the Conformation of *The Abortion*

According to Cicero's design, Evans should have next presented a brief partition giving an overview of what was to come in the conformation. Unlike *The Guide*, it never appears in *The Abortion*.

Evans's conformation, or core arguments, appears in articles 12–18, however, it is actually only in Article 16 that Evans gives a "description of a steam engine on the new principle." Most of the conformation is expended in descriptions of various applications of his high-pressure steam engine:

- the dredging scow completed and operated in the summer of 1805 for the Philadelphia Board of Health (Article 12);
- his 1804 proposal for steam-powered carriages earlier submitted to the Lancaster Turnpike Road Company (Article 12);
- his distillation ideas (Article 14);
- and his patent description for his own screw mill. (Later, mill licenses reprinted the graphic material for the screw mill from this section, just as the flour mill licenses reprinted sections of *The Guide*'s conformation.)

Additionally, unlike *The Guide's* conformation which only included Evans's own work, *The Abortion* includes the work of others (Article XVIII). This placement of material from other inventors diminishes the force of the presentation of Evans's own inventions by wandering off to descriptions of unrelated inventions (a straw cutter, a flour press, and a machine to remove earth).

Also in *The Guide*, Evans devotes some 21% of the book (Part III and Part IV) to a presentation of inventions for license; in *The Principles* Evans devoted 6 of 10 paragraphs to the presentation of his own ideas on high-pressure design and their applications, and in the three and a half pages of his turnpike proposal, he devotes about one and three-quarters pages to his core argument for steam wagons. However, in *The Abortion*, at the very point that Evans should begin to deliver the core of his presentation for his own invention, Evans devotes only 9% of the book to his conformation. Thus, where one would expect Evans to deliver his most significant part of his work in *The Abortion*, he moves very quickly through it. He may have breezed through his conformation because it may actually be the confutation of his opponents that captured his attention in 1805: *The Guide* devoted 22% in Ellicott's Part V to a confutation; *The Abortion* devotes 39%.

Confutation of *The Abortion*

Evans's confutation stands out front and center in *The Abortion*. In his confutation he repeats eleven and a half pages from Dobson's *Encyclopedia* to describe the history of the steam engine, but devotes 30 pages to an emotional exchange of letters between himself and Colonel John Stevens that had just recently appeared in the *Medical Repository and Review of American Publications* in 1805 (vol. 8: 321–327; Turnbull, 1928). Evans's reply in the same journal (vol. 9: pp. 120–127) to Stevens's assault upon his reputation relied heavily on sarcasm:

> Have I been half so dexterous as yourself, who sent Dr. Coxe to view my principles, then in operation and use one year, (publicly exhibited and explained to every one who inquired after the principles) and to put a number of questions to me, which drew in answer, a full explanation of the construction and principles of my invention, and which, when you were in possession of, the improvement became obvious to you, and you went and attempted to take out a patent for, and assumed it to yourself; but herein you have failed for want of a competent knowledge; besides you are not the original inventor.... Are you sure you are competent to assert, that I have assumed very erroneous principles while you show you do not understand them yourself?... Your ignorance of the principles of my invention has caused you to commit and set yourself in the way as an obstacle to the introduction of the most useful improvements ever made on steam engines... (Bathe & Bathe, 1972, pp. 105–106)

In 1805, when these heated letters appeared in the *Medical Repository*, it was the very year of *The Abortion*'s publication. Perhaps Evans's original intentions of presenting scientific information for practitioners and to sell his designs and patents—the intentions he had had for *The Guide*—became confused at the last minute by this heated exchange of letters and by an attack upon his reputation. Could it be that just when his eye should have been glancing through the final pages of his book, editing, amplifying, rewriting source material, and making his attacks more subtle, his eye became fixed on refuting his opponents? Thus, perhaps *The Abortion* becomes abortive both as a result of Evans having his motivation to publish severely hampered by the financial circumstances surrounding his patent affairs, and also because of a last-minute shift of focus caused by a personal attack by Stevens. Thus, where the confutation should operate in support of the conformation, it becomes an end on its own and supplants or, at the very least, confuses Evans authorial intentions. That something went awry at this point in his composition process is primarily suggested by noting that the confutation is larger than his conformation—a very unusual circumstance in classical rhetorical disposition.

Conclusion of *The Abortion*

Evans's conclusion—a part of his rhetoric that amplified his case in *The Guide* to the level of Euclid's *Elements*—collapses under the weight of his indignation and counterattack on Stevens in *The Abortion*:

> I now conclude, and renounce all further pursuit of inventions and discoveries, at least until it shall appear clearly to be my interest…

Thus, at the very point when he should amplify his arguments to excite the passions of his audience, as called for in *The Encyclopaedia*, he does just the opposite and minimizes all his efforts and retreats.

In sum, because of his changed financial circumstances and because of having to shift the focus of his efforts at the last minute to defend his personal reputation, Evans failed in *The Abortion* to craft a persuasive frame for the presentation of his high-pressure steam engine invention:

- by not creating a humble and modest persona in his Preface, his introduction;
- by diminishing the presentation of his own inventions through the inclusion of descriptions of unrelated inventions by others in the central statement of the book, articles XVI to XVIII, his conformation;
- by growing his counterattack upon his critics out of proportion to his central statement, as well as by countering his critics with sarcasm and rudeness in his Appendix, his confutation;
- and by surrendering and minimizing all his own future efforts at invention, rather than amplifying them in the Conclusion.

Two different books reveal differing choices of rhetorical arrangement made by Oliver Evans. His choices reflected different states of mind during his composition period, and produced different self-assessments. The marketplace's judgment was also quite different: *The Guide*, which fulfilled the requirements of a classical arrangement as described by the leading rhetorical handbook of the day, had 15 editions; *The Abortion,* which did not fulfill the requirements and whose structure embodied conflicting purposes, had only three.

One may speculate that Evans was quite aware of the deficiencies of *The Abortion* and, unlike *The Guide*, never stopped tinkering with the text until the day he died. He had a special edition printed that included 214 extra blank leaves amongst the printed pages of the book, and, from 1805 until his death in 1818, Evans proceeded to fill in these blank pages with new ideas, expanded explanations, and even a set of 25 questions and answers "to be answered by a Steam Engineer before he can be entrusted to build a Steam Engine." This edition was one of the prize possessions he bequeathed to his son, Cadwaller, upon his death (Bathe & Bathe, 1972, Appendix A and Appendix T).

Choosing whether or not to fulfill the requirements of rhetorical arrangement can have a dramatic effect on the ability of a writer to control his or her material, as well as the likelihood that audiences will understand and be persuaded by that material.

NOTES

[1] Evans probably intended the heavily nuanced word *abortion* in his title to indicate, as will be explained, the imperfect offspring of an untimely birth. *The Abortion* was once published by the author through Fry & Krammerer of Philadelphia; and the second time in 1826 by H. C. Carey & I. Lea. *The Abortion* was also translated by I. Doolittle into French in 1830. This French edition contains no publication dates and has been variously dated by bibliographic sources. See Bathe & Bathe (1972, p. 345) versus Library of Congress dating of the edition.

[2] A 1788 pamphlet, by Englehart Cruse, attacking the steam boat design of Evans's competitor, James Rumsey, was entitled "The Projector Detected or, Some Strictures on the Plan of Mr. James Rumsey's Steam Boat." The author of the pamphlet suggests the antebellum import of the word "projector" in the following three statements: "If you were lost in your profound researches, it is no more than what has happened to other great men-give you time, and time it seems you must have, and your ingenuity will be displayed with fire and smoke" (p. 7). Or, "Legislatures of any of the States can be induced thereby to patronize or throw away money on such vapoury exploits, they will justly merit the censure of their constituents, and ever man of common sense" (p. 12). Finally, "the world will consider you as a metaphysician, a builder of castles in the air!" (p. 13).

[3] Could Evans have anticipated the use of the book as evidence in later lawsuits? He instructs his lawyer to show *The Guide* "to impress the Court & Jury, with adequate conceptions, of the great exertions I have made, and the expense I have born to disseminate my improvements." ("Newspapers")

[4] Fulton and most others were using low-pressure steam engines. Evans's high-pressure engines were much more powerful—and much more dangerous. Hunter (1985, pp. 129, 136) pointed out the advantages of the high pressure design by noting that it primarily allowed the pressure and power of the engines to increase from 10 psi or about 7 horsepower to higher than 100 pounds per square inch producing 70 horsepower. It greatly simplified engine design in a kind of reverse technology evolution by allowing the designer to get rid of the most complex element of James Watts's invention, the condenser and its equipment; thus, the high pressure design greatly diminished the need for large amounts of cold water to operate the condenser. In a later biographic sketch of Evans (Sellers, 1886, p. 13), the author noted: "In *The Emporium of the Arts and Sciences*, Vol. 2 (1812), we find quite an extended account of the state of the steam engine at that period, and the feeling against the use of high-pressure steam is well illustrated by an account of the explosion of one of Trevethick's boilers with fatal effect. This fear of the power of high-pressure steam dated from the time of Watts, who thought Richard Trevethick ought

to have been hanged for using it, and was a potent factor in the opposition which Evans encountered in his efforts to introduce his engine."

⁵ In an earlier work on Evans, "Oliver Evans and the Weave of Text and Graphics for Antebellum Millwrights," (1998), I suggested another complementary explanation for the differences between *The Guide* and *The Abortion*, and that explanation was based upon the variety in the content of the two books vis à vis Evans's native learning style. In journals, advertisements, and letters to his sons on how to understand technology, Evans repeatedly exhibited himself as a visual learner, and, moreover, *The Guide* shows Evans as an author effectively translating much of his mathematical-verbal source material into visual information for common American mechanics. This must have been an effective manner of communication, for as the *Pittsburgh Gazette*'s April 30, 1819, eulogy mentioned: "A few men such as Franklin, Rumford, Davy and Evans, who bend the highest and most abstract principles of science to the use of man, in facilitating the common operations of life, will be remembered by a nation's gratitude, when the comparatively insignificant herd of metaphysicians and conquerors shall have passed into total oblivion" (qtd. in Bathe & Bathe, op. cit., 1972, p. 273). Thus, my earlier theory suggested that Evans personally needed to translate mathematical-verbal source material into visual information before he was able to control such source material rhetorically. Evans went through such a translation while writing *The Guide*, and the result was judged by himself and the public to be a success. However, similar mathematical-verbal source material for *The Abortion* never underwent a similar translation process, and thus, perhaps, remained only partially understood by Evans. The result of such partial understanding was that the material remained out of his rhetorical control, and thus both he and his public judged the book a loss. The essence of this present article's theory is that while such problems in Evans's own composition process may indeed explain some of the differences between *The Guide* and *The Abortion*, his own disengagement with his audience, arising from dramatically changed financial circumstances on the very eve of *The Abortion's* publication, led Evans to publish source material in an ineffective rhetorical manner.

⁶ In the usual 18th century manner of wholesale reprinting, Dobson reprinted much of the third edition of the *Encyclopaedia Britannica* because the American copyright law of 1790 protected the rights of American publishers, while permitting them to republish British and continental works as their own (Arner, 1991). In its turn, the *Encyclopaedia Britannica* had largely reprinted John Wards's 1759 series of lectures at Gresham College, called *A System of Oratory* ("the most extensive restatement of ancient rhetorical theory in English up to 1759" [Ehninger, 1951, p. 8]). Ward tended to follow the hallowed conventions of Cicero's disposition rather than some of the new, simpler Aristotelian formulas coming into vogue at the time, and such as were used in the composition of the Declaration of Independence (Ward, 1759, vol. 1, lecture 12; Howell, 1971). That it is was Ward's rhetoric transmitted through the two encyclopedia intermediaries that probably provided rhetorical models for Evans corrects an opinion in my earlier article on Evans (Brockmann, 1998, chap.1) that it was Hugh Blair's quite popular rhetoric that had probably provided the models. Not only is it more likely that it was Ward, but if information on rhetoric came to Evans in the pages of Dobson's *Encyclopaedia*, then it is interesting to note one of the very few biographical entries from the *Encyclopaedia Britannica* that Dobson chose not to reprint in his American *Encyclopaedia Supplement* (1798–1803) was one on Blair. Arner suggests Dobson made this decision because of Blair's outspoken opposition to the American Revolution (Arner, pp. 168–169).

⁷One could say that this statement's feigning air of modesty in *The Guide* is the intended effect of his statement in the Postscript to *The Abortion*, where he wrote: "Having read my work in sheets, as it came from the press, I am highly sensible of its deficiencies. It is truly an 'Abortion of the Young Steam Engineer's Guide.'" If the intent in both instances was the same, it would suggest that the word "Abortion" in the title of the second book was a rhetorical feint of modesty by Evans and that he truly had a much higher opinion of the book. However, compare the construction of the two sentences and observe how the first undercuts its self-deprecation, and the second does not. In *The Guide,* Evans undercuts his own declaration of insufficiencies through the use of double negatives—"not safe to conclude," "without error;" an embedded phrase disrupting the flow of the sentence; and the lack of a stated subject (who exactly does the concluding?). However, in *The Abortion,* the self-deprecation is very blunt, direct, and personal, without any intentional undercutting: "I am highly sensible;" "It is truly." That Evans was sincere in the second evaluation finds support in what a life-long friend, Hezekiah Niles, wrote in the July 5, 1834, edition of the *Niles Register*: "Had Mr. Evans not been rendered almost misanthropic by what he, (as we thought, erroneously), believed was the injustice and ingratitude of the public, we are of opinion that a discovery made by him, as to the application of steam power, would have been proclaimed, which, even at this day, would be regarded wonderful."

⁸Gross explained this name dropping and reference to widely-respected books as the use of the rhetorical method of ethos, "the persuasive effect of authority." Gross continued: "Innovation is the *raison d'etre* of the scientific paper; yet in no other place is the structure of scientific authority more clearly revealed. By invoking the authority of past results, the initial sections of scientific papers argue for the importance and relevance of the current investigation; by citing the authority of past procedure, these sections establish the scientist's credibility as an investigator" (1990, p. 13).

⁹"Closely allied to the kind of conclusion that recapitulates or summarizes is the one that generalizes. This is the kind of conclusion that broadens and extends the view of the problem or issue that we have been considering in the body of the discourse, that considers the ultimate consequences" (Corbett, 1990, p. 312).

¹⁰ The Lancaster Turnpike was the first extensive turnpike in America, covering some 62 miles and crossing at least two rivers and many streams by bridge (Bathe & Bathe, 1972).

¹¹ Evans, op. cit., 1805, Postscript. It is difficult to describe exactly how much of Evans's $10,000 would be worth in today's dollars. For example, Derks (1994) does not have data going back far enough. However, prices and salaries of the time can give some sense of the value of $10,000 in 1805. For example, when George Washington wrote to Evans in 1798 requesting help in hiring a miller for the Evans's mill he had constructed at Mount Vernon, Washington wrote that he intended to pay $16 dollars annually in addition to providing the miller with food, firewood, a milking cow, and a cottage (Bathe & Bathe, 1972, p. 62). Without these "benefits," the annual salary for a millwright in 1800, according to Wright (1895), was approximately $260 for 6 days work a week, 10 hours a day. In comparing the cost of objects, the price of *The Guide* with 300 to 400 pages, hardbound binding with lettering on the spine, was $2.00; one of Evans's letters in 1804 mentions that a single bag of feed cost $1.00 (Bathe & Bathe, 1972, p. 98), and, in 1806, when Evans purchased nearly a quarter of an acre lot inside the city limits of Philadelphia, bounded on four sides by streets, he paid $2,000 (Bathe & Bathe, op cit., p. 121).

¹² This entire article was dropped by the publisher from the revised edition of *The Abortion* in 1826.

[13] John Dalton was a famous chemist and secretary of the Literary and Philosophical Society (Musson & Robinson, 1969).

[14] John Leslie was a professor of natural philosophy at Edinburgh University and author of *An Experimental Inquiry into the Nature and Propagation of Heat* (Evans, 1804).

[15] Dr. Black was a chemist at Glasgow and Edinburgh universities (Musson & Robinson, 1969).

REFERENCES

Arner, R. D. (1991). *Dobson's encyclopaedia: The publishers, text, and publication of America's first britannica, 1789–1803*. Philadelphia: University of Pennsylvania Press.

Bathe, D., & Bathe, G. (1972). *Oliver Evans: A chronicle of early American engineering*. New York: Arno Press.

Brockmann, R. J. (1998). *From millwrights to shipwrights to the twenty-first century: Historical considerations of American technical communication*. Cresskill, NJ: Hampton.

Carey, Matthew. (1792, May). Editorial. *American Museum or Universal Magazine*, 225–226.

Corbett, E. P. J. (1990). *Classical rhetoric for the modern student* (3rd ed.). New York: Oxford University Press.

Cruse, E. (1788). *The projector detected or, some strictures on the plan of Mr. James Rumsey's steam boat*. Baltimore: John Hayes.

Derks, S. (Ed.). (1994). *The value of a dollar, 1860–1989*. Detroit, MI: Gale Research.

Dobson, T. (Ed.). (1790–1798). *The encyclopedia; or a dictionary of arts, sciences, and miscellaneous literature* (Vols. 1–18). Philadelphia: Thomas Dobson.

Dobson, T. (Ed.). (1798–1803). *Encyclopedia supplement*. Philadelphia: Thomas Dobson.

Ehninger, D. (1951). John Ward and his rhetoric. *Speech Monographs, 17*, 1–16.

Emporium of the arts and sciences (Vol. 2). (1812). Philadelphia: J. Delaplaine.

Evans, O. (1794). Letter of September 12 to John Nicholson. Copy of letter in Grenville Bathe papers in the National Museum of American History, Smithsonian Institution; original letter was owned by Simon Gratz.

Evans, O. (1795). *The young mill-wrights and millers guide*. Wallingford, PA: The Oliver Evans Press.

Evans, O. (1804). The principles of steam engines. In *Early American imprints* (Second Series no. 6265).

Evans, O. (1805). *The abortion of the young steam engineers guide*. Wallingford, PA: The Oliver Evans Press. (Reprinted in facsimile, 1990)

Evans, O. (1813). *Patent right oppression exposed, or knavery detected. By P. N. I. Elisah, Esq*. Philadelphia: R. Folwell.

Evans, O. (1817). *Oliver Evans to his counsel who are engaged in defense of his patent rights*. Philadelphia.

Evans, O. (1824). *The young steam engineers guide*. Philadelphia: H. C. Carey & I. Lea.

Evans, O. (1826). *The young mill-wrights and millers guide* (T. Jones, Ed.) (5th ed.). Philadelphia: M. Carey & Sons.

Ferguson, E. S. (Ed.). (1965). The Philadelphia of Oliver Evans. In *Early engineering reminiscences (1815–40) of George Escol Sellers* (p. 38). Washington, DC: Smithsonian Institute.
Ferguson, E. S. (1980). *Oliver Evans, inventive genius of the American industrial revolution.* Greenville, DE: Hagley Museum.
Ferguson, J. (1772). *Lectures on select subjects.* London: W. Strahan and F. Rivington.
Franklin, B. (1990). *Autobiography* (D. Aaron & J. A. L. Lemay, Eds.). New York: Vintage Books.
Gross, A. G. (1990). *The rhetoric of science.* Cambridge, MA: Harvard University Press.
Howell, W. S. (1971). John Ward's lectures at Gresham College. In *Eighteenth-century British logic and rhetoric* (pp. 83–124). Princeton, NJ: Princeton University Press.
Hunter, L. C. (1979). *A history of industrial power in the United States, 1780–1930, volume one: Waterpower in the century of the steam engine.* Charlottesville, VA: University Press of Virginia for the Eleutherian Mills-Hagley Foundation.
Hunter, L. C. (1985). *A history of industrial power in the United States, 1780–1930, volume two: Steam power.* Charlottesville, VA: University Press of Virginia for the Eleutherian Mills–Hagley Foundation.
Martin, B. (1771). *Philosophia Britannica.* London: John Franic, Charles Rivington, etc.
Musson, A. E., & Robinson, E. (1969). *Science and technology in the industrial revolution.* Manchester, England: Manchester University Press.
Newspapers & Advertisements Certificate Etc. from James L. Woods's collection of Evans's papers, Franklin Institute Library.
Niles, H. (1834, July 5). Editorial. *Niles Register, 50.*
Odell, L., Gotswami, D., Herrington, A., & Quick, D. (1983). Studying writing in non-academic settings. In P. Anderson, R. J. Brockmann, & C. Miller (Eds.), *New essays in scientific and technical communication: Research, theory, and practice* (pp. 17–40). Farmingdale, NY: Baywood Publishing.
Review of the book *The Young Mill-Wright and Millers Guide.* (1795, September). *American Monthly Review, 9,* 1–5.
Sellers, C. (1886). Oliver Evans and his inventions. *Journal of the Franklin Institute,* 122.
Smeaton, J. (1794). *Experimental enquiry concerning the natural powers of wind and water to turn mills and other machines.* London: I & J Taylor.
Tredgold, T. (1827). *The steam engine, comprising an account of its invention and progressive development.* London: John Weale.
Turnbull, A. D. (1928). *John Stevens: An American record.* New York: The Century.
Ward, J. (1759). *A system of oratory delivered in a course of lectures publicly read at Gresham College, London* (Vols. 1–2). London: Georg Olms Verlag. (Reprinted in facsimile from 1759 ed., Hildesheim, Germany).
Waring, E. (1793). Observations on the theory of water mills. In *Transactions of the American Philosophical Society* (pp. 144–149). Philadelphia: Aitken & Son.
Wright, C. D. (1895). *Industrial evolution of the United States.* Meadville, NY: Flood & Vincent.

4

Sada A. Harbarger's Contribution to Technical Communication in the 1920s*

Teresa Kynell
Northern Michigan University

As Jo Allen (1994) suggested in her piece "Gender Issues in Technical Communication Studies," while some scholarship has appeared on gender studies and technical communication, "other issues of gender have found virtually no place in our studies" (p. 161). One area that warrants further investigation is the historical role of women in the evolution of technical communication as a distinct discipline in this country. Certainly interest in the history of technical communication is evident in works like Russell's *Writing in the Academic Disciplines* (1991), Kynell's *Writing in a Milieu of Utility* (1996), and Tebeaux's *Emergence of a Tradition* (1996). Each, too, highlights the contributions of some important individuals. Tebeaux, in fact, evaluates the contributions of women during the English Renaissance and the 17th century to demonstrate that women were active technical writers, involved in the running of their households and writing how-to books for other women involved in the same enterprise. Thus, a redefinition of workplace—that is, home—reveals a thriving arena for women (see *Technical Communication Quarterly*, Summer 1998). What is striking about these first women technical writers is both the voice and format of their writings. Women wrote about medicinal potions, home cures, giving birth, canning and preparing foods, etc. They wrote, then, as women for

---

* Many thanks to the Martha Kinney Cooper Ohioana Library for providing me with copies of S. A. Harbarger's biographical data and publishing history.

other women. The research of Tebeaux and others, as a result, opens up a vast area of further inquiry for future scholars. Another important, though perhaps less-researched area, involves the contributions of women who changed the direction of technical communication pedagogy in an environment dominated by a primarily masculine scientific hegemony.

One such individual, Sada A. Harbarger, is noteworthy for her contributions to the field of technical communication in the 1920s, indeed, perhaps influential in her elevation of the discipline as it evolved from a predominately composition-related pedagogy to an engineering-centered foundation for writing. As Sue Ellen Holbrook (1991) noted in "Women's Work: The Feminizing of Composition," the long-standing association between the feminine and writing has been well documented (p. 201). As this study will point out, Harbarger found herself initially involved in teaching composition (a service course routinely taught by women), but embraced and advanced the shifts that led to an engineering English curriculum. Her contribution, thus, was not necessarily a "feminization" of the discipline—though her obituary described her in nurturant terms—but an adherence to the principles of the discipline she came to understand and indeed even define. Tracing her contributions permits a better understanding of a "woman's place" in higher education and how one woman reconceived that place not only for herself but for the development of the discipline as well.

Because Harbarger spent the bulk of her career in an engineering environment, some background on the discipline of engineering from roughly 1850 to 1862 will help establish ways in which English played a role in an engineer's education. Most engineers in that time period were trained (as opposed to educated) by either apprenticeships or through a random engineering or science course within a standard baccalaureate curriculum. Thus, anyone working as an engineer during or prior to this period was typically perceived by members of the "professional" class as "vocational." By 1862, when the Morrill Act was enacted to establish a permanent endowment of acreage and funding to promote both liberal and practical education, land grant colleges emerged as places where engineering would become a curricular alternative. By 1870, though, engineers were concerned about the lingering perception of their field as vocational. Engineering educators embarked on a series of curricular revisions as a means both to address the vocational issue and to elevate the social status of engineers. English instruction emerged as one curricular cure to the status concerns (literature) and the growing concern over the near illiteracy of many graduating engineers (writing).

Thus, writing instruction became part of a necessary means to resolve remaining status concerns and solve illiteracy problems. Engineering students were required to take, as a result, a composition course—a current traditional pedagogy of drills, practices and exercises—and, in some cases, a literature course. By the end of the 19th century, however, the students themselves were beginning to rebel against extra coursework in an already crowded curriculum; they could find little purpose in either composition or literature since both seemed far removed from both the engineer's purpose and the generally practical nature of the engineering curriculum. Indeed, one faculty member, Professor H. L. Creek (1939), speaking at the Society for the Promotion of Engineering Education, noted that students often perceived their English teachers as "not masculine." "One cannot," he continued, "simply glide into a classroom and greet a class of engineers with a sweet schoolgirl smile and my is not this a beautiful spring morning" (p. 301). The association of the composition requirement with that which is feminine is not surprising. Holbrook (1991) postulated that the connection of writing and civilizing skills with the female or nurturant figure led to a concentration of women "in the lower ranks in certain subjects (including the Humanities)" (p. 204).

Compounding the problem was the lack of English faculty—and certainly women—involved in any national dialogue on the role of English in an engineering curriculum. Then too, because teachers of English typically specialized in literature, few of the English faculty teaching composition in an engineering environment embraced the job. Those that did, though, truly charted a course very different than the one in which they had been trained to face. In fact, if composition already held the relatively low status of "service" course, imagine the status accorded *engineering English*.

A great irony in the evolution of technical communication in an engineering curriculum was the virtual second class status imposed on the discipline by those who taught it. Perceived as a service course in much the same manner as composition, faculty were naturally reluctant to teach a course for which they either were not trained or which offered little chance for professional advancement. Indeed as Holbrook (1991) noted in her piece, "composition teaching began positioned in the lower status" (p. 207).

Many educators who taught English for engineers perceived themselves to be looked down upon by colleagues who were teaching what virtually all English faculty were trained to do—teach literature. As a result, teachers of engineering English were, as Russell (1991) noted, "beneath the teaching of literature [and] beneath the teaching of engineering" (p. 122). With little opportunity for professional development, such faculty often felt stranded, teaching a service course that meant, as Connors (1982) noted, "no glory and no real chance for professional advancement" (p.335). Quoted in the English Committee's minutes of the Society for the Promotion of Engineering Education, W. T. Magruder, a faculty member, called English "the scullery maid of our engineering college household" (1916, p. 185). The connection of the English requirement with the female figure of relative low status was not unintentional. In this professional atmosphere, English faculty certainly did not embrace teaching students who perceived them negatively, and certainly engineering faculty were, at the least, uncertain of the role of English in the curriculum of their students.

In this educational and cultural milieu, Harbarger's contributions are even more striking. She emerged during a period of professional uncertainty, yet she embraced the discipline and arguably shaped it in her unfailing devotion to its dictates. She approached technical communication in much the same manner as a colleague, J. Raleigh Nelson. Chair of the 1922 Society for the Promotion of Engineering Education Conference of Teachers of English, Nelson proclaimed:

> I teach the course in engineering reports to seniors. I do not believe anybody was ever born to begin more poorly adapted to do this work than I, because I had no technical training, or no particular taste for engineering. It was a very big cross to me. I took it up as a consecrated cross and I bore it bravely. I made myself think it was necessary. I know I could not read those papers as well as an engineer could have done it but I have been conscientious about it and I am reaping my reward. (1922, p. 264)

Harbarger's dedication to the discipline was no less sincere. Indeed, her enthusiasm was infectious, as evidenced in her obituary, published along with her writings in a collection titled, *Sada A. Harbarger, 1885–1942* in 1942. A colleague at Ohio State University remembered her as a teacher of agriculture and engineering students, a woman who had an unusual skill in dealing with young men and helping them in the practical mastery of English. (He continued that although she dealt primarily with male students, she retained all her womanly traits.) The unknown writer of her "Consummation," included in her memoir, continued:

> There must be literally hundreds of men working in these fields [agriculture and engineering] to whom her passing brought genuine pangs of regret. Continuing the sort of work she had begun here, as an alumna coming home, the same unusual skill in dealing with young men, the knack of getting on with them, counseling them, helping them in the practical mastery of English, interesting herself in an almost maternal way not only during their student life, but keeping track of them long after they had left the University. (p. ii)

Her obituaries reflect not only the lack of women connected to the profession of engineering, but the general inability of the men writing those obituaries to separate her professional contributions from her gender. Thus, she worked with men but remained "womanly," and interested herself in the development of male students in an almost "maternal" fashion.

The Ohio State Lantern (1942), on the other hand, remembered her for working in one of the most difficult fields of English, that of providing instruction for professional and technical students. The *Journal of Engineering Education* (1942) noted that she was widely respected by men in the engineering profession (p. 776), and the Society for the Promotion of Engineering Education remembered her as one of the first teachers to have taken an active part in developing the work of the English Section. Her work with engineering students, the SPEE noted, was at first accidental. As a young instructor in the English Department at the University of Illinois, she took over some sections of engineers that none of the other instructors wanted. "In doing so," C. W. Park, Chairman of the English Division, continued, "she turned 'an educational chore into a career' " (1942, p. 365).

Indeed, Harbarger found herself almost completely in the company of men throughout her career[1]; however, the enormous respect implicit in the comments of male colleagues for Harbarger points to the tremendous commitment she made to the teaching of writing for engineers. First, and in many ways foremost, Harbarger was known and is still known for her 1923 textbook, *English for Engineers*, a text which, unlike other texts of that period, espoused a kind of workplace, or "real world," philosophy; an appeal, if you will, to engineering students that they embrace the relevance of writing instruction as key to their ultimate professional success. Harbarger's book, unlike T.A. Rickard's 1920 book, *Technical Writing* (then in its second edition) and Trelease and Yule's text *The Preparation of Scientific and Technical Papers* (1925), was a departure for a variety of reasons. First, both texts reflected a decidedly academic, almost distancing tone. Harbarger, whose first chapter is entitled "Professional Prestige and English," very clearly links job-related success with English skills, which is a much more direct attempt to pull the student into the study of English.

Harbarger, in fact, spent three chapters convincing students that good English skills—and specifically good technical writing skills—would make them more marketable. While this may, on the surface, appear to be an appeal to the most vulnerable aspect of a student's psyche (achieving financial success), Harbarger was really reflecting the consensus of the period. English had to be practical and work-related to be relevant to an already very busy student body.[2] In fact, before Harbarger moved on to discussions of letters and reports, she made very clear to students that ultimately they were the salesmen of their own ideas, and English represented the best means to sell them. In her *English for Engineers* (1923), she wrote:

> To the employer of engineers in the large industries, skill in English is one of the specifications for the employees who are, in one way or another, to be salesman whether they are in the sales department or the factory. (p. 5)

Harbarger's book, and her teaching philosophy, endured through several more editions of *English for Engineers*, including the fourth edition, published after her death and updated by Ohio State alumni Anne B. Whitmer and Robert Price (1943).

As significant as her textbook was for teachers, her other considerable contributions are, in many ways, more important. In fact, her overall contributions fall into three definable categories. First, Harbarger was a true pioneer. Her work in technical writing for engineers was original and thus played a role in the evolution of the discipline in the United States. Second, though Harbarger promoted the practical applications of technical writing to her students, she was also a proponent of teaching rhetoric as the primary base for technical writing. Last, as a woman who climbed to national rank in many important organizations, she played a key role in guiding the future of technical writing.

PIONEERING THE PEDAGOGY

Though Harbarger found herself teaching English to engineers "by accident," she wrote often about the special skills necessary for an Engineering English teacher.

Her articles, many of which detailed the qualities of a good technical writing teacher, are straightforward and uncompromising. Harbarger believed, as she wrote in her 1919 SPEE article "Some Unconsidered Factors," that English teachers who work with engineers should be eager, enthusiastic, and interested in the principles of science and engineering. Interested, she continued, but not necessarily an engineer or technician because "a general knowledge of the sciences and a broad view of the technical course means less danger of making a humanistic subject too highly specialized and too technical in its leaning" (p. 280). Harbarger was one of the first to advocate a unique kind of English teacher for engineering students, one who had not only an interest in the topic (and a passionate one), but an education in the humanities. Thus, Harbarger appealed to those English teachers dealing with the inequities of a service course (usually called *engineering English or technical exposition*) for engineers by demonstrating not only the importance of the course, but also by illustrating the reciprocity of English and more technical undertakings. Indeed, as she noted in her 1938 article, "Aims of the Applied English Courses," English courses for engineers have two focuses:

1. Utilitarian: to prepare students to meet acceptably the writing demands of later professional courses, and still later active professional practice.
2. Cultural: to add and enrich the practical, and to inculcate a point of view toward general reading for later adult study and enjoyment. (p. 340)

Her belief that the course in technical communication could "enrich the practical" came at a time in the history of engineering education when curriculum reformers were concerned about status-related issues, professionalism and lingering perceptions of vocationalism connected to the discipline. Harbarger believed that the technical and humanistic need not be artificially separated to provide the engineer experience in practical writing; instead, she believed that the cultural was implicit in the activity of technical communication itself.

Harbarger continued in "Some Unconsidered Factors," "the engineer is perceived as little more than an earth-bound being whose spirit is easily stirred only the most crass materialism. No one will deny that he is practical; that he demands tangible results from every project upon which he expends his thought and energy" (1919, p. 279). In addition, Harbarger, in a noteworthy 1920 SPEE article, "The Qualifications of the Teacher of English for Engineering Students," engineering students will work hard for an English teacher if that teacher "realizes that the engineer needs English as a tool" (p. 301). More importantly, though, she emphasized that the English teacher "is free from the *culture obsession*...[and] associates English, therefore, with reality" (p. 302).

The *culture obsession* reference is directly attributable first to the concerns of engineering educators implicit in students' greater exposure to culture (usually in the form of literature) as an increase in status, status enjoyed by other "professionals" like doctors and lawyers and second, to the training of virtually all university-level English faculty—in literature. In other words, English faculty in an engineering environment often found themselves teaching literature as a means to ennoble the engineer. The results were largely disastrous for both faculty and students. Students needed a reading/writing experience grounded in the practical. English faculty, frustrated by negative student attitudes, saw little chance for advancement in such an environment. Student attitudes were sometimes so negative, notes Connors (1982), that students regularly regarded their English teachers as "effeminate," with one even calling his teacher a "budding pinko" (p. 337).

The role of the engineering English teacher, according to Harbarger, was three-fold: to reinforce engineering principles through English instruction, to connect English to the future professional life of the engineer, and to view English as the link to professional and social success. She was not, however, arguing that literature or writing in isolation was a means to those ends. Instead, Harbarger believed that engineering and English had to band together in order to make the course in technical English more relevant to engineering students. In an article, "Better English," co-authored with William H. Hildreth, Harbarger (1934) argued convincingly that placing the entire burden for proficient writing upon the English department presupposed the separation between writing-related interests and engineering interests. She, very much in keeping with trends of the period toward greater cooperation, worked with the College of Agriculture as a cooperative effort.

> It has been co-operation in the true sense. Each not only has recognized individual responsibility, but has had a particular part to contribute toward sustaining and maintaining the common aim and objective—acceptable English in all written work. (p. 148)

Distinctive features of the Ohio State University plan included: an English consultant, a whole-college approach to the initiative, and involvement of administration in mounting the effort (p. 149).

It should be noted that the success of Harbarger's Ohio State University experiment in cooperation as a means to increase the viability and visibility of technical communication was, in some ways, unique. Experiments in cooperative efforts at other institutions often resulted in a model of instruction that deemed the engineer qualified to assess the technical aspects of the paper, whereas the English teacher was left to assess only the mechanical correctness of the piece. An SPEE English Department survey (1925) on cooperative models revealed the negativity of some involved.[3] One English faculty member perceived that "instructors in English are merely clerks to mark spelling and grammar" (p. 331). Harbarger, in the conclusion of her article, seemed aware of the potential problems in cooperation, noting that "no attempt at a co-operative project in English...should be undertaken until there is evidence of unity of objective both on the part of the college and the English Department" (1934, p. 154). Harbarger's model, not unlike those tried successfully at the University of Cincinnati and the University of North Carolina,[4] played a role in the development of a technical communication curriculum in this country, as engineering educators slowly recognized the unique features of an English course that reflected the principle purpose of the practical, utilitarian discipline of engineering.

In addition to creating an atmosphere where experiments in technical communication could succeed, Harbarger called for English teachers to write articles for technical journals (as she did), in the belief that the reputation of English instructors with their students would rise if their work were printed alongside the work of engineers. "The English instructor," she wrote in the article, "Qualifications of the Teacher of English," "has a splendid chance to demonstrate that he can practice what he preaches" (1920, p. 303). Although Harbarger encouraged English faculty to write for technical journals in order to increase students' estimation of them, there seems little doubt that she also meant that perhaps engineering faculty as well might regard the English instructor differently if, rather than emphasizing the ways in which the humanities and sciences were different, similarities between the two were instead highlighted. Students and engineering faculty, stressed Harbarger, needed to see what English had in common with the sciences, not what it could do to improve the sciences.

A FOUNDATION IN RHETORIC

Interestingly, Harbarger felt that one way to demonstrate what English had in common with the sciences was to teach rhetoric. To view rhetorical principles as foundational to effective communication skills means establishing the fundamental, unique qualities of technical communication—coding, decoding, and translating information for the user. To recommend the course as she did in her 1916 article, "A Defense of Rhetoric," printed in a highly technical journal, *The Technograph*, was to look ahead to the ultimate goals and needs of the technical student. Harbarger felt the value of rhetoric for an engineering student was implicit in what she felt were the practical goals of rhetoric. Rhetoric, she wrote, "first equips the engineer to express satisfactorily his ideas," so it therefore is a tool that can be put to immediate use. Second, rhetoric, she argued, held a service-related value. "Though two men," she noted, "may possess an equal amount of technical information, the man who is capable of using his mother tongue accurately and effectively can gain recognition where the one unskilled in rhetoric cannot." Finally, she added, rhetoric helps the engineer to explain the technical so that the uninformed, the disinterested will not only admire mechanical perfection but also appreciate the possibilities; therefore, "rhetoric is as essential to a technical course as theoretical and applied mechanics." Harbarger truly believed that engineers and their students failed to recognize the value in rhetoric, particularly the value in helping technical students to understand better the intricacies and practical uses of the language in the real world.

In fact, as she noted in the conclusion of "A Defense of Rhetoric,"

Someone has said: "Tell me what a man reads, and I'll tell you what he is." I should like to adapt this quotation thus: "Tell me an engineer's final attitude toward rhetoric and I'll tell you what kind of an engineer he is." (n.p.)

Even as early in her career as 1916, Harbarger was bringing together the pedagogical strands—in rhetoric and in its practical application—that would allow her later experiments in technical communication to help shape the discipline as we know it today.

I should add that although Harbarger advocated a rhetorical foundation for her students, she also taught (along with other early technical writing faculty like Samuel Chandler Earle at Tufts University) recognizable forms. In a 1933 book titled, *Service Studies in Higher Education*, Harbarger wrote Chapter 11, "Courses in Applied English." In it, Harbarger put together a year-long plan of study for engineering English faculty, beginning with the mandatory freshman writing experience, moving to a second quarter professional writing experience, and culminating in a third quarter researched effort drawing upon the "most frequently used forms of professional writing of the preceding quarters" (p. 160). Harbarger believed that the package she put together for her engineering students not only prepared them for the workplace, but made their chosen field of study ultimately more relevant.

For example, in the second quarter of Harbarger's proposed year-long curriculum for the engineer, she advocated technical description, technical definition, writing of specifications, and a series of memos and a long report reflecting observation, inspection, or interview (p. 271). These assignments reflecting the move to written discourse in technical communication are easily recognizable today and highlight not only Harbarger's contribution to a then still-evolving curriculum in technical writing, but as well her commitment "to the use of English in advancing professional prestige" (1938, p. 157). Throughout her career and during her varying experiments in cooperation and English curricular advances, Harbarger consistently perceived the lingering status-related concerns *vis-à-vis* engineering and, as a response, helped shape the direction of technical communication by focusing attention on discourse skills as a bridge between the practical or utilitarian and the humanistic and enculturating.

NATIONAL SERVICE CONTRIBUTIONS

So passionate was Harbarger's belief in the place and virtues of a solid preparation in technical writing, that she not only published considerably in a field dominated by engineers, but she went on to national rank in the one organization that oversaw curricular changes in engineering education in this country—the Society for the Promotion of Engineering Education. By 1934, the date, in fact, of the publication of the third edition of her book *English for Engineers*, Harbarger had an impressive national record. She served as a member of the Committee on English for the Society for the Promotion of Engineering Education from 1918 to 1928. In 1928, she became the Chair of the Committee on English. She was on the Guiding Council of the SPEE from 1926 to 1929, and in 1932 became the local director for the SPEE's summer institute.

In the Report of the Committee on English in June 1928, for example, Harbarger (still a member of the committee and about to become chair) is cited for a large-scale study of English education for engineers in American programs. Her study, interestingly, assured the committee that there was no "alarm as to the likelihood of the professional aims of engineering students being overstressed or the broadening aspects ignored in the first year of English" (p. 325). The emphasis of her study, the committee report continued, was "very strongly upon the supreme importance of teachers with right ideals" (p. 325).

Not surprisingly, when Harbarger subsequently became Chair of the Committee on English, the issue of the "teacher" remained paramount. She noted in the June 1929 report, her first, the need to put together materials and outlines from experienced teachers "for the guidance of the young and inexperienced teachers of English to engineering students" (p. 203). She concluded her report by noting that the contribution of the English teacher is dependent upon

1) the personality of the teacher, which obviously affects,
2) the presentation of the material, or the project, and
3) the cooperation of the instructors of the technical subjects through their handling of their class material. (p. 204)

It is not surprising that Harbarger's first report would reflect such emphasis on the teacher, for the qualifications of the teacher of English—the title of her 1920 article—remained her great concern. Harbarger was a woman who devised an early, if rudimentary, technical writing curriculum, worked with engineering faculty to make the curriculum work for a specific kind of student, and ultimately came to passionately appreciate, the topic she was virtually relegated to as a new English teacher.

Thus, Harbarger's contributions cannot be overlooked as we in the discipline continue to trace our own evolution. From her popular textbook, to her considerable publications, to her national rank in an engineering society, S. A. Harbarger is truly a woman, a teacher, and a technical writer, worthy of (re)consideration and further exploration.

NOTES

[1] Connors (1982) describes Harbarger as "tough-minded and professionally determined." He also suggests that her decision to use the initials S. A. (rather than her full name, Sada Annis) in her published material was "perhaps because the publisher felt that many readers

might resent a woman claiming to be able to teach technical writing" (pg. 335). It should be noted, however, that she regularly used her first name in the Proceedings of the Society for the Promotion of Engineering Education.

[2] Engineering educators were very concerned about the rigors of an engineering curriculum and the ability of students to complete a degree in four years. Any additions to the curriculum, therefore, had to be relevant and inextricably linked to the work of an engineer.

[3] For example, although 91% of respondents answered "yes" when asked if "cooperatively teaching English would promote the habitual use of correct English," only 43% answered "yes" when asked if cooperation was feasible.

[4] For more information on the two models of integrated cooperation, see Kynell, 1996, chap. 4.

REFERENCES

Allen, J. (1994). Gender issues in technical communication studies: An overview of the implications for the profession, research, and pedagogy. In P. M. Dombrowski (Ed.), *Humanistic aspects of technical communication* (pp. 161–179). Amityville, NY: Baywood.

Connors, R. J. (1982). The rise of technical writing instruction in America. *Journal of Technical Writing and Communication, 12,* 329–352.

Creek, H. L. (1939). Teachers of English in engineering colleges: Selection and training. *Proceedings of the Society for the Promotion of Engineering Education, 47,* 300–313.

English Department, The. (1925). *Proceedings of the Society for the Promotion of Engineering Education, 33,* 324–332.

Harbarger, S. A. (1916). A defense of rhetoric. *The Technograph, 1,* n.p.

Harbarger, S. A. (1919). Some unconsidered factors. *Proceedings of the Society for the Promotion of Engineering Education, 5,* 278–281.

Harbarger, S. A. (1920). The qualifications of the teacher of English for engineering students. *Proceedings of the Society for the Promotion of Engineering Education, 28,* 299–308.

Harbarger, S. A. (1928). *English for engineers* (2nd ed.). New York: McGraw Hill.

Harbarger, S. A. (1933). Courses in applied English. In *Service studies in higher education.* (Bureau of Educational Research Monograph #15). Columbus, OH: Ohio State University Press.

Harbarger, S. A. (1934). *English for engineers* (3rd ed.). New York: McGraw Hill.

Harbarger, S. A. (1938). The aims of the applied English courses. *American Journal of Pharmaceutical Education,* 339–344.

Harbarger, S. A. (1942). *Sada A. Harbarger: 1885–1942.* Columbus, Ohio: Ohio State University Press.

Harbarger, S. A., & Hildreth, W. H. (1934). Better English. *The Journal of Higher Education, 4,* 148–154.

Harbarger, S. A., Whitmer, A. B., & Price, R. (1943). *English for engineers* (4th ed.). New York: McGraw Hill.

Holbrook, S. E. (1991). Women's work: The feminizing of composition. *Rhetoric Review, 9,* 201–228.

Kynell, T. C. (1996). *Writing in a milieu of utility: The move to technical communication in American engineering programs, 1850–1950.* Norwood, NJ: Ablex.

Magruder, W. T. (1916). Discussion. *Society for the Promotion of Engineering Education, 24,* 183–193.

Nelson, J. R. (1922). Conference of teachers of English. *Proceedings of the Society for the Promotion of Engineering Education, 30,* 255–282.

Obituary. (1942, May). *The Journal of Engineering Education, XXXLL,* 776–777.

Park, C. W. (1942). Chairman of the English Division. *Proceedings of the Society for the Promotion of Engineering Education, 33,* 362–369.

Professor Harbarger succumbs. (1942, April 24). *The Ohio State Lantern.*

Report of Committee No. 12 - English. (1928). *Proceedings of the Society for the Promotion of Engineering Education, 30,* 324–326.

Report of Committee No. 12 - English (1929). *Proceedings of the Society for the Promotion of Engineering Education, 20,* 202–204.

Rickard, T. A. (1931). *Technical writing* (3rd ed.). New York: Wiley.

Russell, D. R. (1991). *Writing in the academic disciplines: 1870–1990: A curricular history.* Carbondale, IL: Southern Illinois State University.

Tebeaux, E. (1996). *Emergence of a tradition.* Amityville, NY: Baywood.

Trelease, S. F., & Yule, E. S. (1937). *The preparation of scientific and technical papers* (3rd ed.). Baltimore: Williams and Wilkins.

part II
Key European Movements in the History of Technical Communication

5

The Emergence of Women Technical Writers in the 17th Century: Changing Voices Within a Changing Milieu

Elizabeth Tebeaux

In 1985, Michael Moran wrote that the history of technical writing has not been written, but, as William Rivers reported in 1994, segments of technical communication's history are now being explored. The increase of interest in the history of technical communication has come at a time when studies of women's contributions to technical communication are also emerging. However, most studies of technical communication's history and women's contributions have focused on the 19th and early 20th centuries and on individual women technical writers in those periods. In this essay, I want to move to an earlier period, the English Renaissance (1475–1640) and then the 17th century (1641–1700), to examine the first published technical books and documents by women, the context in which these books emerged, and the changes in women's voices that surfaced in these texts.

APPROACHING THE HISTORY OF TECHNICAL WRITING FOR AND BY WOMEN IN THE ENGLISH RENAISSANCE

In attempting to define technical communication as it emerged in the English Renaissance, I have surveyed technical writing during the period of 1475 to 1640—the nature and characteristics of technical writing in this period, examples of technical books, and the audiences for these how-to books—in *The Emergence of a Tradition* (1997). In pursuing these goals, I have relied heavily on the *Short Title Catalogue of Books Printed in England, Scotland, & Ireland and of English Books Printed Abroad 1475–1640* (Pollard & Redgrave, 1976–1986). This bibliography, a standard tool of scholars and bibliographers, has served since 1976 as testimony to the growth of knowledge during the Renaissance, the impact of print technology on the dissemination of knowledge, the increase of literacy, and the reading interests of English readers.

If we examine the titles of the *STC*, interspersed among hundreds of titles about religion, history, politics, and literature are a variety of what we can call "technical books." These little books, many cheaply printed and designed to be easily carried in the saddle bags or pockets, explain to their readers how to perform tasks in farming, gardening, beekeeping, household management, estate management, home medicine, navigation, and military science—to name a few of the common topics. However, within this array of how-to books that describe methods for performing common tasks in everyday English Renaissance life, I found how-to books written specifically for women—books on growing and harvesting silkworms, gardening, cooking, household medicine, household management, and care of farm animals. The four most popular books on gardening, farming, and estate management contained sections for both men and women (Tebeaux, 1990). These were Thomas Tusser's *Five Hundred Points of Good Husbandry* (1580; 21 editions); Gervase Markham's *Countrey Contentments* (1615; 10 editions);[1] John Fitzherbert's *Booke of Husbandrie* (1598; 11 editions); and William Lawson's *A New Orchard and Garden* (1618; 5 editions).

These books, and other similar ones written specifically for women (see Hull, 1982), tell us that middle-class men and women had similar literacy levels. Furthermore, if we define "literacy" as the ability to read printed material, these books do not suggest that their writers—who were apparently all men—believed that women readers required a different kind of presentation, based on a lower ability to read. As I noted in 1993, books and book sections for women did not include Greek and Latin phrases because women were generally prohibited from attending grammar school, where Greek and Latin were a major part of the curriculum. Aside from that difference, sentence structures and diction in the sections for men and in the sections for women show no differences. Women and men assumed specific roles in running estates and households, and these books help us understand the spheres in which women and men lived and worked. Women and men did have different responsibilities, and differences in instructions for carrying out those responsibilities guide the content of these how-to books. While tasks differed, the reading comprehension level of men and women was assumed to be the same. Each of the four books previously mentioned was extremely popular, as suggested by the number of editions of each book. As Louis Wright (1958) reminded us, Renaissance printers couldn't afford to provide books for which there wasn't a market. Thus, we may assume that these popular books met the needs of an increasingly literate audience.

As I explained in an earlier essay entitled, "Books of Secrets" (1990), these technical books show that Renaissance technical writers were aware of the needs of their readers. Writers, such as Gervase Markham, who wrote technical books as well as dream visions, were fully capable of a wide range of styles. Markham, for example, uses a Latinate style for his dream visions but a decidedly modern technical style for his books on farming and estate management. Sir Thomas Elyott, best known for *The Boke named the Governour* (1531–1580), directed toward readers of the ruling class, used a Latinate style in this work and a plain style in his highly popular self-help medical book, *The Castle of Health* (1537–1610; 16 editions).[2] Books on military science, directed toward men in the higher social strata, often contained extensive Latin references, as well as Latin and Greek notation and allusions. Barnabe Rich, a prolific writer, wrote books on military science, directed toward male readers, and also six titles directed specifically to women, or for both men and women. Comparing the styles in these books does not suggest that Rich believed that women were less literate than men.

However, within this diverse group of how-to books *for* women, I have been unable to find any written *by* women. In examining books of all categories written by women and published during the period of 1475 to 1640, the *STC* records a small number of books by women, of approximately 60 titles, according to Patricia Gartenberg and Nadine Whittemore (1977). These published texts, many which enjoyed multiple editions, are mostly translations, poems, advice books for children, moral admonitions, or devotions penned by well-educated, upper-class women, genres that were apparently the acceptable outlets for women writers. The prefaces to these books suggest that their writers were aware that their authoring these books was not a standard, acceptable activity for women, and they asked for forgiveness for their "forwardness" in presenting their messages in published form. A typical example is *Miscelanea Meditations Memoratives*, by Elizabeth Grymeston (1604). This slim work is divided into 14 chapters (11 are prose meditations; the remaining three, poetic meditations). Ms. Grymeston dedicated the work to her son Bernye. The tone of the preface is typical of those found in Renaissance women's books: Because she is ill with consumption, she states that she writes these meditations for Bernye's spiritual edification:

> [I] must yeeld to this languishing consumption to which it hath brought me: I resolued to breake the barren soile of my fruitlesse braine, to dictate something for thy direction; the rather for that as I am now a dead woman among the liuing. (n.p.)

We see a similar treatment in *The Mothers Blessing*, by Dorthoy Leigh (1616). The dedication explains the purpose of her book:

> My chilren, God hauing taken your Father out of this vale of teares, to his euerlasting mercie in Christ, my selfe not onely knowing what a care hee had in his life time, that you should bee brought vp godly, but also at his death beeing charged in his Will...to see you well instructed and brought vp in knowledge, I could not choose but seeke...to fulfill his will in all things, desiging me greater comfort in the world, then to see you grow in godlinesse. (n.p.)

In Chapter 2, she then states:

> Bvt lest you should maruell, my children, why does not according to the vsual custome of women, exhort you by words and admonitions, rather then by writings; a thing so vnusuall among vs,...know therefore that it was the motherly affection that I bare vento you all, which made mee now (as it often hath done heretofore) forget my selfe in regard of you. (pp. 3–4)

WOMEN WRITERS OF THE 17TH CENTURY

If the number of published books by women was minimal before 1640, the situation abruptly changed after 1640. As Crawford (1985) noted, between 1600 and 1640, the number of first editions of books published by women ranked between one and six for each five-year period. However, from 1641 to 1700, the numbers of first editions by women increased dramatically, to an average of over 50 per five-year period. Even though women's published works accounted for only a maximum of 1.5% of the estimated total published material between 1600 and 1700, the increase of works by women was substantial. For example, during 1686–1690, 67 first editions by women appeared! And, as Smith and Cardinale (1990) reported, approximately 650 titles by women appeared during the period of 1641 to 1700. The titles of these books suggest that after 1640, women no longer felt constrained about topics on which they could write. While Renaissance women like Elizabeth Grymeston, Dorothy Leigh, and several dozen other well educated women had written and published within the genres approved for women (Tebeaux & Lay, 1995), post-1640 female writers published plays, biographies, poetry, fiction, and natural philosophy. Religious writings continued to flourish, particularly among Quaker women. This was the period of Aphra Behn, who sought to earn a living as a writer and published 70 separate titles (not including the total number of editions of each of these works) by 1700.

As the following works will illustrate, the increase in works by women was accompanied by a change not only in topics but also in tone and voice. It is during this period that the first technical works by women appeared.

THE FIRST TECHNICAL BOOKS WRITTEN BY ENGLISH WOMEN

The First Domestic Medicine Books Written by Women

During the Renaissance, domestic medicine books and books of cookery, which contained sections on domestic medicine, were the most popular technical books written for women before 1640. The fact proves to be significant in forecasting the direction that women's technical writing would take after 1650, when the first technical books by women emerged. As Lynette Hunter (1997) has reported, the first published technical books by women were books on domestic medicine. While these books were written between 1602 and 1620, none of these were published until after 1650: Elizabeth Grey, *A Choice Manual of Rare and Select Secrets* (1653); Queen Henrietta Maria, *The Queen's Closet Opened* (1655); and Alethea Howard Talbot (sister of Elizabeth Grey), *Natura Exenterata* (1655).[3] Elizabeth Grey's book is particularly important, as it serves as a harbinger of changes in women's writing that would occur during the last half of the 17th century.

Like the women authors of the English Renaissance, these women were members of the privileged classes. Elizabeth Grey and Alethea Talbot were sisters, both politically powerful aristocrats, who were close friends with Henrietta Maria, wife to Charles I. Yet, because of their status, they were responsible for the health and well being of their families, their servants on their estates, and their communities, a role they assumed with commitment. As how-to books for women written in the Renaissance—by Thomas Dawson, John Partridge, Hugh Platt, Gervase Markham, plus several anonymous books, like *A Closet for Ladies and Gentlewomen*—reveal, women were the main providers of home medicine in Renaissance England.

Hunter (1997) also notes that the technical expertise required to prepare and administer home medicine was substantial, and this knowledge was passed from one generation of women caregivers to another. Home medicine continued to be the domain of women because the formal education afforded to men in the grammar schools and in the university curriculum included classics, arithmetic, history, and geography, but not science or technology. Education in chemical technology and medicinal preparations was acquired by observation, as women observed their mothers, apothecaries, and other adults. Thus, home medicine books, though sparsely phrased, imply a knowledge of techniques by both writer and reader, acquired not only by reading but also observation and apprenticeship. In examining the shift from orality to textuality, we can see that home medicine books assumed that textualized medicinal receipts extended, rather than replaced, knowledge learned by oral instruction.

What is particularly interesting about these first technical books—and Elizabeth Grey's work (1653) serves as a useful example—is the length and complexity of the work. In the domestic books by Partridge, Platt, Markham, and Murrel, receipts for cooking, cleaning, preserving, and personal hygiene are combined with receipts for medicinal preparation. The repetition of the combinations and the popularity of these books indicates that women were responsible for preparing as well as administering these receipts. No book, directed to women and published before 1640, was as expansive as Elizabeth Grey's *Choice Manual*, not published until 1653. Approximately 175 pages long, this small octavo has an alphabetized table of contents that provides advice about preparing medicines and treating disorders, from coughs and simple wounds to the plague. The style, however, is quite similar to that used by Partridge, Markham, and Dawson:

> A very good Medicine for a Consumption, and Cough of the Lungs.
>
> Take a pont of the best Honey as you can get, and dissolve it in a Pipkin, then take it off the fire, and put in two penniworth of flower of Brimstone, and two Penniworth of Pouder of Elecampana, and two penniworth of the flower of Liquirice, and two penniworth of red Rosewater, and so stir them together, till htye be all compounded together, and put it into a gallie pot, and when you use it, take a Liquorish stick beaten at one end, and take up with it as much almost as half a Walnut at night when you go to bed, and in the morning, bathing, or at any time in the night when you are troubled with the Cough, and so let it melt down in your mouth by degrees. (pp. 1–2)

The receipts that comprise Talbot's book are much more sophisticated than those in the books by Partridge, Platt, and Markham and reflect the movement from a herbal-based Galenic medicine to a balance of herbal and chemical medicine. In the index, the chemical receipts are not listed in a different section, but are added separately to the end of each alphabetical listing of herbal remedies. This work is particularly important in viewing women's roles as technical writers and as domestic medicine experts who continued to develop their techniques long before their works were published.

First, as Hunter notes, Talbot's book has much in common with Grey's book, specifically, a number of common receipts. Hunter, who has viewed the manuscript of Talbot's book, states that the printed book begins with Talbot's receipts in order of preparation techniques. This section is followed by experimental receipts of the new chemistry, accompanied by chemical symbols.

> Then comes Ann Dacre's receipts from the 1570s–80s, thirty to forty years earlier than Alethea's initial writing and a good eighty years before the publication, which include herbal preparations, planting and techniques for distillation. Accompanying the earlier work by Ann Dacre are medical receipts by her contemporaries, including letters of advice on specific ailments. (p. 11)[4]

The reason that these books were not published until after 1650 is uncertain, but as Patricia Crawford (1985) explained, the effect of the English civil wars on the sudden increase of women's published writing cannot be underestimated. As Hunter (1997) also has suggested, the monetary value of publication cannot be ignored, as neither Elizabeth Grey nor Alethea Talbot had social and financial positions that were secure as a result of the political upheaval of the Civil War years. However, as Ezell (1984) has shown, even though women did not often publish their works, the failure to publish does not mean that women were not active intellectually. As studies of English Renaissance women have concluded, many women were heavily involved in their family's business affairs and were in charge of estates during the time their husbands were away on business (Hogrefe, 1975). Once the Civil War began, women were forced to assume greater responsibility than before, and for a longer time. As a result, women entered political debates, controlled the financial transactions for the family's business or estate, and continued their domestic responsibilities. Following the civil wars, they had no intention of reassuming their former roles. Speaking out in print can be seen as one indication of the change in women's social roles, and the voices they adopted to announce their involvement in the post-Renaissance social and political order.

This trend, as manifested in other technical books written by women after 1641, can be seen in works by Hannah Woolley, who published eight how-to books during 1661 to 1677. Woolley's 1675 work, *The Gentlewomans Companion Or, a Guide to the Female Sex*, shows how books like Markham's *The Countrey House-Wife* (1631) were transformed when they were written by women. Woolley's *Guide*, like many domestic books for women written a century earlier, focused on "the virtuous and good Education of young Ladies and Gentlewomen" (n.p.) along with "instruction in the most considerable matters of Physick and Chyrurgery, Candying, Preserving and Distilling."

Woolley opened with a five-page narrative resume of her qualifications for writing her book. From the opening passages of this work, we can see that the delicate, self-effacing tone found in the writing of Dorothy Leigh and Elizabeth Grymeston has been replaced with a thinly masked defiant tone that clearly exudes unmistakable self-confidence in her abilities:

> I would not presume to trouble you with any passages of my life, or relate my innate qualifications, or acquired, were it not in obedience to a Person of Honour, who engag'd me so to do, if for no other reason than to stop the mouths of such who may be so maliciously censorious as to believe I pretend what I cannot perform.(1675, n.p.)

Yet, even in 1677, Woolley recognized that women who sought publication were taking a bold step and that women had to justify their expertise. Like other women writers, Woolley showed contempt for women who pursue lives of leisure rather than usefulness:

> It is no ambitious design of gaining a name in print (a thing as rare for a Women to endeavor, as obtain) that put me on this bold undertaking; but the meer pity I have entertain'd for such Ladies, Gentlewomen, and others, as have not received the benefits of the tythe of the ensuring Accomplishments. These ten years and upwards, I have studies how to repair their loss of time, by making publick those gifts which God hath bestow'd upon me. To be useful in our Generation is partly the intent of our Creation; I shall then arrive to the top of the Pyramid of my Contentment, if any shall profit by this following Discourse. If any question the truth of what I can perform, their trial of me I doubt not but will convince their infidelity. (1675, p. 10)

This entire narrative resume is well worth reading, as it illustrates Ms. Woolley's desire to move beyond the roles and social behavior expected of women, while asserting her mastery of skills that English society had decreed she must master. Yet, this resume, as well as the sections that follow, also reveal another important point: Hannah Woolley's command of English sentence structures equals that of Partridge, Markham, Murrell, and Dawson; her narrative technique is superior to that of Partridge and Murrell, but certainly parallels Markham's: Rather than a terse, direct style, she adopts a conversational style in the opening segment on "physicke." The following excerpt from "An Introduction to Physick and Chyrurgery," though lengthy, will show her way of establishing a relationship with her readers:

> As it is a very commendable quality in Gentlewomen, whether young or old, to visit the sick; so it is impossible to do it with the charity some stand in need of, without some knowledg in Physick, and the several operations of Herbs and Spices: But since it will take up too much room to insert here what may make you a compleat herbalists I shall refer you to such who have largely treated on that Subject, *viz. Mr. Gerhard*, and *Mr. Parkinson*, with many more expert in the knowledg of Vegetables. Wherefore, since the knowlg of sundry sort of Spices is very requisite both for person diseased, and in health, I shall begin with them.
>
> Pepper is a spice of the most common use, not and dry to the fourth degree almost. The black is that which is generally coveted; but inconsiderately by the younger sorts of people, it being hurtful to them, though comfortable to old Age. When you use it, beat it not too small for fear of inflaming the blood, otherwise it cutteth gross flegm, dispelleth Crudities, and helpeth Digestion.
>
> The next thing, which is hotter than Pepper, is Ginger; not that it is really so, but because the biting heat of Ginger is more lasting and durable. This spice is not so much used in dressing meat, as the other; however it is very good for concoction, and opens obstructions, and is very expedient for the expulsion of Wind. Green ginger in the *Indies* preserved, is excellent good for a watry and windy stomack, if taken tasting; the better sort is unfleaky, and so clear you may almost see through it; but there is little good made in *England*. (1675, pp. 161–162)

Her receipts for medicinals follow the succinct linear instructions common to domestic medicine books published throughout the Renaissance. Yet, even these often have a conversational, personal quality absent from staccato, direct medicinal instructions of Markham or Partridge:

> Aginst a Stinking-breath
>
> To prevent a Stinking-breath you ought to keep your teeth very clean by rubbing them every morning with sater and salt, which will also cure the Scurvy, you may if you please try Mr. *Turners* Dentrifices, which are every-where much cryed up. But if you breath be trainted, proceeding from some other cause, take Rosemary-leaves with the blossoms, if to be had, and seeth them in White-wine, with a little Myrrh, and Cinammon, and you will find the effect to answer your desires if you use it often. (1675, p. 120)

The style of books by Woolley and Grey illustrate one important quality of women's technical writing: the sense that they are talking with their readers rather than simply providing objective, succinct information. In short, as Mary Lay and I (1995) observed, many of the qualities of women's writing that were first delineated in the 1950s are present in the earliest published writing by English women.

The First English Midwifery Book Written by a Woman

During the Renaissance, the most popular book for midwives was *The Birth of Mankinde* (1540–1604) by Eucharius Roesslin, which enjoyed 10 editions during the Renaissance. Another book on midwifery, *Child-birth or, The Happy Deliverie of Women*, a translation of Jacques Guillemeau's French treatise, was first published in 1612, but never achieved the popularity of *The Birth of Mankinde*. The Preface to *The Birth* states that the audience is women midwives, while *The Happy Deliverie of Women* was directed toward physicians. The large number of pharmaceuticals may have made the work less useful for midwives than *The Birth of Mankinde*. While a number of instructions for midwives were never published and remained in manuscript form in private collections, the first published book by an English woman was *The Midwives Book; Or the whole Art of Midwivery*, by Jane Sharp (1671), which enjoyed four editions by the turn of the century.

The Midwives Book was divided into six parts: (1) male and female gynecological descriptions, (2) the process of procreation, (3) causes of infertility (4) labor, (5) childbirth complications, and (6) postpartum diseases and illnesses in young children. Sharp's work shows a mature style, and like so many other books by women, begins with a justification of the book, which is historically significant. In her preface, Sharp acknowledges that many midwives lack needed skills. She explains that she wrote her book to provide essential information, which she has gleaned from books in French, Dutch, and Italian, and from her own experience.

By the closing decades of the 17th century, the College of Physicians was establishing its official stance against women midwives, based on their lack of formal medical education. Women were beginning to resent and openly oppose their exclusion from medical training and from midwifery, but Sharp makes her case from a historical perspective: women had, since Biblical times, practiced midwifery.

> There being no so much as one word concerning Men-midwives mentioned there that we can find, it being the natural propriety of women to be much seeing into that Art; and though nature be not along sufficient to the perfection of it, yet farther knowledge may be gain'd by a long and diligent practice, and be communicated to other of our own sex....where there is not Men of Learning, the women are sufficient to perform this duty. (1671, p. 3)

The current charge by physicians, that women who have not had medical training do not have the knowledge to be midwives, is also a point about which Sharp disagrees. The midwife's work has nothing to do with her understanding medical nomenclature:

> It is not hard words that perform the work, as if none understood the Art that cannot understand the Greek. Words are but the shell, that we of times break out Teeth with them to come at the kernal, I mean our brains to know what is the meaning of them; but to have the same in our mother tongue would save us a great deal of needless labor. (1671, p. 4)

In short, if medical books could be translated into English (which they had been since the mid-16th century), then these works on midwifery should also be made available in a form in which midwives could read and understand.

The specific segments of the book present detailed discussion, based on the current knowledge about pregnancy, even though the number and quality of female anatomical descriptions does not begin to approach those in *The Birth of Mankinde*. Sharp's style is decidedly modern, and like Woolley's, conversational, despite the medical quackery of much of its advice:

> Young women especially of their first Child, are so ignorant commonly, that they cannot tell whether they have conceived or not, and not one of twenty almost keeps a just account, else they would be better provided against the time of their lying in, and no so suddenly be surprised as many of them are.
>
> Wherefore divers Physicians have laid down rules whereby to know when a women hath conceived with Child, and these rules are drawn from almost all parts of the body. The rules are too general to be certainly proved in all women, yet some of them seldom fail in any.
>
> First, if when the seed is case into the womb, she feel the womb shut close, and a shivering or trembling to run through every part of her body, and that is by reason of the heat that draws inward to keep the conception, and so leaves the outward parts cold and chill. (1671, p. 103)

A work that offers an interesting comparison to Jane Sharp's work is *The Compleat Midwifes' Practice* (1663), collaboratively written by a group of English physicians. The book's purpose, as stated in the preface, is to correct erroneous knowledge used by many midwives. The authors state their method of compiling sources in several languages and material from Madam Loug Bourgeo, late midwife to the Queen of France, whose picture appears with the title page. While the writers do not discredit women midwives directly, they do denounce them for relying on works that do not include reference to state-of-the-art information about gynecology.

In comparison to *The Birth of Mankinde* (1598), *The Complet Midwifes' Practice* contains more information, more focus on anatomy, but fewer gynecological illustrations. The quantity of folklore surrounding pregnancy has been replaced with medical information about safe delivery. The presentation of information is, however, similar to that used by Jane Sharp, although Ms. Sharp eliminated Latin medical terms. The physician writers use Latin medical terms, although they define each term in context. If we compare these two books, we see a similarity in style and presentation suggesting that female readers and writers could handle written English as well as male writers.

The First Proposals Written by English Women

The education of midwives was the subject of the first proposal written by a woman: *A Scheme for the Foundation of a Royal Hspital, and Raisng a Revenue of Five or Six-thousand Pounds a year, by, and for the Maintenance of a Corporation of skilful Midwifes*, by Elizabeth Cellier (1687). As the title suggests, Cellier proposed a college for educating midwives and caring for homeless children. This proposal followed a contemporary proposal arrangement: rationale for the proposal, the proposal itself, tasks to be performed in realizing the college, implementation procedures, and costs. Ms. Cellier's opening salvo—that midwives need more education—echoes Jane Sharp's concern that women midwives were under attack by physicians because the midwives lacked proper medical education. Cellier's proposal, although unsuccessful, would have helped correct the problem. Her argument, as well as her style, is vigorous and direct, as the opening passage of the proposal reveals:

> That within the Space of twenty Years last past, above Six-thousand Women have died in Chaild-bed, more than Thirteen-thousand Children have been born abortive, and above Five-thousand chrysome Infants have been buried within the weekly Bills of Mortality, above two Thirds of which, amounting to Sixteen-thousand Souls, have in all Probability perished, for Want of due Skill and Care, in those Women who practice the Art of Midwifry.
> . . . .
>
> To remedy which it is humbly proposed, that your majesty will be graciously pleased, by your Royal Authority, to united the whole Number of skilful Midwifes, now practicing within the Limits of the weekly Bills of Mortality, into a Corporation, under the Government of a certain Number of the most able and matron-like Women among them, subject to the Visitation of such Person or Persons, as your Majesty shall appoint; and such Rules for their good Government, Instruction, Direction, and Administration, as are hereunto annexed, or may, upon more mature Consideration, be thought fit to be annexed. (p. 241)

The proposal immediately moves to what might be called the work plan section, which states the specifics of the proposal. Each statement emphasizes the need for the College as well as the means of supporting its operations. Elizabeth Cellier was a realist who combined the rationale for the College with the necessity for ongoing financial support and the need for good management. For example,

> That for the better Maintenance and Encouragement of so necessary and royal a foundation of Charity, it is humbly proposed that by your Majesty's royal Authority, one fifth Part of the voluntary Charity, collected or bestowed in any of the Parishes within the Limits of the weekly Bills of Mortality, may be annexed for ever to the same, other than such Money taxed for the Maintenance of the Parish Poor, or collected on Briefs by the royal Authority for any particular charitable Use.
>
> That likewise, by your Majesty's royal Authority, the said Hospital may have Leave, to set up in every Church, Chapel, or public Place of Divine Service of any Religion whatsoever within the Limits aforesaid one Chest or Box, to receive the Charity of all well-minded People, who may put money into the same, to be employed for the Uses aforesaid.
> . . . .
>
> That such Hospital may be allowed to establish twelve lesser convenient Houses, in twelve of the greatest parishes, each to be governed by one of the twelve Matrons, Assistants to the Corporation of Midwifes, which Houses may be for the Taking in, Delivery, and Month's Maintenance, at a Price certain of any Woman, that any of the Parishes, within the Limits aforesaid, shall by the Overseers of the Poor place in them, such Women being to be subject, with the Children born to them, to the future Care of that Parish, whose Overseers place them there to be delivered.... (1687, pp. 244–245)

This proposal is one of approximately 18 extant documents written by Elizabeth Cellier. Three were written in defense of midwifery. Through her writing, we can trace the efforts of English physicians to regulate the practice of midwifery and confine obstetrics to physicians. Elizabeth Cellier pursued an ongoing argument in print with English physicians who attacked the competence of midwives. In one earlier polemic, *The Mid-wives just Petition* (1643), Ms. Cellier blamed men's historical pursuit of war for the lack of medical practitioners. Thus, women have had to assume the role of midwives throughout history and, most recently, during the English civil wars. Women throughout history have had to assume and excel in this role, for without them many children and mothers would have died for lack of proper care. She patiently argued that midwifery has preserved mankind and the English Commonwealth.

In *To Dr.-----An Answer to his Queries, concerning the Colledg of Midwives* (1688), Ms. Cellier's voice has become impatient and exudes a strident tone absent from *The Petition*. In answering the physician, who had accused her of arrogance in proposing the college, she delineates the historical roots of midwifery in Jewish, Greek, and Egyptian history.

> And such a favour from God of *building Houses for them*, we do not read the Physicians ever received; nor was Physick then a regular Study, nor brought under

Government in that Learned Nation of *Egypt*, in Herodotus his time which put together, proves the Antiquity of the Midwives Government so much antienter than that of the Doctors. (p. 2)

And, now, Doctor, let me put you in mind, that tho you have often Laughted at me, and some Doctors hav accounted me a Mad Woman these last four Years, for saying her Majesty was full of Children, and that the Bath would assist her Breeding: 'Tis now provded so true, that I have cause to hope my self may live to praise God, not only for a Prince of Wales, and a Duke of York but for many other Royal Babes by Her.... (p. 7)

Other Published Technical Books by Women After 1641

Other technical books written and published by women during the period of 1641 to 1700 included numerous cookbooks and the first collaboratively written book by women: *The Ladies Companion, or a Table furnished with sundry Sorts of Pies and Tarts, Gracefull at a Feast, with many receipts. By Persons of Quality whose names are mentioned* (1653–1654). Mary Tillinghast's 1690 book, *Rare Excellent Receipts, Experienc'd, & Taught by Mrs. Mary Tillinghast*, is likely the first cookbook with instructions for cooks in training.

Two other proposals argued for improved education for women: Mary Astell, *A Serious Proposal to the Ladies* (1694), argued for a women's college dedicated to the pursuit of learning and good works. Echoing Hannah Woolley's reasons for writing her books, Astell attacked the frivolous life of women of the leisure classes and encouraged all women to develop their intellects. Bathsua Makin proposed a girls' school offering foreign language, science, math, and philosophy.

Bathsua Makin's proposal, like Cellier's, used a decidedly modern approach in delivering its message. Makin's 43-page proposal uses document design—headings, deliberate italics, and listings—and begins with a rationale that would have appealed to a male audience: Because women are naturally inclined to sin, education is essential to counteract their nature. In her rationale section, she surveys the history of women's education in Biblical, Greek, and Roman literature and gives more current Renaissance education practices available during the reign of Elizabeth I. In the rationale, each topic is introduced by a centered, italicized sentence heading—e.g., "Women have formerly been educated in Arts & Tongues;" "Women educated in Arts & Tongues have been eminent in theology;" "Women have been good Linguists;" "Women have been profound Philosophers." In these sections, she shows recent and classical historical precedence for affording women education in linguistics, oratory, logic, philosophy, mathematics, and poetics. Her examples show that she had an impressive command of history.

Thus, her proposal, "Care ought to be taken by us to Educate Women in Learning," is argued from a pragmatic perspective: women will be better people, as education is a "hedge against heresies." In addition, women can lead more productive lives; widows can manage their own affairs; married women can help their husbands in business and trades; educated women can improve their children's learning; educated nurses become better teachers. Makin then enumerated likely objections from men and offered rebuttals to these objections. Her tone here is decidedly more acrimonious than it is in the rationale. For example:

> Object. *Women are ill of Natures, and will abuse their Education: They will be proud, and not obey their Husbands; they will be pragmatick, and boast of their Parts and Improvements. The Ill Nature that it is in them, will become more wicked, the more wits you furnish them with.*
> *Answ.* This is the killing Objection, and every thick-skull'd Fellow that babbles this out, thinks no Billingsgate Women can Answer it. I shall take this Objection in pieces.
> 1. *They will abuse Learning.* So do men; he is egregiously simple, that argues against the use of necessary or very convenient thing from the abuse of it. By this Argument no men should be liberally brought up. strong Drinks should never be used any more in the World, and a hundred such things.
> 2. *They are ill Natures.* This is impudent calumny; as if the whole Sex of Women, or the greatest part of them, had that malice infused into their very Natures and constitutions, that they are ordinanarily made worse by the Education that makes Men generally better.
> -------- *Ingenus didicisse fideliter artes*
> *Emollit mores, nee finit esse feros.*
> The Heathen found, that Arts wrought upon Men, the rougher Sex. Surely it is want of fidelity in the Instructer, if it have not the like effect upon softer and finer Materials. (1673, p. 32)

She concluded her proposal by outlining a specific curriculum for schools to show how her proposed subjects could be covered during a typical school day. Half the day would be spent on religion, dancing, music, singing, writing, and keeping accounts. "The other half would be imployed in gaining the *Latin* and *French* Tongues; and those that Please, may learn *Greek* and *Hebrew*, the *Italian* and *Spanish*" (1673, p. 42).

She also addressed cost:

> The Rate certain shal be *20 l.per annum*:: But if a copetent improvement be made in the Tongues, and the other things aforementioned, as shall be agreed upon, then something more will be expected. But Parents shall judge what shall be deserved by the Undertaker.
> Those that thik these Things Iprobable or Impracticable, may have further account every *Tuesday* at Mr. *Masons* Cossee-House in *Cornhil*, near the royal Exchange; and *Thursdayes* at the Bolt and tun in *Fleetstreet*, between the hours of three and six in the Afternooons, by some Person whom Mris. *Makin* shall appoint. (1673, p. 43)

Technical Books by Women—A Microcosm of Social Change

The growth of published works by women during the period of 1641 to 1700, the variety of topics addressed by women shows that women were assuming more active social roles. Women's voices were becoming more strident, more confident, more knowledgeable, more argumentative and logical, and more demanding of their right to education and opportunities appropriate to women. The first technical books written by English women stand as arguments for women's rights to practice midwivery better and safer. These first proposals and the apologetic tracks supporting them, such as those by Elizabeth Cellier, directly announce that women wanted and deserved access to education, to formal schools that would allow them more opportunity to develop intellectually, and to study subjects that for centuries had been the province of men. Studies by Anglin (1980), Camden (1975), Campbell (1983), Clark (1982), Cressy (1980), Houston (1988), Laqueur (1976), Masek (1979), McMullen (1977), Ong (1982), Prior (1985), Schofield (1968), Simon (1966), Stone (1964), Thomas (1986), Wright (1931, 1958), and Wrightson (1982) provide essential historical framework for understanding changes in literacy and educational opportunity throughout the Renaissance and 17th century, background which space precludes from discussion here. Against the milieu, the first technical documents written for and by women provide a microcosm for illuminating these changes and their effects on the evolution of technical writing and women's contribution to this genre.

NOTES

[1] The section for Women, "The Countrey Huswife," was also published separately and in other works by Markham, who was the most prolific technical writer of the English Renaissance. By 1700, "The Countrey Huswife" had appeared in 30 editions, a testimony to its popularity, as perceived by printers and by Markham himself.

[2] While Elyott has received the most attention for his views on education, articulated in *The Boke named the Governour* and *The Castle of Health*, was much more popular in the Renaissance. The STC records 8 editions of *The Governour* (1530–1580), but 18 editions of *The Castle* (1537–1610). *The Castle* became the prototype for the extremely popular self-help books published throughout the Renaissance.

[3] Neither *The Queen's Closet Opened* or *Natura Exenterata* are recorded in Wing (1945). Both books are held in special collections at the University Of Leeds.

[4] Historical material about Elizabeth Grey, Alethea Talbot, and Queen Henrietta Maria originally appeared in Lynette Hunter's lecture, "Women and domestic Medicine: Lady Experimenters 1570–1620," one of four lectures given at Gresham College in October 1995. These essays, along with several others, were published as "Women and 17th Century Medicine Science" to honor the 500th anniversary of Gresham College.

REFERENCES

Anglin, J. P. (1980). The expansion of literacy: Opportunities for the study of the three Rs in the London diocese of Elizabeth I. *Guildhall Studies in London History, 3*, 63–74.

Anonymous. (1608). *A closet for ladies and gentlewomen, or, the art of preseuing, conseruing, and candying.* London. Short-Title Catalogue 5434.

Astell, M. (1694). *A serious proposal to the ladies, for the advancement of their true and greatest interest.* London. Early English Books 9.6 and 1375: 24, 25.
Camden, C. C. (1975). *The Elizabethan woman.* Mamaroneck, NY: Paul P. Appel.
Campbell, M. (1983). *The English yeoman under Elizabeth and the early Stuarts.* London: The Merlin Press.
Cellier, E. (1643). *The Mid-wives just petition: Or, a complaint of divers good gentlewomen of that faculty.* London. Early English Books 242: E86, No. 14.
Cellier, E. (1687). *A scheme for the foundation of a royal hspital and raisng a revenue of five or six-thousand pounds a year, by, and for the maintenance of a corporation of skilful midwifes* (Fourth Series. #8325.0–6. Vol. 2, pp. 243–249). London.
Cellier, E. (1688). *To Dr.----- An answer to his queries, concerning the colledg of midwives.* London. Early English Books 1457:3.
Clark, A. (1982). *Working lfe of wmen in the sventeenth cntury.* London: Routledge and Kegan Paul.
Crawford, P. (1985). Women's published writings 1600–1700. In M. Prior (Ed.), *Women in English society 1500–1800* (pp. 211–274). London: Methuen.
Cressy, D. (1980). *Literacy and the social order, reading and writing in Tudor and Stuart England.* Cambridge, England: Cambridge University Press.
Dawson, T. (1598). *The second part of the good hus-wifes Jewell.* London. Short-Title Catalogue 6394.
Early English books 1641–1700 (Vols. 1–9) [Microfilm]. (1900). Ann Arbor, MI: University Microfilms.
Elyott, T. (1531–1580). *The boke named the Governour.* London.
Elyott, T. (1537–1610). *The castle of health.* London.
Ezell, M. M. (1984). *The patriarch's wife: Literary evidence and the history of the family.* Chapel Hill, NC: The University of North Carolina Press.
Fitzherbert, J. (1598). *Booke of husbandrie.* London: The English Experience 926; Amsterdam and New York: Theatrvm Orbis Terrarvm Ltd. 1979. Short-Title Catalogue 11004.
Gartenberg, P., & Whittemore, N. T. (1977). A checklist of English women in print, 1475–1640. *Bulletin of Bibliography, 34*(1), 1–13.
Grey, E. (1653). *The choice manual of rare and select secrets in physick and shyrurgepy.* London. Early English Books 152.3 and 1573:1.
Grymeston, E. (1604). *Miscelanea, meditations, memoratives.* London: The English Experience, No. 933; Amsterdam and New York: Theatrvm Orbis Terrarvm. 1979. Short-Title Catalogue 12407.
Guillemeau, J. (1612). *Child-birth or, the happy deliverie of women.* London. Short-Title Catalogue 12496.
Hogrefe, P. (1975). *Tudor women: Commoners and queens.* Ames, IA: Iowa State University.
Houston, R. A. (1988). *Literacy in early modern Europe.* New York: Longman.
Hull, S. (1982). *Chaste silent & obedient, English books for women 1475–1640.* San Marino, CA: Huntington.
Hunter, L. (1997). Women and domestic medicine: Lady experimenters 1570–1640. In L. Hunter & S. Hutton (Eds.), *Women, medicine and science 1500–1700* (pp. 29–62). Gloucestershire, England: Sutton Publishing.
The ladies companion, or a table furnished with sundry sorts of pies and tarts, gracefull at a feast, with many receipts. By persons of quality whose names are mentioned. (1653–1654). London.

Laqueur, T. (1976). The cultural origins of popular literacy in England, 1500–1850. *Oxford Review of Education, 2,* 255–275.
Lawson, W. (1618). *A new orchard and garden.* London. Short-Title Catalogue 15329.
Leigh, D. (1616). *The Mothers blessing. Or the godly counsaile of a gentle-woman.* London. Short-Title Catalogue 15402.
Makin, B. (1673). *An essay to revive the ancient education of gentlewomen in religion, manners, arts & tongues.* London. Early English Books 697.2.
Maria, Q. H. (1655). *The queens closet opened. Incomparable secrets. Physick, chirugery, preserving, candying, and cookery; As they were presented to the Queen by the most experienced persons of our times.* Cornhill.
Markham, G. (1615). *Countrey contentments.* London: The English Experience 613; Amsterdam and New York: Theatrvm Orbis Terrarvm Ltd. 1973. Short-Title Catalogue 17342.
Markham, G. (1631). *The countrey huswife.* London. Short-Title Catalogue 17353.
Masek, R. (1979). Women in the age of transition: 1485–1714. In B. Kanner (Ed.), *The women of England, from Anglo-Saxon times to the present* (pp. 138–162). Hamden, CT: Archon Books.
McMullen, N. (1977). The education of English gentlewoman, 1540–1640. *History of Education, 6,* 87–101.
Moran, M. (1985). The history of technical and scientific writing. In M. G. Moran & D. Journed (Eds.), *Research in technical communication: A bibliographic sourcebook* (pp. 25–38). Westport, CT: Greenwood.
Murrel, J. (1617). *A daily exercise for ladies and gentlewomen.* London. Short-Title Catalogue 18301.
Ong, W. J. (1982). *Orality, and literacy: The technologizing of the word.* London: Methuen.
Partridge, J. (1573). *The treasurie of commodious conceits, and hidden secrets. and may be called, the huswives closet of healthfull provision.* London. Short-Title Catalogue 19425.
Partridge, J. (1595). *The widdowes Treasure, plentifully furnished with sundry precious and a approved secrets in physicke and chirurgery, for the health and pleasure of mankinde.* London. Short-Title Catalogue 19434.
Platt, H. (1602). *Delights for ladies, to adorne their persons, tables, closets and distillatories.* London. Short-Title Catalogue 19978.
Pollard, A. W., & Redgrave, G. R. (1976–1986). *A short-title catalogue of books printed in England, Scotland, & Ireland and of English books printed abroad 1475–1640* (2nd ed., Vols. 1–2) (W. A. Jackson, F. S. Ferguson, & K. F. Panzer, Eds.). London: Bibliographical Society.
Prior, M. (1985). Women in the urban economy: Oxford 1500–1800. In M. Prior (Ed.), *Women in English society, 1500–1800* (pp. 93–117). London: Methuen.
Rivers, W. E. (1994). Studies in the history of business and technical writing: A bibliographical essay. *Journal of Business and Technical Communication, 8*(1), 6–57.
Roesslin, E. (1598). *The birth of mankinde, otherwyse named the womans booke* (T. Raynalde, Trans.). London: Richarde Watkins. Short-Title Catalogue 21160.
Schofield, R. S. (1968). The measurement of literacy in pre-industrial England. In J. Goody (Ed.), *Literacy in traditional societies* (pp. 96–119). Cambridge, England: Cambridge University Press.

Sharp, J. (1985). The midwives book. Or the whole art of midwifry. In R. Trumback (Ed.), *Marriage, sex and the family in England 1660–1800* (a 44-vol. facsimile series). New York: Garland. (Original work published 1671)
Simon, J. (1966). *Education and society in Tudor England*. Cambridge, England: Cambridge University Press.
Smith, H. L., & Cardinale, S. (1990). *Women and the literature of the seventeenth century: An annotated bibliography based on Wing's short-title catalogue*. Westport, CT: Greenwood.
Stone, L. (1964). The educational revolution in England, 1560–1640. *Past & Present, 28*, 41–80.
Talbot, A. H. (1655). *Natura Exenterata*. London.
Tebeaux, E. (1990). Books of secrets—Authors and their perception of audience in procedure writing of the English Renaissance. *Issues in Writing, 3*, 41–67.
Tebeaux, E. (1993). Technical writing for women of the English Renaissance: Technology, literacy, and the emergence of a genre. *Written Communication, 10*(2), 164–199.
Tebeaux, E. (1997). *The emergence of a tradition: English Renaissance technical writing 1475–1640*. Farmingdale, NY: Baywood.
Tebeaux, E., & Lay, M. M. (1995). The emergence of the feminine voice. *Journal of Advanced Composition, 15*(1), 53–81.
Thomas, K. (1986). The meaning of literacy in early modern England. In G. Baumann (Ed.), *The written word: Literacy in transition* (pp. 97–131). Oxford, England: Clarendon Press.
Tillinghast, M. (1690). *Rare excellent receipts, experienc'd, & taught by Mrs. Mary Tillinghast*. London.
Tusser, T. (1984). *Five hundred points of good husbandry* (G. Griegson, Ed.). Oxford, England: Oxford University Press. (Original work published 1580)
Wing, D. (1945). *Catalogue of books printed in England, Scotland, Ireland, Wales and British America and of English books printed in other countries 1641–1700*. New York: Columbia University Press.
Woolley, H. (1675). *The gentlewomans companion or, a guide to the female sex*. London. Early English Books 1602:55.
Wright, L. B. (1931). The reading of Renaissance English women. *Studies in Philology 28*, 139–156.
Wright, L. B. (1958). *Middle-class culture in Elizabethan England*. Ithaca, NY: Cornell University Press.
Wrightson, K. (1982). *English society 1580–1680*. New Brunswick, NJ: Rutgers University Press.

6

Landmark Essay: The Plain Style in Scientific and Technical Writing[*]

Merrill D. Whitburn
Rensselaer Polytechnic Institute

Marijane Davis

Sharon Higgins

Linsey Oates

Kristene Spurgeon

Plainness is perceived as a stylistic ideal in scientific and technical writing today. In his successful textbook, *Basic Technical Writing*, Herman Weisman accurately states: "By tradition technical style is plain..." (1974, p. 1). Weisman rightfully finds the source of this tradition in the scientific revolution of the 17th century. While the plain style may have emerged as a corrective to certain stylistic abuses, however, it may itself be creating problems for today's writers. In this paper, I shall

---

[*] For the student quotations in this paper I am indebted to Guinn Hubbard, Thomas Cool, Kristene Spurgeon, and Lucinda Irwin, respectively.

explore the emergence of plainness as a stylistic ideal in our field, the difficulties it might create, and some cautious experiments I began in a graduate course, Technical Writing for Publication, to engage these problems.

ORNATE STYLE

English prose style before the scientific revolution has been ably described by Richard Foster Jones in his noted article "Science and English Prose Style in the Third Quarter of the Seventeenth Century" as "characterized by various rhetorical devices which use figures, tropes, metaphors, and similes, or similitudes, to use a term of the period. The sentences are long, often obscurely involved, and rhythmical, developing in writers like Sir Thomas Browne a stately cadence, which, in the studied effect of inversions, is the prose counterpart of Milton's blank verse. The penchant for interlarding a work with Latin and Greek quotations is also apparent. The diction reveals a host of exotic words, many Latinisms..." (1951, p. 2). The stylistic ideal was copiousness and ornamentation. Writers were often so caught up in verbal exuberance, so delighted with word play, that they seemed more intrigued with rhetorical devices than the search for truth.

Perhaps the work which best exemplifies the ornate style of the late 16th and early 17th centuries is John Lyley's *Euphues: The Anatomy of Wit*. A sentence from *Euphues* suggests the extent to which the work is imbued with rhetorical devices. Lyley writes: "As therefore the sweetest rose hath his prickle, the finest velvet his brack, the fairest flour his bran, so the sharpest wit hath his wanton will and the holiest head his wicked way" (1967, p. 127). The whole sentence is a simile, comparing the flaws in roses, velvet, and flour with the imperfections of man. The three phrases "the sweetest rose hath his prickle, the finest velvet his bracket, the fairest flour his bran" are parallel with an ellipsis, an omission of *hath*, in the second and third. Lyley also uses an example of asyndeton, an omission of the conjunction *and* between the last two phrases. In the remainder of the sentence, "so the sharpest wit hath his wanton will and the holiest head his wicked way," the words *wit* and *head* are examples of synecdoche, a figure in which the part stands for the whole, and the words "wanton will" and "wicked way" are examples of alliteration. Numerous other examples of rhetorical devices can be found in sentence after sentence of *Euphues*. Such devices were catalogued in works on rhetoric toward the end of the 16th century. Among these works are Thomas Wilson's *The Arte of Rhetorique* (1553); Henry Peacham's *The Garden of Eloquence, Conteyning the Figures of Grammar and Rhetoric* (1577); and George Puttenham's *The Arte of Englishe Poesie* (1589).

RISE OF THE PLAIN STYLE

The nature of the scientific revolution in the 17th century predisposed it against rhetorical devices. Pre-revolutionary scientists acquired knowledge through books, and philological training involving Latin and Greek was essential. However, post-revolutionary scientists acquired their knowledge through the observation of nature, and words, associated with the rejected science of the past, became suspect. This skepticism toward language was strengthened by the belief that traditional philosophy—usually linked to Aristotle—tended to concern itself with words rather than realities. The new scientists wanted words to approximate mathematical symbols. Such symbols were to possess no virtue in themselves but stand for quantities and relationships. Nothing was to exist between the mind and its true object; rhetorical devices were not to be an obstruction between observation and description.

Advocates of the new science in the 17th century were remarkably sensitive to the problem of style. In his book, Jones writes: "We may say without exaggeration that their program called for stylistic reform as loudly as for reformation in philosophy" (1951, p. 2). Francis Bacon is quoted as attacking rhetorical devices because he believes they lead men to "study words and not matter" (p. 2). John Wilkins rejects them because they cause obscurity; he describes the ideal style as "plain and naturall, not being darkned with... Rhetoricall flourishes. Obscurity in the discourse is an argument of ignorance in the mind. The greatest learning is to be seen in the greatest plainnesse..." (quoted in Jones, p. 78). John Webster suggests that subjects like rhetoric might be all right "if there were not too much affection towards them, and too much precious time spent about them, while more excellent and necessary learning lies neglected and passed by..." (quoted in Jones, p. 82). In his *History of the Royal Society,* Thomas Sprat attacks rhetorical ornamentation as being "in open defiance against Reason: professing not to hold much correspondence with that; but with its Slaves, the Passions..." (1667, p. 4). Sprat indicates that members of the society are resolved "to reject the amplifications, digressions, and swellings of style: to return back to the primitive purity, and shortness, when men deliver'd so many things, almost in an equal number of words. They have exacted from all their members, a close, naked, natural way of speaking; positive expressions, clear senses; a native easiness; bringing all things as near the Mathematical plainness, as they can..." (pp. 85–86). These goals became part of the statutes of the Royal Society when they were published in 1728. In Chapter V, Article IV, we find: "In all Reports of Experiments to be brought into the Society, the Matter of Fact shall be barely stated, without any Prefaces, Apologies, or Rhetorical Flourishes, and entered so into the Register-Book, by order of the Society" (p. 84).

PLAINNESS—AN IDEAL TODAY?

Nobody would deny the importance of the stylistic reforms encouraged by the advocates of the new science. Too many men of that time did seem to be caught up in verbal games rather than the pursuit of truth. All too often, rhetorical devices obscured the communication of matters they should have been clarifying. And too many previous thinkers did tend to concern themselves with words rather than realities. But a number of factors suggest that the stylistic ideal of plainness ought not to continue as unchallenged as in the past. Revolutions are typically reactions against excesses, and the reactions are often as excessive as the original abuses. We need to explore the extent to which the attempt of the new scientists to overcome stylistic excesses of the past resulted in excesses of their own. Numerous changes in science and technology since 1700 should encourage us to question whether a stylistic ideal appropriate to the 17th century is still viable. Finally, the writing crisis we now confront should lead us to explore all of our traditionally unquestioned assumptions.

A standard approach to style in current textbooks focuses on the avoidance of error. For instance, excessive words in one sentence are eliminated in a corrected version that follows. Such an approach is extremely useful, in that students who grasp its lessons will be better able to revise their work and avoid stylistic problems in the future. But their posture will remain essentially defensive. The advocates of the new science in the 17th century so reacted against the excesses of stylistic artistry that a reluctance to use any artistry at all seems to have prevailed ever since. Scientific or technical students rarely confront a writing task armed with sufficient stylistic tools to shape their compositions aggressively. They tend to be more concerned with what *not* to do than what *to* do in their writing. Scientific and technical writing textbooks tend not to drill students in even such basic stylistic techniques as antithesis, climax, parenthesis, and apposition. On the contrary, students are expected to muddle their way through by stylistic instinct. Such an approach promotes the myth of the born writer.

Numerous changes in science and technology should also encourage us to question approaches characteristic of the stylistic ideal of plainness. These changes have often involved an increase in complexity, and compounding this complexity has been the extraordinary development of scientific and technical jargon. Yet coincident with these developments has been the growing need by scientists and technologists to address such nonexperts as government representatives, foundation representatives, supervisors, colleagues from other fields, laborers, and the general public. My own experience in industry has convinced me that audience adaptation is the most serious problem in our discipline. Scientists and technologists need considerable practice in developing alternative ways of expressing the same materials. Such practice was common for students before the scientific revolution of the 17th century. They often learned to vary a theme hundreds of different ways. They strove to amplify their writing through comparison, example, description, repetition, paraphrasis, and digression. But the resulting excesses of expansion led advocates of the new science to condemn techniques of amplification as working against plainness, more specifically, brevity. Such a condemnation seems to have been to the detriment of modern scientific and technical writing style. Practice in expressing the same content in alternative ways does not necessarily preclude brevity; as Erasmus suggested in the Renaissance, no one can better achieve brevity in his style than he who knows what words and figures to choose from among a great variety. Modern scientists and technologists have simply not had the practice to enable them to attain the ideals of economy and simplicity possible through training in techniques of amplification.

Lastly, the plain style also militates against rhetorical devices that might better enable writers to communicate their personalities. In a previous paper, *Personality in Scientific and Technical Writing* (1976), I stress the importance of the personal touch. Many of the students in our technical writing classes will be largely confronting scientific and technical communication that involves the adaptation of a message for an audience relatively unfamiliar with the subject being presented. They will be educating students, helping to create public awareness, persuading a supervisor, or addressing a group of employees. These are serious activities, and we should make our students aware that a personal touch can often help them attain their ends. A personal touch arouses interest, and interest sharpens awareness and understanding. Rhetorical devices are one means of attaining the right personal touch. For instance, such devices as irony, hyperbole—the use of mock exaggeration, and litotes—the use of mock understatement—might be used to humorous effect in a non-specialist communication. Again, one seeks in vain for information about such devices in technical writing textbooks, and the ideal of the plain style seems at fault.

SOME RECENT EXPERIMENTS

For some time, then, I have been troubled by the inadequacy of the stylistic tools I have been providing my students. Their work tends to lack artistry in its smaller units, and their resulting stance toward writing tends to be overly passive. They have extraordinary difficulty adapting abstruse information so that nonspecialists can understand it. Few of them know how to take advantage of stylistic techniques to help them let their personalities through. In an effort to begin engaging these problems, I have tried some cautious experiments in a graduate course, Technical Writing for Publication. The students in this course come from various fields with graduate programs in the university—oceanography, chemistry, architecture, agricultural economics, wildlife and fisheries, parks and recreation, and education. In addition to presenting them with the standard materials in scientific and technical communication, I have been giving them a few exercises involving various rhetorical devices. I frankly inform the students that our exercises are experimental.

The results have been encouraging. A few examples can serve to illustrate the kinds of things the students are doing. For instance, almost all textbooks in scientific and technical writing have a section on parallelism. In line with the tradition of the plain style, however, the stance tends to be defensive. If students find themselves with a series of some kind, they should be careful to make the elements of one member of the series parallel with the elements of the other members of the series. I have tried to change this procedure by giving my students exercises in which they aggressively work toward parallelism. As part of one such exercise, a student in oceanography wrote the following passage:

> The Polychaeta exhibit every feeding type known to animals. Some burrow through the sediments ingesting all material in their path and digesting any available organic matter; these are the nonselective deposit feeders. Some burrow through the sediments testing the material before them and selecting only those choice morsels which meet their criteria as food; these are the selective deposit feeders. Some lie in holes, burrows, or tubes and strain particles of organic material from the water using tentacles and/or mucous nets; these are the suspension or filter feeders. Some aggressively attack and devour prey, often even their own kind; these are the predatory carnivores. Some eat only the remains of dead animals; these are the scavengers. Some may combine many or all of the previously mentioned methods, depending upon circumstances; these are the omnivores.

This passage may not be as economical as it should be, but the gains in clarity, emphasis, smoothness, and reader interest strike me as substantial.

If economy is a virtue, a student ought to practice the techniques that promote it. The stylistic device of ellipsis, the omission of a word or words implied by the context, would certainly be one such technique. Yet I have never encountered any mention of this technique in a modern textbook on scientific and technical writing. A student writing about astronomy composed the following two sentences as part of one of his exercises:

> Neptune is recognized by a series of alternating, vertically-oriented belts colored green and gray; and Uranus by a sequence of horizontally-oriented green and gray zones.
> Galileo was more concerned with the laws of physics here on earth and their relation to planetary movements; Copernicus with the motions of planets relative to one another.

The student effectively takes advantage of the parallelism in two parts of a sentence to omit elements in the second part.

A student, writing on the topic of education, developed the following passage:

> We—like our students—must realize that there is more to college than lectures and notes and assignments and tests. We—like our students—must realize that the true test is being and becoming, not remembering and reviewing. We—like our students—must realize that learning is a journey, not a destination.

Only close examination reveals the extent to which rhetorical devices are responsible for the effectiveness of the above passage. Within the context of the article in which this passage appears, the parallelism of "We—like our students—must realize that" is not only inherently pleasing but also climactic. The parentheses, the interjections of "like our students," distinguish these sentences from others in the article and so add variety. The parentheses also serve to isolate and emphasize "We," which reinforces the meaning of the sentences. The polysyndeton, the repetition of the word *and* in the first sentence, effectively conveys an impression of multiplicity and boredom. The antithesis in the second sentence, "being and becoming, not remembering and reviewing," strengthens the distinction being made. Furthermore, it is itself strengthened by alliteration. Climax, parenthesis, polysyndeton, antithesis, and alliteration—none of these techniques tend to be found in current textbooks.

Occasionally, a student will experiment to the point of heresy. One of the absolute laws found in textbook after textbook is the necessity of subject-verb-object or subject-verb-complement sentence order, yet a student from parks and recreation included the following inverted sentence in one of her exercises:

> Through the smell of good food cooking over a wood fire or the freshness of new mown hay, the sound of an Indian tribal dance or a hoedown in an old barn, the sight of wind filling the sails of a tall sailing ship or wooly balls of fluff being spun to thread on a spinning wheel, the feel of a sheep's kinky wool or the soft smoothness of a ceramic pot, the taste of first run maple syrup cooled on snow or the unusual taste of sassafras tea, living history interpretation communicates with all of the physical senses.

The reaction of the class to this passage was mixed. One student seemed to like the structure of the five senses, the move from smell to sound, sight, feel, and taste. Some felt that the parallelism was overdone; others liked it. A few specifically mentioned that they liked the way sound reinforced sense; the alliteration of S's in "the sight of wind filling the sails of a tall sailing ship" does suggest the sound of a whistling wind. None of the students found the sentence obscure as a result of the inversion. In fact, one member of the class believed that the presence of concrete images before the abstraction made the abstraction more clear. It may be that we have become too dogmatic in some of our dictates about style.

THE VALUE OF RHETORICAL DEVICES

My experiment is far from complete. I have been working with exercises involving stylistic devices over three successive semesters, but much more evidence needs to be accumulated. Some things have already become clear, however. Rhetorical devices do promote such traditional ideals in technical writing as economy and clarity. They encourage students to be more aggressive writers and let their personalities through in nonspecialist communication. They strengthen the belief that writing is an art that can be learned and not a talent inherent in the genes. Rhetorical devices help with invention, the discovery of content. They arouse reader interest. Perhaps I have been most delighted with the fun so many of my students have been having. Most of them attack the exercises with an exuberance that renews my faith in the magic of words. In many facets of our lives these days we seem to be reaching back to the past. For instance, we are beginning to run short of energy and are turning to windmills again. I have come to the conviction that the past also holds stylistic riches for the modern practice of scientific and technical writing. Adjustments need to be made in the ideal of the plain style passed on to us by the scientific revolution of the 17th century.

REFERENCES

Jones, R. F. (1951). *The seventeenth century: Studies in the history of English thought and literature from Bacon to Pope.* Stanford, CA: Stanford University Press.
Lyley, J. (1967). Ephues: The anatomy of wit. In M. Lawlis (Ed.), *Elizabethan prose fiction* (pp. 112–188). New York: Odyssey.
Sprat, T. (1667). *History of the Royal Society.* London.
Weisman, H. W. (1974). *Basic technical writing* (3rd. ed.). Columbus, OH: Charles E. Merrill.
Whitburn, M. D. (1976). Personality in scientific writing. *Journal of Technical Writing and Communication, 6,* 299–306.

7

Deconstructing Depression: A Historical Study of the Metaphorical Aspects of an Illness*

Henrietta Nickels Shirk
University of North Texas

INTRODUCTION: TEXTS AND TERMS

The historical development of the rhetorical uses of metaphors in biomedical communication can be traced by examining written descriptions relating to a particular illness over several centuries.[1] In this study, I focus on the uses of metaphor to describe the mental illness of depression in selected published texts spanning the period from the Hippocratic writings of the 5th century B.C. to the current *Diagnostic and Statistical Manual of Mental Disorders* published by the American Psychiatric Association (1994).

The approach is that of deconstructing portions of selected texts in terms of the metaphors that their authors use to describe and respond to the mental illness that we today identify as depression, and which in earlier times was identified as melancholia. It is not my intent to provide a comprehensive history of the different concepts of depression, but rather to employ a deconstructive reading of selected major

---

* I wish to express my appreciation to the Department of English at the University of Central Florida, Orlando, for creating a stimulating and encouraging environment in which to research and write the first draft of this essay while in residence there as a visiting professor during the Spring of 1997.

texts on this disease entity in order to demonstrate historical, as well as conceptually evolving, changes in the metaphors used to describe melancholia/depression.

The assumption is that, even though translations are used for some of the texts covered here, metaphors are an aspect of style that is usually translated literally, and their meanings are therefore typically not lost in the translation process. Although this current selection of texts is limited, it is nonetheless representative of the predominant medical views on depression during each major period of history. The texts are those that can be classified as nonfiction—written for both experts and laypersons—in order to communicate information about the nature of melancholia/depression.

Finally, a specifically focused historical study such as this one provides a basis for drawing several conclusions about the uses of metaphor as a rhetorical device by today's technical communicators, especially those working in biomedical fields. It also provides an opportunity to reflect on the nature of metaphors in general by observing their historical evolution through a single example. The informing methodology, or strategic device, of this analysis is that of deconstruction as articulated by the French philosopher and rhetorician, Jacques Derrida. The metaphors in the selected texts are subjected to an analysis that assumes several characteristics about texts.

THE "DECONSTRUCTIVE" WAY OF READING TEXTS

While it is difficult, if not impossible, to articulate a complete set of rules and procedures for reading texts from the perspective of deconstruction, the following fundamental maxims are generally agreed upon (Aune, 1990):

> (1) The very nature of language prevents meaning from ever being fully present in any act of communication.
> (2) Texts (like the larger philosophies or ideologies which they profess to re-present) work by establishing pairs of opposed terms in which one term has priority.
> (3) These paired terms tend to collapse into one another when one examines their interplay within a text. (p. 259)

According to Aune (1990), deconstruction provides a useful way for reading the history of rhetoric in the West. A deconstructive reading of texts demonstrates that any discourse is caught up in webs of signification that precede it, over which the writer and reader have little control, and which prove to be logically contradictory upon rigorous examination. Crowley (1989) states:

> Deconstruction exposes the dissemination of textual meaning beyond what are aberrations, or inconsistencies in a given text. It does this because it is aware that language, especially written language, is reflexive rather than representative; it folds back in on itself in very interesting and complex ways which produce meanings that proliferate beyond an author's conscious control. (pp. 7–8)

In terms of using metaphors effectively, the deconstructive assumption is that, in order to consider webs of signification, one must become aware of the particular metaphor's historical background. According to Derrida, a deconstructive reading of a text does not try to aim or turn the text toward some overarching system of meaning that would "make sense of it," but rather, to look for places in texts where the author mis-speaks, loses control of intention, or says what is not "meant" to be said. Derrida's method is to select loaded metaphors from texts and to show how they work to support a whole powerful structure of presuppositions (Norris, 1991, pp. 27–28).

As Derrida states, "the reading must always aim at a certain relationship, unperceived by the writer, between what he commands and what he does not command of the patterns of language that he uses" (1976). Deconstruction has been called "infectious, corrosive, and irrepressible" in that it threatens all forms of convention (Leitch, 1983, p. x). The notion that there are aspects of our texts that we do not command is unsettling to many writers, as well as readers. A historical analysis can show us how texts do indeed exhibit lives beyond themselves, and have not only pasts, but also futures often unknown to their writers and readers.

MAKING MEANING OF METAPHORS

What then is a metaphor? It is a figure of speech in which one thing is thought or expressed in terms of another, oftentimes unrelated, thing. Generally, metaphors use some word or phrase in a new sense in order to remedy a gap in vocabulary and one's understanding of a concept (in this case, an illness). A metaphorical expression gives meaning that transforms literal statements from their normal literal meanings by providing insight, usually insight that the literal expression would fail to give (Black, 1962, 1979).

It has been suggested that "our ordinary conceptual system, in terms of which we both think and act, is fundamentally metaphorical in nature" (Lakoff & Johnson, 1980, p. 3; Lakoff, 1987). If these characteristics of metaphor are true, then it follows that the metaphorical language used to describe a particular illness will also provide a more comprehensive understanding of that illness. Such a study of language should also enable readers, and possibly writers, to use metaphorical language more effectively.

For deconstructionists, all language is metaphorical. Although deconstructionism has been applied as a critical approach primarily to philosophical and literary texts, it also has relevance for the texts of technical communication as well. Specifically, it can be applied to the use of metaphor as a rhetorical technique in descriptions of illnesses. Depression/melancholia has been recognized as a problem for human beings since the beginning of time. How authors have written about depression provides a way of focusing on the uses of metaphors to describe this disease in particular and to provide some conclusions about metaphorical language for technical communication in general.

It has been suggested that, when deconstruction concentrates on the metaphors in a text, "it is in order to dramatize, through these contingent associations, connections that repeat themselves in various guises and contribute to a paradoxical logic" (Culler, 1982, p. 60). Indeed, the concept of melancholia and depression is surrounded by numerous examples of paradoxical logic and multiple interpretations. My goal is to deconstruct the play of concept and metaphor behind the notions of melancholia and its evolution into the disease today known as depression.

THE "BLACK BILE" OF THE ANCIENTS

There is a certain descriptive, and constrictive, consistency among the ancients who wrote about melancholia, considered to be one form of madness. The scattered fragments of the Hippocratic writings of the 4th and 5th centuries B.C. are the first known written works to indicate that the key causative factor for melancholia was an excess of "black bile" (Tellenbach, p. 1). The English term *black bile* was derived from the Greek *melaina chole*, which was translated into Latin as *atra bilis*. As one of the four humors in the humoral theory prevalent in classical times, black bile was called the melancholic humor, and it was thought to be the essential factor in causing melancholia, as well as various other disorders that came to be referred to as melancholic diseases.

In terms of the humoral theory of illness, melancholia was initially associated with the season of autumn (when people were more inclined to be susceptible to it) and its corresponding qualities of coldness, dryness, and viscosity. Probably as early as the Pythagoreans, the four seasons were also matched with the Four Ages of Man, making autumn the time of human decline preceding old age. The Hippocratic writers referred to melancholia as being both the humor of black bile and the mental and physical state caused by an excess of this humor, when it was unbalanced in the body and therefore affected the brain (Hippocrates, trans. 1923–1931, vols. II & IV).

Galen (131–201) devotes a chapter in his book, *On the Affected Parts* (trans. 1976), to melancholia. Essentially, he elaborates on the Hippocratic humoral view and suggests that melancholic patients exhibit either fear or despondency. Galen describes these patients:

> Because of this despondency patients hate everyone whom they see, are constantly sullen and appear terrified, like children or uneducated adults in deepest darkness. As external darkness renders almost all persons fearful, with the exception of a few naturally audacious ones or those who were specially trained, thus the color of the black humor induces fear when its darkness throws a shadow over the area of thought [in the brain]. (p. 93)

For approximately 2,000 years, the humoral theory remained the major explanatory scheme for dealing with all diseases. In terms of melancholia, Hippocrates, Galen and others introduced the major metaphorical motifs that are to this day associated with it—the notion of being in a condition characterized by darkness and shadows.

Black bile itself is thus a metaphor for describing melancholia. Indeed, because the term *melancholia* literally means "black bile," to say that it is caused by black bile is a logical tautology. It is not only a metaphorical inadequacy, but it is an example of an interplay of terminology in which both terms tend to merge into one another, in that the cause for the disease is the disease itself, and vice versa. Hippocrates' blackness and imbalance and Galen's darkness and shadows are both collapsed metaphors, or perhaps we could more accurately say "as-yet unfolded" metaphors. It was in the following centuries that melancholia came to be viewed not as a punishment that happened to people, but rather as a sign of evil, something to *be* punished.

THE SIN OF ACEDIA IN THE MIDDLE AGES

While the Middle Ages continued to widely use the metaphors for melancholia introduced by the classical writers and Galen, by the late 4th century A.D. the Christian church introduced the then-familiar term of *acedia* to designate a constellation of feelings and behaviors that were related to melancholia. According to John Cassian (360–435 A.D.), the characteristics of acedia were a "weariness or distress of heart," "akin to dejection," and "especially trying to solitaries" (trans. 1955, vol. 11: pp. 266–267). The gradual development of acedia's relation to melancholia resulted from the initial experiences of monks and religious anchorites who lived in isolation from the rest of society and often found such isolation difficult. The notion of acedia was eventually applied to nonreligious life as well. Indeed, the medieval relationship of acedia to states of sorrow and dejection is a complex one, and it is filled with several contrasting elements.

A major concern was how acedia fit into the medieval worldview of cardinal sins, a list that frequently changed over the centuries. In the early Middle Ages, acedia was associated with *tristitia* (dejection, sadness, or sorrow), and there were frequent references to *desperatio* (despair) in writings about acedia. By the later Middle Ages, acedia was associated directly with *melancholia*. In any case, as a sin acedia was dealt with in terms of doing required, and sometimes harsh, penance, while as a medical condition, it was dealt with as a disease that required compassion and curing. The latter view began to predominate with the developing humanism of the Renaissance.

Additionally, two main groupings of themes surrounding acedia evolved during the Middle Ages—that of sorrow-dejection-despair, and that of neglect-idleness-indolence. How one thought of acedia depended on the emphasis one placed on the total condition. Sorrow or dejection could be a state of mind, and slothful behavior could be the external manifestation of this state of mind. As Stanley Jackson (1986) explains:

> Whereas traditionally acedia had implied...a combination of inner state and mode of behavior, through the eleventh century there had been a trend toward more emphasis on the physical phenomenon of idleness and drowsiness. During the following century, however, spiritual authors laid more emphasis on the inner phenomenon of spiritual slackness, weariness and boredom with religious exercises, and a dejected state of mind. (pp. 72–73)

Eventually, it was the neglect-idleness-indolence aspect that was brought more into focus and was linked to melancholia as a disease entity, as the cardinal sins became less significant in subsequent and more secularly-oriented centuries.

For the purposes of this analysis, a significant contribution of the Middle Ages is Cassian's sustained medical metaphor as a way of treating acedia. When discussing the question of remedy, Cassian views the treatment for acedia as being administered by someone with characteristics like St. Paul, who was "like a skillful and excellent physician" providing "the healing medicines of his directions" (trans. 1955, vol. 11: pp. 267–268). This comparison of spiritual remedy to being like a medical remedy is an interesting perspective that has overtones for consideration of the moralistic aspects of disease.

Representative of a 20th-century viewpoint on illness in general, Susan Sontag (1978), in her essay on current perceptions of the disease of cancer, reminds us that:

> Nothing is more punitive than to give a disease meaning—that meaning being invariably a moralistic one. Any important disease whose causality is murky, and for which treatment is ineffectual, tends to be awash in significance.... Feelings about evil are projected onto a disease. And the disease (so enriched with meanings) is projected onto the world. (p. 58)

The fact that the disease of acedia was considered as a sin during the Middle Ages is a kind of metaphor in itself. Acedia, in its association with melancholia, is indeed "awash in significance" as the religious emphasis of the Middle Ages merges with the physical emphasis of earlier times. It is as though the notion of melancholia has simply been given an additional layer of metaphorical meaning to carry along with it. As Sontag (1978) reminds us, "disease metaphors are never innocent" (p. 84). Just as the characteristics of melancholia (i.e., acedia) were given religious metaphors in order to explain them, so too did the Renaissance add its own value-laden images to this illness.

MELANCHOLIA IN THE RENAISSANCE

For the purposes of medical history, the period known as the Renaissance has generally been associated with the 16th century and the early decades of the 17th century. In spite of new discoveries in anatomy, physiology, and therapeutics, Galenic medicine was still dominant during most of the Renaissance, and most medical texts on melancholia were essentially elaborated versions of Galen's views and metaphors. Two of these texts provide representative Renaissance examples.

Timothie Bright (1550?–1615) published *A Treatise of Melancholie* in 1586. In this work, he makes extensive use of the metaphors of darkness, cloudiness, and heaviness. These things obscure the functions of the mind, with "melancholie vapours rising from that pudle of the splene" to obscure the normal "clearenes, which our spirites are endued with" (pp. 102–109). He goes on to describe at length the "monstrous terrors of feare and heaviness without cause" that came from the effects of "the unnaturall melancholie rising by adjustion" and that led to various excited, agitated, and frightened states, which he indicates were "of another nature far disagreeing from the other, & by an unproper speech called melancholy" (pp. 110–116).

Robert Burton (1557–1640) published *The Anatomy of Melancholy* in 1621. This work is a comprehensive compilation of Renaissance views on melancholia. Burton uses the term *melancholy* for the disease known as "melancholia," whether referred to in Latin or one of the vernacular languages. He also frequently uses the term to refer to the melancholy humor (black choler, black bile) and sometimes to the melancholy temperament. Burton defines the disease of melancholy/melancholia as "a kind of dotage without a fever, having for his ordinary companions fear and sadness, without any apparent occasion." Dotage meant that "some one principal faculty of the mind, as imagination, or reason, is corrupted, as all melancholy persons have." He emphasized that "*Fear & Sorrow* are the true characters and inseparable companions, of most melancholy" (pp. 332–338). For Burton, then, melancholia is a kind of inward corruption or disintegration, but without any overt cause. And he compares the characteristic attitudes of melancholia (fear and sorrow) to "inseparable companions" (pp. 148–149). Burton's metaphors relating to the personal burdens caused by melancholia express the Latin meaning of depression, from *depressus*, to press down or the act of pressing down.

In addition to the traditional views of Bright and Burton, which defined melancholia either as one of the humors or as a disease, there is a third perspective on melancholia that was prevalent in the common, vernacular language practices during the Renaissance, where there was no distinction between the melancholy temperament and the disease of melancholia. Jackson (1986) describes this universe of meaning and discourse:

> In somewhat simplified terms, this melancholy disposition was thought to be the basis for intellectual and imaginative accomplishments, to be the wellspring from which came great wit, poetic creations, deep religious insights, meaningful prophecies, and profound philosophic considerations; and yet, at the same time, those so disposed lived at a certain risk that their melancholic temperament might lead them into melancholia the disease. (p. 99)

Actually, this view of melancholia was first articulated by Aristotle and later brought to popularity in the early Renaissance by Marsilio Ficino (1433–1499), with far-reaching effects on both his own time and later times. Ficino was a philosopher, physician and priest, whose writings fused the notion of Platonic divine inspiration with the Aristotelian superior melancholic disposition. Ficino's view was that "the highly gifted melancholic—who suffered under Saturn, in so far as the latter tormented the body and the lower faculties with grief, fear and depression—might save himself by the very act of turning voluntarily towards that very same Saturn," to a life of creative contemplation (qtd. in Klibansky, Panofsky, & Saxl, 1964, p. 271).

The developing humanism of the Renaissance is apparent in the view of the melancholic person somehow having greater insight and wisdom than the ordinary unafflicted person. During this period of history, it is no longer a sin to be sad, fearful, and depressed, but rather a virtue associated with creative superiority. Like the Medieval religious perspective, the humanistic Renaissance perspective here also becomes part of the metaphorical territory associated with the illness of melancholia.

This relationship between disease and insight has persisted until the present. As current essayist Susan Sontag (1978) reminds us: "In the twentieth century, the repellent, harrowing disease that is made the index of superior sensitivity, the vehicle of 'spiritual' feelings and 'critical' discontent, is insanity" (p. 35). Of course, in its extreme forms, what we now call depression is one of many different kinds of insanity or mental illness. However, our current views of depression come to us with the additional metaphorical layers of meaning provided by the intervening centuries.

CHEMICAL EXPLANATIONS IN THE 17TH CENTURY

The Galenic humoral theory of melancholia did not begin to wane until the second half of the 17th century, when it was replaced by chemical theories of causation. These developments in the concepts about melancholia undoubtedly resulted from developments in the scientific field. Again, a brief look at two representative texts will elucidate the focus of this century.

Richard Napier (1559–1634) was a physician and clergyman in rural England. Among Napier's mentally troubled patients were frequently those who demonstrated, as he writes, "melancholy," "took grief," and "grieving," "simply sad," and "all sadness symptoms" (qtd. in MacDonald, 1981, p. 117). In Napier's medical notebooks are references to standard medical ideas about melancholy during his time. For example, letters from his patients refer to the fact that the person is "Desirous to have something to avoid the fumes arising from the spleen." And another states that he has "Deep melancholy fearfulness, almost of every object. Fumes ascending from the stomach, distempering the brain" (pp. 152–153). Napier's view is that melancholy is caused by improper chemicals (fumes) in the body. The shadows of earlier times are now expressed in more scientific terms as "fumes," themselves as metaphorical as their shadowy predecessors.

Thomas Willis (1621–1675) was a proponent of the chemical basis for melancholy, and he argued against the traditional humoral theory. In his *Practice of Physick*, he favored the five "Principles of Chymists" (1684, pp. 3–8), which meant "affirming all Bodies to consist of Spirit, Sulphur, Salt, Water, and Earth, and from the diverse motion, and proportion of these, inmixt things, the beginnings and endings of things, and chiefly the reasons, and varieties of fermentation, are to be sought" (p. 2). Willis explains that when "Melancholick people talk idly, it proceeds from the vice or fault of the Brain, and the inordination of the Animal Spirits dwelling in it" (p. 188). He further explains that, while the animal spirits would ordinarily have been "transparent, subtle, and lucid," they "become in *Melancholy* obscure, thick, and dark, so that they represent the Images of things, as it were in a shadow, or covered with darkness." He then suggests that the animal spirits, "with the Vehicle to which they cleave" (namely, the blood), were analogous "to some *Chymical* Liquors, drawn forth by distillation form natural mixtures" (p. 189). Chemical notions thus become the basis by analogy for the formation of the animal spirits and their pathological alternations in the mental disorder of melancholia.

Both Napier and Willis present their figurative language about "fumes" and "animal spirits" on the basis of extended analogy. In accordance with the developing scientific knowledge of their time, they have both seized upon a metaphor (that of chemistry) to talk about human life, including the notion of depression. Their concern in looking for causes of melancholy is in keeping with the developing attitudes of scientific inquiry during their time. No longer are texts about melancholia merely descriptive, or attributed to a sin or to Saturn, but there is a "scientific" reason for the existence of this disease entity. In metaphorical terms, both Napier and Willis are "victims" of the metaphorical images (biases) of their century, but perhaps no more so than other writers on the same topic during the following centuries.

As the 17th century drew to an end, however, the chemical theories used to explain melancholia by Napier and Willis were short-lived. By the end of the century, mechanical notions were already displacing them and functioning to completely replace the earlier humoral explanations.

MECHANICAL EXPLANATIONS IN THE 18TH CENTURY

During the 18th century, mechanical explanations gradually became a significant feature in science, and they also began to displace the chemical explanations for melancholia. William Harvey's establishment of the circulation of the blood gave encouragement to the use of hydrodynamic concepts to explain the operation of the human body. These mechanical views were initially articulated by medical writers such as Friedrich Hoffmann (trans. 1971) and Hereman Boerhaave (1735), and given broad acceptance by the work of Richard Meade (1751, 1762).

In his *Fundamental Medicinae* (1783), Hoffman conceived of the human body as being "like a machine which is composed of solid and fluid particles, disposed and arranged in varying order and position" (pp. 6–10). Hoffman describes melancholia in terms of the nervous system and the flow therein of animal spirits. He states: "In melancholics the spirits are indistinct and fixed, and approach a sort of acid nature. They not only leave enduring fixed ideas in the brain pores, but promptly uncover similar traces, of ideas of sadness, terror, fear, and so on" (pp. 71–72). "With this body fluid turning acidic and fixed, its motions became sluggish, and the person became slow, timid, and sad" (pp. 12, 70).

In his *Aphorisms* (1735), Boerhaave also puts forth a mechanistic explanation of the body and of melancholia. Although he maintains that "this Disease arises from that Malignancy of the Blood and Humors, which the Ancients have called *Black Choler*," he is not using the traditional humoral theory, and implying that this is an excess of black bile. Rather, this is a pathogenic material that develops from malfunctioning of certain mechanical or hydrodynamic actions, when "the most moveable Parts of all the Blood be dissipated and have left the less moveable united, then will the Blood become thick, black, fat and earthy. And this Defect will be call'd by the name of an *Atrabiliar Humor*, or *Melancholy Juice*" (pp. 312–313). Boerhaave sees melancholia resulting from a pathogenic sequence that included "a long continued preceding Sorrowfulness, in which the Vessels of the abdominal Bowels create a Stagnation, Alternation, and Accumulation of black Choler Which insensibly increaseth, though the Body was very healthful but a little before: And also that the same black Choler, when bred from bodily Causes, doth produce that Delirium" (p. 320).

With Hoffman and Boerhaave, the conceptual matrix surrounding the humoral theory was lost. As Jackson (1986) explains:

> As with others before him, Boerhaave had abandoned the traditional humoral theory, but he used the term *humor* to refer to the various bodily fluids and secretions. Black bile was no longer a normal component in a physiological theory that, in excess, might have pathological results. It was now a darkened substance that separated out from the blood on standing or on slowing, a thickening mass with an obstructive potential rather than a traditional humor with the qualities of coldness or dryness. Boerhaave was introducing a then modern meaning for a notion that had gradually lost its traditional conceptual matrix. (p. 120)

This change in metaphorical direction of the 18th century demonstrates the way old terms can be wrapped in the new "clothing" of emerging metaphors. The term *humor* does not disappear, but was merely appropriated as terminology to explain objectively observable bodily fluids like the blood. As metaphors are used to describe disease entities, they seem to subsume their meanings from prior historical times and merge, or collapse, them into the current predominant views of the time.

Machines were of great interest and importance to the scientific and industrial communities of the 18th century. In his *Medical Works* (1762), Richard Meade states that the human body "ought to be considered as a hydraulick machine...in which there are numberless tubes properly adjusted and disposed for the conveyance of fluids by different kinds" (p. 455). He uses the concept of a kind of nerve juice, a subtle fluid that is brought to the brain in the blood; the brain, functioning as "a large gland," separated it from the blood; the nerves served as excretory ducts for this "thin volatile liquor, of great force and elasticity," and, "lodged in the fibres of the nerve" (p. xxi). What he calls "melancholy disorders," a kind of madness, are brought about by alterations in this active liquor, or "animal spirits." In his *Medical Precepts and Cautions* (1751), Meade further defines these "diseases of the head" as being caused when "images present themselves to the mind...and excite certain affections or passions of the soul, which are instantly followed by suitable motions in the body"; these passions cause "alterations in the body, by raising commotions in the blood and humours" (p. 77). He then adds that "the instrument of all these motions, both of mind and body, is that extremely subtle fluid of the nerves, commonly called *animal spirits*" (p. 78).

Meade's *animal spirits* are not the animal spirits (fumes) of the prior century mentioned by writers like Napier and Willis. Rather, they are examples of figurative language used to describe the motion of fluids in the blood, acting upon the brain and the body—and able to cause melancholia. They are part of an extended metaphor used to describe mechanical motion.

It is notable that these 18th century mechanical metaphors for melancholia had little lasting influence on the medical understanding of the disease itself. In spite of all their influence on terminology and theoretical formulations, they left the clinical syndrome essentially unchanged. As Jackson (1986) correctly observes:

> While mechanical ideas sometimes colored the theoretical rationale for a treatment endeavor, the therapeutics of melancholia continued much the same, or at least relatively unchanged by any influence from such ideas. As much as anything else, under the influence of a series of leading medical authors and teachers, melancholia was kept terminologically and conceptually in fashionable dress. And each explanatory scheme offered its own metaphors for referring to the signs and symptoms of melancholia. (p. 132)

It was during the following centuries that more layers of metaphorical clothing were added to melancholia's "fashionable dress."

MELANCHOLIA-MANIA CONNECTIONS IN THE 19TH CENTURY

A strong descriptive tradition dominated medical writing about melancholia during the 19th century. Jackson explains this trend: "During the nineteenth century these descriptive contributions increasingly entailed careful attention to the course of a disorder, perhaps reflecting the growing populations in psychiatric institutions and the concomitant increase in opportunities to follow the course of these illnesses over longer periods" (1986, p. 185). In terms of metaphorical language, this descriptive tradition resulted in several changes in the perception of melancholia, as it began to be thought of as functioning within a continuum of characteristics that included its opposite, or mania. Three 19th-century medical texts represent typical views.

Benjamin Rush (1745–1813), who is known as the father of American psychiatry, believed that all diseases, including melancholia, or what he called *tristimania*, were essentially fevers and were thus variants of one basic condition. His view is that within the vascular system was a central and disordered excitement, and therein is the crucial explanatory factor. In his book, *Medical Inquiries and Observations upon the Diseases of the Mind* (1827), Rush outlines his physiological orientation:

> All the operations of the mind are the effects of motions previously excited in the brain, and every idea and thought appears to depend upon a motion peculiar to itself. In a sound state of mind these motions are regular, and succeed impressions upon the brain with the same certainty and uniformity that perceptions succeed impressions upon the senses in their sound state. (p. 9)

He goes on to explain that "the cause of madness is seated primarily in the blood-vessels of the brain, and that it depends upon the same kind of morbid and irregular actions that constitutes other arterial diseases. There is nothing specific in these actions. They are a part of the unity of disease, particularly of fever; of which madness is a chronic form, affecting that part of the brain which is the seat of the mind" (1827, pp. 15–16). For Rush, melancholia is thus a kind of persistent fever of the brain. Rush's view is typically a turn-of-the-century unitarian one—that is, he looks for *one* cause to explain all forms of mental disease. His metaphor for explaining how the human body works becomes, in essence, his metaphor for explaining all forms of mental illness.

This unitarian view is further expressed by Jean Christian Heinroth (1773–1843). Heinroth introduced a detailed classification system for melancholia and depression (genus and species). He characterized melancholia in a unique way:

> If we now consider the main forms of disturbances of the disposition, insanity and melancholia, we find that they are distinguished by altogether different characters; in melancholia the disposition has lost its world, and becomes an empty, hollow Ego which gnaws at itself, while the insane disposition is torn and removed from itself and flutters among the dream images and airy figures of the imagination. We find here signs of two opposite physical principles: the centripetal or contractive force, that is, a tendency to lose oneself in one central point and thus gradually fade out into nothing; and the centrifugal, or expansive force, that is, a tendency to expand without limit and thus also fade out into nothing. (1818/1975, pp. 221–222)

Heinroth presents the concept of symptom patterns along a continuum of severity, with "two opposite physical principals" (p. 221), and he describes how some cases of melancholia might gradually worsen and be transformed into cases of mania. He explains that there are a variety of "intermediate forms through which this transition from melancholia passes into maniacal excitement" (1818/1975, p. 271). He refers to such cases as "morbid states in which the patients are *out of themselves*," in contrast to the turned-inward, self-preoccupied tendencies of melancholic patients (p. 273). With Heinroth, we have the beginnings of descriptions about manic-depressive disorders, perhaps a result of the prevailing interest in classifying and schematizing scientific information.

Richard von Krafft-Ebing's (1840–1902) *Text-Book of Insanity* was a widely influential work on psychiatry that had far-reaching influence in the latter portion of the 19th century. Among the "forms of insanity" covered in his famous textbook were melancholia and mania. He said: "The fundamental phenomenon in melancholia consists of the painful emotional depression, which has no external, or insufficient external, cause, and general inhibition of the mental activities, which may be entirely arrested" (1879/1904, p. 286). Von Krafft-Ebing's descriptions of melancholia exhibit his extended metaphor of nutrition. As he explains it:

> Melancholia, from a comprehensive, unprejudiced point of view, may be defined as an abnormal condition of the psychic organ dependent upon a disturbance of nutrition, characterized, on the one hand, by a psychic painful emotional state and manner of reaction of the whole consciousness (psychic neuralgia), and, on the other hand, by inhibition of the psychic activities, feelings, intellect, and will, which may go to the extent of arrest. (p. 286)

Von Krafft-Ebing continues to describe "the depressed state" as "a painful, depressed condition of feeling (psychalgia, phrenalgia), that has arisen spontaneously and exists independently [and] is the fundamental phenomenon in the melancholic states of insanity" (p. 49). In addition, he says, "Here we have a phenomenon analogous to that which occurs in a sensory nerve as the result of disturbance of nutrition, in the form of neuralgia. Disturbance of nutrition in the cerebral cortex leads to mental pain (psychic neuralgia)" (1879/1904, p. 49).

This nutritional theme to explain melancholia resulted perhaps from the 19th-century emphasis on providing detailed information in describing the clinical picture of the patient, and as the result of increased scientific understanding of the importance of nutrition in human development. Likewise, the melancholia-mania connection became prominent as medical writers saw a continuum of severity in a single disease, an insight that perhaps resulted from their desire for a unifying concept to give meaning to observable differences in mental disorders and to find the causes of disorders. Metaphorically, the language used to describe melancholia in the 19th century typically reflects the concerns and interests of that century's worldview.

DEPRESSION IN THE 20TH CENTURY

At the beginning of the 20th century, melancholia gradually became identified as a symptom of depression, rather than depression being a symptom of melancholia (as it had been in earlier centuries). This inversion of meaning was enhanced by further—and still ongoing—attempts to classify various mental diseases, and especially to systematize various depressive disorders and dichotomies.

In 1917, the first official classification system for mental disorders was formulated by the American Medico-Psychological Association (later to become the American Psychiatric Association). This publication is known as the *Diagnostic and Statistical Manual of Mental Disorders* (or *DSM*). It is now in its fourth edition (*DSM-IV*, 1994). The 1917 edition covered only two mood disorders: manic-depressive psychosis and involutional melancholia. This information was based primarily on the work of Emile Kraepelin (1856–1926), whose influence continued into the 20th century. Kraepelin was responsible for defining the type of melancholia that came to be known as "involutional melancholia." This melancholia, he writes in his famous *Lehrbuch* (1902),

> represents two groups of cases which are characterized by uniform depression with fear, various delusions of self-accusation, of persecution, and of a hypochondriacal nature, with moderate clouding of consciousness and disturbance of the train of thought, leading in the greater number of cases after a prolonged course to moderate mental deterioration. (pp. 254–255)

Kraepelin describes the depressive forms of manic-depressive insanity as also including a "great clouding of consciousness" (p. 258). Kraepelin's "clouding" metaphor contains reminders of Galen's "dark shadows" metaphors used to describe melancholia.

Gradually, however, the term *depression* has came to be the noun to which the descriptive term *melancholy* was attached as a kind of adjectival metaphor. Three examples from 20th-century medical literature, which follow, demonstrate the movement toward this significant change in perspective, as well as focus on the ongoing concern for accurate nosological systems.

Adolf Meyer (1866–1950): "In its current use melancholia applies to all abnormal conditions dominated by depression.... It is obvious that many dissimilar conditions are thus brought under one heading, simply because they are dominated by depression" (1957, p. 566).

Eugen Bleuler (1857–1939): When describing "manic-depressive insanity"—"melancholia," or "the depressive phase colors all experiences painfully" (1924, p. 472).

Oskar Diethelm (1897–): "Depressions are characterized by the depressed mood which is expressed by various individuals..., by the attack form of the illness, and by accompanying physical symptoms..." (1950, pp. 205–206).

Diethelm's statement reminds us of the ancients' view that melancholia was caused by the melancholic bile. However, the term *melancholia* was not easily put aside as a disease entity, and it has continued to thrive in various 20th century contexts.

Sigmund Freud (1856–1939), the originator of psychoanalysis, wrote an influential paper on "Mourning and Melancholia" (1917/1957) that has had wide impact on 20th-century thinking on depression. In this essay, Freud draws an extended analogy between the dejected state connected with mourning the death of a loved one and the characteristic symptoms of melancholia. Freud observes that in grief and mourning "it is the world which has become poor and empty" due to the quite apparent loss, whereas "in melancholia it is the ego itself" which is lost (p. 246). In mourning, the sufferer slowly withdraws emotional attachment from the now-lost object and gradually frees him/herself from this painful state. In melancholia, however, matters are much more subtle, and the nature of the loss much less clear, or totally unclear. As Freud states:

> The difference is that the inhibition of the melancholic seems puzzling to us because we cannot see what it is that is absorbing him so entirely. The melancholic displays something else besides which is lacking in mourning—an extraordinary diminution of his self-regard, an impoverishment of his ego on a grand scale.... The patient represents his ego to us as worthless, incapable of any achievement and morally despicable; he reproaches himself, vilifies himself and expects to be cast out and punished.... He is not of the opinion that a change has taken place in him, but extends his self-criticism back over the past; he declares that he was never any better. (p. 245)

The series of themes introduced by Freud in "Mourning and Melancholia" have continued to influence subsequent considerations of melancholia and depression.

Freud's emphasizes something he called "introjection." He describes the melancholic person:

> If one listens patiently to a melancholic's many and various self-accusations, one cannot in the end avoid the impression that often the most violent of them are hardly at all applicable to the patient himself, but that with insignificant modifications they do fit someone else, someone whom the patient loves or has loved or should love.... So we find the key to the clinical picture: we perceive that the self-reproaches against a loved object which have been shifted away from it on to the patient's own ego. (1917/1957, p. 248)

This lost "loved object," or aspects of that person, had been "taken into" the melancholic person by a process of introjection, and an identification with that person had occurred, that is, aspects of the lost object had become internalized aspects of the person who has now become the melancholic sufferer. "The shadow of the object" had fallen "upon the ego," and a critical inner agency judged the ego as though it were "an object, the forsaken object" (p. 249). Thus, the self-reproaches of the melancholic amount to both overt attacks on himself and unconscious attacks on the lost "loved object" (pp. 248–249). Such relationships, according to Freud, also involves significant ambivalence, which tends to manifest itself in the form of hatred and sadism toward the ego (and the introjected object) in the melancholic state (pp. 250–252). In concluding, Freud states that "the three preconditions of melancholia" are "loss of the object, ambivalence, and regression of libido into the ego" (p. 258). All three factors are deemed necessary, but only the regression of emotional investment was considered specific to melancholia.

Freud's system and its attendant figurative language become themselves a kind of metaphorical system for addressing mental disease. From the perspective of melancholia and depression, Freud's extended analogy highlights the sadness, diminishment, and lack among those suffering from depression.

Phoenix-like, the term *melancholia* has been resurrected as a new synonym and a descriptive adjective for certain kinds of depression. Indeed, 20th century biomedical communication on depression contains many metaphorical and conceptual characteristics from earlier centuries. A recent booklet titled *Depression in Primary Care: Detection, Diagnosis, and Treatment*, published in 1993 by the U.S. Department of Health and Human Services as a "Quick Reference Guide for Clinicians," provides one example of metaphorical accumulations of influences from past centuries.

In words reminiscent of the 4th century B.C., the booklet's authors describe a person with a "major depressive disorder" as suffering from "intense mental, emotional, and physical anguish, and substantial disability" (USDHHS, 1993, p. 1). One is reminded of the "fear and despondency" exemplified by the "darkness" and "shadow" of Galen's description of melancholia.

In words reminiscent of the Middle Ages' moralistic viewpoint about the sin of acedia, the booklet's authors caution:

> Depression is viewed by many patients and the lay public as evidence of a character defect or lack of will power. Thus, those with major depressive disorder must endure the additional burden of having an illness that society views as the reflection of an inherent personal weakness or fault. (USDHHS, 1993, p. 1)

Our 20th century society's view of this disease has indeed not moved conceptually very far beyond the judgmental views of the Middle Ages.

In words reminiscent of the "reduced animal spirits" of the view of the 17th and 18th centuries, the reader of the booklet is told that "common complaints of patients in primary care settings with major depressive disorder include...low energy—excessive tiredness, lack of energy or a reduced capacity for pleasure and enjoyment," "a mood of apathy," and "symptoms of fatigue, malaise,...or sadness" (USDHHS, 1993, p. 2). This description has hidden roots in the chemical and mechanical metaphorical attributes of melancholia expressed by the biomedical writers of the 17th and 18th centuries.

METAPHORICAL ASPECTS OF AN ILLNESS 147

In words reminiscent of 19th century classification systems that placed melancholia and mania on a continuum, the authors of the booklet report that a small percentage of patients will have bipolar disorder. They offer a checklist and a flowchart to be used in determining whether the manic aspects of the disease are present (USDHHS, 1993, p. 4; see Figure 7.1 below). The flowchart itself provides a continuum of focused symptomatic considerations that visually demonstrate the metaphorical continuum for manic-depressive disorder established in earlier centuries. Perhaps indicative of our 20th century viewpoint, the term *melancholia* is not once directly mentioned in the booklet.

In fact, the term *melancholia* is not needed in this publication, because it is an underlying assumption that has meaning equal to depression. Although Rorty's essay (1982) on Derrida's notion of "written-ness" refers to philosophy, it could equally apply here:

> Writing is an unfortunate necessity; what is really wanted is to show, to demonstrate, to point out, to exhibit, to make one's interlocutor stand and gaze before the world.... In a mature science, the words in which the investigator "writes up" his results should be as few and transparent as possible. (p. 94)

The term *depression* has indeed come to mean the same as melancholia in the fields of psychology and psychiatry—or almost the same. However, perhaps the 20th century has not quite yet been able to give up melancholia's metaphorical meanings, as the *DSM-IV* reveals.

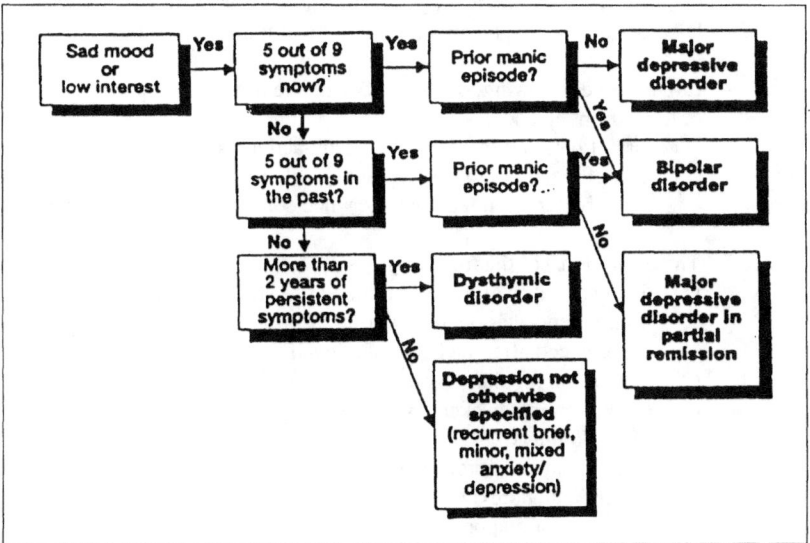

FIGURE 7.1. Differential Diagnosis of Primary Mood Disorders (from USDHHS, *Depression in Primary Care*, 1993, p. 4).

The *DSM-IV* (1994) does use the words "melancholic." In the section on depressive disorders, it provides a classification of major depressive disorders, and it suggests applying a "Melancholic Features Specifier" that is included in the manual (see Figure 7.2 below). This specifier exhibits the influence of Freud's essay on "Mourning and Melancholy" in its distinction between the depressed mood being "experienced as distinctly different from the kind of feeling experienced after the death of a loved one" (p. 384). It is meaningful that the word *specifier* is used to list the "melancholic features." It implies that these characteristics, much like the element of comparison in a metaphorical figure of speech, are used to make the original term more specific and understandable. They have become adjectival in that they lend details and "written-ness" to depression.

Indeed, it seems that metaphors related to an illness like melancholia/depression are often inherent in the illness itself. Melancholia's presence in the 20th century comes laden with "webs of signification" from all the previous centuries in which the term has been used and in which it has evolved. Perhaps it is appropriate that it has come to be employed in its adjectival form, where its multiple descriptive qualities can be employed more effectively than in its noun forms.

■ **Criteria for Melancholic Features Specifier**

Specify if:
    With Melancholic Features (can be applied to the current or most recent Major Depressive Episode in Major Depressive Disorder and to a Major Depressive Episode in Bipolar I or Bipolar II Disorder only if it is the most recent type of mood episode)

  A. Either of the following, occurring during the most severe period of the current episode:
    (1) loss of pleasure in all, or almost all, activities
    (2) lack of reactivity to usually pleasurable stimuli (does feel much better, even temporarily, when something good happens)

  B. Three (or more) of the following:
    (1) distinct quality of depressed mood (i.e., the depressed mood is experienced as distinctly different from the kind of feeling experienced after the death of a loved one)
    (2) depression regularly worse in the morning
    (3) early morning awakening (at least 2 hours before usual time of awakening)
    (4) marked psychomotor retardation or agitation
    (5) significant anorexia or weight loss
    (6) excessive or inappropriate guilt

FIGURE 7.2. Criteria for Melancholic Features Specifier (from USDHHS, *Depression in Primary Care*, 1993, p. 384).

CONCLUSION: A SUGGESTED METAPHORICAL STANCE

What, then, is the benefit of using metaphors, since they change so quickly and shift in concept and emphasis so easily? Derrida says that "all the concepts which have played a part in the delimitation of metaphor always have an origin and a force which are themselves 'metaphorical'" (1982, p. 301). As we have seen, the metaphors associated with melancholia/depression are really clusters of metaphors, each with a long historical past. These metaphors have many faces and many cumulative and even contradictory meanings.

Paul De Man, in his essay entitled "The Epistemology of Metaphor" (1978), maintains that attempts to control metaphor cannot extract themselves from metaphor and that in each case a crucial distinction between the literal and the metaphorical breaks down. "The resulting undecidability is due to the asymmetry of the binary model" that opposes the figural to the literal (p. 28). The literal is the opposite of the figurative, but a literal expression is also a metaphor whose figurality has been forgotten.

Metaphors are not simplistic mechanisms of comparison, significant only at the peripheral areas of texts, but rather they are figures of speech fraught with multiple and powerful meanings. Michel Foucault's statement about the manic-depressive cycle in *Madness and Civilization* (1965) is appropriate to all considerations of metaphorical language:

> The essential thing is that the enterprise did not proceed from observation to the construction of explanatory images; that on the contrary, the images assured the initial role of synthesis, that their organizing force made possible a structure of perception, in which at last the symptoms could attain their significant value and be organized as the visible presence of the truth. (p. 135)

Often what is considered unimportant and actually unknown by earlier centuries, and therefore marginalized, becomes central to the meaning of a metaphor for a particular time in history. This continual rewriting of knowledge produced the varying metaphors for describing melancholia/depression. Derrida's notion of "differance" has been aptly explained by Crowley:

> What a culture, or an individual, "knows" at any given moment is available only because its configurations differ from, and yet depend on, what preceded it.... In short, the dream that objects or events can be isolated from their contexts, that lines or borders can be drawn around them, is another metaphysical fiction. (1989, pp. 10–11)

A deconstructive reading of texts, such as those analyzed here, shows how the rhetoric of a metaphor is not necessarily compatible with its explicit meaning. Metaphors are always tainted by their historical and cultural conventions. However, this incompatibility can be systematically and historically explored in order to provide knowledge of the current status and influence—the figurative accouterments—of the metaphor itself.

Technical communicators can effectively apply deconstructive concepts to both the reading and writing of texts. Terry Eagleton (1983) reminds us that all language, for Derrida, displays a certain "surplus" over exact meaning; it is always threatening to outrun and escape the sense which tries to contain it. Although this happens most frequently in literary discourse, it applies equally effectively to nonliterary discourse. Written products, then, become the results of particular systems of meaning, as we have observed in terms of the metaphors of melancholia/depression. They are defined most effectively by what they exclude, as well as by what they include. Historical studies of metaphorical usages can serve to make us more alert to the cumulative and varying meanings associated with metaphorical language. They can also make us more aware of the truly restrictive nature of the historicity of metaphors.

As the metaphors connected with melancholia/depression demonstrate, they are sometimes misleading because of their historical parochialism. However, this does not mean that technical communicators should avoid using metaphors. Sontag's (1978) view that "illness is not a metaphor, and that the most truthful way of regarding illness—and the healthiest way of being ill—is one most purified of, most resistant to, metaphoric thinking" (p. 3) is a position that contradicts the deconstructionist viewpoint. Derrida's method of selecting loaded metaphors for analysis is an approach to explicating how language folds back on itself. This means that the multi-faceted historical aspects of metaphors require that technical communicators use metaphors carefully and with as much knowledge of each metaphor's explicit and implicit meanings as possible.

We need to begin "deconstructing" the field of technical communication by looking at its marginal areas—for example, places where rhetorical figures of speech such as metaphors are incorporated into texts. Figurative language is an important feature in the descriptive writing that is an essential part of technical communication. Additional historical studies of metaphors will undoubtedly reveal some ways that these figures of speech are the results of their historical structures of thought and are embedded in the cultures and often unrelated scientific discoveries of their times. History, however, should always make us aware of our future.

In conclusion, when using metaphors to describe diseases or anything else, technical communicators need to be aware of the webs of signification that accompany all metaphors they might use. Metaphors have pasts; they also have histories that are moving and meaningful. As I have demonstrated, the history of a disease entity like depression has powerful metaphors connected with it. Understanding the past does not always make the present completely intelligible, but it certainly helps.

NOTES

[1] For historic references from Hippocrates to Freud, I am indebted to the historical compilation in Stanley W. Jackson's *Melancholia and Depression: From Hippocratic*

Times to Modern Times (1986). Jackson's perspective is from the history of medicine, while my perspective is from the history of rhetoric as expressed through metaphorical language.

REFERENCES

American Psychiatric Association. (1994.). *DSM-IV: Diagnostic and statistical manual of mental disorders* (4th ed.). Washington, DC: American Psychiatric Association.
Aune, J. A. (1990). Rhetoric after deconstruction. In R. A. Cherwitz (Ed.), *Rhetoric and philosophy* (pp. 253–273). Hillsdale, NJ: Lawrence Erlbaum.
Black, M. (1962). *Models and metaphors: Studies in language and philosophy.* Ithaca, NY: Cornell University Press.
Black, M. (1979). How metaphors work: A reply to Donald Davidson. In S. Sacks (Ed.), *On metaphor* (pp. 181–192). Chicago: University of Chicago Press.
Bleuler, E. (1924). *Textbook of psychiatry* (A. A. Brill, Trans.). New York: Macmillan.
Boerhaave, H. (1735). *Boerhaave's aphorisms: Concerning the knowledge and cure of diseases.* London: W. and J. Innys.
Bright, T. (1586). *A treatise of melancholie.* London: Thomas Vautrollier.
Burton, R. (1948). *The anatomy of melancholy* (F. Dell & P. Jordan-Smith, Eds.). New York: Tudor. (Original work published 1621)
Cassian, J. (1955). *The twelve books on the institutes of the Coenobia* (E. C. S. Gibson, Ed. and Trans.). In P. Schaff, & H. Wace (Eds.), *A select library of the Nicene and post-Nicene fathers of the Christian church: Vol. 11* (2nd ser., pp. 266–267). Grand Rapids, MI: Wm. B. Eerdmans.
Crowley, S. (1989). *A teacher's introduction to deconstruction.* Urbana, IL: National Council of Teachers of English.
Culler, J. (1982). *On deconstruction: Theory and criticism after structuralism.* Ithaca, NY: Cornell University Press.
De Man, P. (1978). The epistemology of metaphor. *Critical Inquiry 5,* 13–30.
Derrida, J. (1976). *Of grammatology* (G. C. Spivak, Trans.). Baltimore: Johns Hopkins University Press.
Derrida, J. (1982). *Margins of philosophy* (A. Bass, Trans.). Chicago: University of Chicago Press.
Diethelm, O. (1950). *Treatment in psychiatry* (2nd ed.). Springfield, IL: Charles C. Thomas.
Eagleton, T. (1983). *Literary theory: An introduction.* Minneapolis, MN: University of Minnesota Press.
Foucault, M. (1965). *Madness and civilization: A history of insanity in the age of reason.* New York: Random House.
Freud, S. (1957). Mourning and melancholia. In J. Strachey (Ed.), *The standard edition of the complete psychological works of Sigmund Freud: Vol. XIV* (pp. 239–260). London: The Hogarth Press and the Institute of Psycho-Analysis. (Original work published 1917)
Galen. (1976). *On the affected parts* (R. E. Siegel, Ed. and Trans.). Basel, Switzerland: S. Karger.
Heinroth, J. C. (1975). *Textbook of disturbances of mental life: Or disturbances of the soul and their treatment* (J. Schmorak, Trans.). (Vols 1–2.). Baltimore: Johns Hopkins University Press. (Original work published 1818)

Hippocrates. (1923–1931). *Works of Hippocrates: Vol. 1* (W. H. S. Jones & E. T. Withington, Eds. and Trans.). Cambridge, MA: Harvard University Press.

Hippocrates. (1923–1931). *Works of Hippocrates: Vol. 2* (W. H. S. Jones & E. T. Withington, Eds. and Trans.). Cambridge, MA: Harvard University Press.

Hoffman, F. (1971). *Fundamenta medicinae* (L. S. King, Trans.). London: MacDonald. (Original work published 1783)

Jackson, S. W. (1986). *Melancholia and depression: From Hippocratic times to modern times.* New Haven, CT: Yale University Press.

Klibansky, R., Panofsky, E., & Saxl, F. (1964). *Saturn and melancholy: Studies in the history of natural philosophy, religion, and art.* New York: Basic Books.

Kraepelin, E. (1902). *Psychiatrie. Ein lehrbuch fur studirende und aerzte* (6th ed.). In A. R. Defendorf (Trans.), *Clinical psychiatry: A text-book for students and physicians* (6th ed.). New York: Macmillan.

von Krafft-Ebing, R. (1904). *Text-book of insanity* (C. G. Chaddock, Trans.). Philadelphia: F. A. Davis. (Original work published 1879)

Lakoff, G. (1987). *Women, fire, and dangerous things: What categories reveal about the mind.* Chicago: The University of Chicago Press.

Lakoff, G., & Johnson, M. (1980). *Metaphors we live by.* Chicago: The University of Chicago Press.

Leitch, V. B. (1983). *Deconstructive criticism: An advanced introduction.* New York: Columbia University Press.

MacDonald, M. (1981). *Mystical bedlam: Madness, anxiety, and healing in seventeenth-century England.* Cambridge, England: Cambridge University Press.

Meade, R. (1751). *Medical precepts and cautions* (T. Stack, Trans.). London: J. Brindley.

Meade, R. (1762). *The medical works of Richard Mead, M.D.* London: C. Hitch.

Meyer, A. (1957). *Psychobiology: A science of man* (E. E. Winters & A. M. Bowers, Eds.). Springfield, IL: Charles C. Thomas.

Napier, R. (1981). *Notebooks.* In M. MacDonald, *Mystical bedlam: Madness, anxiety, and healing in seventeenth-century England.* Cambridge, England: Cambridge University Press. (Original work published 1640)

Norris, C. (1991). *Deconstruction theory and practice* (Rev. ed.). New York: Routledge Press.

Rorty, R. (1982). *Consequences of pragmatism.* Minneapolis, MN: University of Minnesota Press.

Rush, B. (1827). *Medical inquiries and observations upon the diseases of the mind* (3rd ed.). Philadelphia: J. Grigg.

Sontag, S. (1978). *Illness as metaphor.* New York: Doubleday.

Tellenbach, H. (1980). *Melancholy: History of the problem, endogeneity, typology, pathogenesis, clincial considerations.* Pittsburgh, PA: Duquesne University Press.

U.S. Department of Health and Human Services. (1994). *Depression in primary care: Detection, diagnosis, and treatment* (No. 5). Washington, DC: U.S. Government Printing Office.

Willis, T. (1684). *Dr. Willis's practice of physick, being the whole works of that renowned and famous physician* (S. Pordage, Trans.). London: T. Dring, C. Harper, & J. Leigh.

8

Renaissance Surveying Techniques and the 1590 Hariot-White-de Bry Map of Virginia*

Michael G. Moran
University of Georgia

As Rivers notes in his recent bibliography on the history of business and technical communication, little work has been done on the history of visuals in technical communication (1994, p. 19). This fact is not surprising given that technical communication, as a field, has tended to privilege verbal over visual communication while largely ignoring the rhetorical function of the latter. According to Kostelnick (1994), for instance, until the late 19th century, business writing textbooks separated visual and verbal rhetoric. Visual rhetoric was characterized by the "dress metaphor": visuals were seen as "the mere external wrapping that clothed the linguistic text" (p. 96). As Barton and Barton argue of more recent practice, technical communicators too often mistakenly hold a "product view of visuals," which assumes writing comes first and visuals merely support the finished written document

---

*This research was supported by a 1989 National Endowment for the Humanities Summer Institute for College and University Faculty Grant and a 1989 Newberry Library/Columbian Quincentennial Fellowship for Work in Residence, National Endowment for the Humanities, Newberry Library. A preliminary version of this essay appeared as Newberry Library Slide Set Number 19 in 1990. I wish to thank the Newberry for its kind support and its permission to use the illustrations of Renaissance survey methods and instruments in its collection.

"visual communication is an integral part of communication activity, rather than a discrete component" (p. 135). Other current theorists have expanded this position. Killingsworth and Gilbertson (1992) have developed the "principle of complementarity," an argument that visual and verbal signs, though different, should function together to create documents with integrated sign systems (p. 45). In addition, Shriver (1997) argues that the field of document design should be "concerned with creating texts (broadly speaking) that integrate words and pictures in ways that help people to achieve their specific goals in using texts" (pp. 10–11). In other words, technical communicators should view visuals as central to the communication process. Williams agrees that images and words should be effectively integrated in documents. Both words and images, he argues, are signs that substitute for some entity, a referent, that is usually not immediately present. However, the visual sign has some advantages over the verbal. While language transforms fleeting thought into propositions, which readers process sequentially, visuals present directly large amounts of information that the viewer can process simultaneously (1993, p. 674). Bertin (1983) attributes this advantage of visuals to the fact that special systems, unlike linear systems such as the printed word, "communicate the relationships among three variables...the variation of marks [on the page] and the two dimensions [vertical and horizontal] of the plane" (p. 3). A map, for instance, provides a quicker, more accessible overview of a region and the relationships among that region's natural features than a verbal description can. Finally, technical communicators sometimes forget that visual communication, like verbal communication, functions rhetorically. As Read argues, visual communication attempts to "influence the receiver in some way" (1972, p. 252). The sender uses the visual to achieve a purpose—or a mixture of purposes—"to inform, motivate, persuade, instruct, or entertain" a viewer (p. 252). In short, technical communicators should view visuals as an important channel of communication, and specialists interested in the history of engineering, business, and scientific writing should examine the development of visual communication for technical purposes. Tebeaux (1991a, 1991b, 1997) has begun this work with Renaissance texts by analyzing the format and page design that made early how-to books useful to readers. This chapter attempts to add to our understanding of this development by examining the White-Hariot-de Bry 1590 map of Virginia, which was the first widely published map of any area of the New World based on a careful survey.

HISTORICAL BACKGROUND

The map was one of the important results of Sir Walter Raleigh's attempt in 1585 to establish a permanent English colony in North America. After receiving a patent dated March 25, 1584, from Queen Elizabeth I to settle all lands in the mid-Atlantic region not possessed by a Christian prince, Raleigh financed a successful 1584 reconnaissance voyage to the Outer Banks of what is now North Carolina. Raleigh's next move was to organize a fleet of colonists under the governorship of Ralph

Lane, which left England on April 9, 1585, to settle a colony in what was then called Virginia on Roanoke Island near the site of present day Manteo, North Carolina (see Stick, 1983; Kupperman, 1984; and Quinn, 1985, for discussions of the 1585 colony). One of the goals of this colony was to survey and map Raleigh's new domain. To carry out this survey, Raleigh sent one of his closest advisors, Thomas Hariot (sometimes spelled "Harriot"), who was one of Renaissance England's greatest mathematicians and astronomers, and John White, a relatively unknown artist (see Moran, 1990, for White's artistic background). This team combined their talents to create the first maps of North America based on a detailed survey.

These maps set a new standard for New World cartography because of their technical accuracy (Cumming, 1988), and, on account of this, they represent an important development in the history of Renaissance technical communication. The maps are technical in two ways. First, they are the products of the developing technologies of surveying made possible by the many instruments of navigation and survey that the Renaissance produced. White and Hariot used these technologies to construct maps that possessed an accuracy never before achieved in North American cartography. Second, the Hariot-White-de Bry maps, along with other Renaissance technical communication, shared an emphasis on "usability" (Tebeaux, 1997, p. 1). The maps were designed to be used by various groups interested in the English colonization movement. Raleigh and his circle used them to entice potential colonists and investors to participate in Raleigh's scheme because the maps made Virginia seem real and desirable. Those colonists and inventors used the maps to make decisions about whether or not to participate in the movement and what to expect when they arrived at the colony. Navigators used them to negotiate the treacherous shallows around the Outer Banks. In short, the utility of the maps grew directly from their accuracy.

To understand how Hariot and White achieved this accuracy, we must examine the most important instruments and techniques that the two cartographers used to conduct their survey and the widely-distributed Hariot-White-de Bry map that resulted from this work. Any account of their techniques must be conjectural, as no instructions for the survey have survived. However, instructions, written by an unknown hand, do exist from another expedition surveyor, Thomas Bavin, who was to have sailed on a voyage to America in 1582 or 1583 that never materialized. These instructions indicate exactly what Bavin was to accomplish and how he was to conduct his survey. Also extant is the list of surveying instruments that Martin Frobisher carried on his 1576 expedition to North America. These included a meridian compass, a universal ring dial, an astronomer's ring, an astrolabe, a cross staff, and a *holometre*, or plain table (Taylor, 1954, p. 35). Both of these sources give us a good idea of the instruments and techniques that Hariot and White would have used on the 1585 voyage. The most direct evidence that the two men had surveying instruments appears in Hariot's *A Briefe and True Report of the New Found Land of Virginia* (1588/1972). In this report on the 1585 expedition, Hariot comments on the Native Americans' curiosity about his surveying and navigational equipment, which included "Mathematical instruments, sea Compasses, the vertue of the loadstone in drawing yron, a perspective glasse..., burning glasses,...[and] spring clockes that seeme to goe of themselves..." (1972, p. 70). The general category of "mathematical instruments" probably included various kinds of angle-measuring devices, such as the cross staff, sextant, astrolabe, universal ring dial, plane table, and theodolite, that 16th-century surveyors commonly used to determine latitude and to measure land by means of triangulation.

It is impossible to say with any accuracy what experience Hariot and White had as surveyors before 1585. Hariot was definitely familiar with the instruments and mathematics of Renaissance surveying. Raleigh hired him in about 1579 as his personal tutor in mathematics. Hariot later established and taught in a school to train Raleigh's sea captains in the developing art of open sea navigation, a discipline that used many of the instruments also used then by English surveyors. Hariot's duties included collecting state-of-the-art instruments and teaching the captains to use them and the basic mathematics needed for transatlantic navigation (Shirley, 1983). To accomplish these goals, he thoroughly studied modern navigation techniques, mathematics, and instruments and wrote the *Arcticon*, a lost navigational manual. In preparation for the 1585 voyage, he probably collected and studied the newest books, instruments, and techniques connected with surveying.

White most likely had experience in surveying, but his must have been more practical than that of Hariot. His drawings of the temporary encampment that Governor Lane constructed in Puerto Rico and Salina during the voyage to the Outer Banks suggest that he had a background in estate and, perhaps, military surveying (see Hulton, 1984, pp. 41–42). In England at the time, artists often helped draw and color elaborate estate and house plans for wealthy builders. These illustrations were often decorated and colored to emphasize special features of the manor house and its lands, then used as wall hangings.

Although little direct evidence exists about Hariot and White's particular training as surveyors, it is clear that the two worked at a time when surveying and mapmaking were becoming professionalized in 16th-century England. This growth of the profession was due to four influences. One of these was the development of the science of artillery, with its need to measure accurately distance and height on both land and sea. In 1578, William Bourne published a translation of Tartaglia's standard work on gunnery under the English title, *The Arte of Shooting in Great Ordanaunce*, in which he discussed the application of survey instruments and techniques in firing artillery (Taylor, 1954). Another influence was the development of the art of navigation, which demanded accurate sea charts and instruments for measuring angles to determine nautical position and direction. Yet another influence was the growth of cartography, which in England was heavily influenced by the military's need for accurate maps from which to plan battles. Some evidence exists, for instance, that Christopher Saxton, who produced the first surveyed maps of England during the 1570s, learned his trade as a surveyor in the Irish campaigns. Perhaps most importantly, surveyors served a central function in the enclosure movement that began in England during the 15th and 16th centuries (Richeson,

1966). As England shifted from the feudal system of communal and hereditary ownership to private ownership of property, the need arose for professionals to survey the land to protect both buyer and seller by determining property lines and by evaluating the value of property. Because of this social need for a new class of professional surveyors, numerous books appeared during the period that discussed instruments and techniques, of which Hariot might well have collected and read in preparation for his survey of Virginia.

RENAISSANCE SURVEYING INSTRUMENTS AND TECHNIQUES

One of the first operations that Hariot and White probably performed in Virginia was to determine their location on the earth's surface. To accomplish this, they needed to establish their latitude (the distance north of the Equator) and their longitude (the distance west of a prime meridian, probably near London).

Establishing Latitude

To determine their latitude, they might well have employed one of the oldest and most common Renaissance instruments of survey and navigation, the cross staff, which mariners and surveyors used to measure vertical angles. Made of wood, the staff consisted of two major parts: a long, graduated staff and a transom or crossbar that slid up and down the staff. The staff had scales marked on each of its four sides, and these scales corresponded to up to four transoms, each of a different length. Renaissance surveyors and navigators used the cross staff to measure the angle of the sun or another star above the horizon, the height of the celestial body determining the length of transom to use. As Figure 8.1 shows, the cross staff could also be used to determine the height of a building, mountain, or other vertical object.

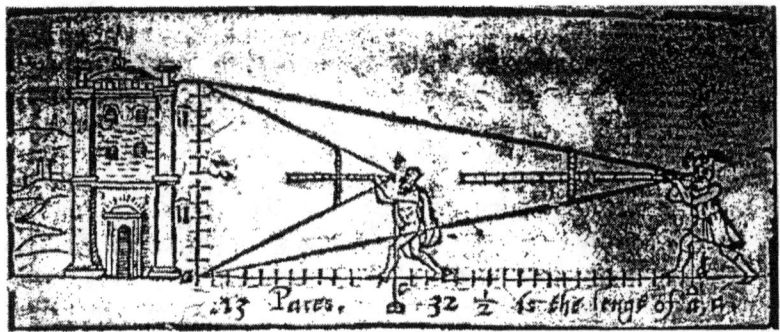

FIGURE 8.1. Use of the cross staff (from Digge, 1637). Photo courtesy of the Newberry Library, Chicago.

To use the cross staff, surveyors and navigators began by estimating the length of the crossbar needed by the height of a celestial body, such as the sun. After sliding the proper bar onto the staff, they would point the staff in the direction of the sun, placing the sighting end of the staff against their eye socket or cheek. They then adjusted the crossbar by sliding it up and down the staff until one end of the bar was on the horizon and the other was on the sun. After adjusting the transom correctly so that it fell across the proper degrees marked on the staff, they would then read off these degrees to determine the angle of the sun above the horizon. To determine latitude required them to take an additional step in their calculations. They had to sight the sun at its meridian, its highest point in the sky, which was at exactly noon. They used figures from Ephemerides charts (which Bavin was to have taken with him on his voyage) that gave figures of the sun's ecliptic, determined by the date, to calculate their latitude. The charts were necessary because they contained information on the irregularity of the calendar year, the oblique track of the sun across the sky, and celestial changes caused by the precession of the equinoxes (Brown, 1977), all of which needed to be calculated in to determine exact latitude.

Proof that Hariot possessed expert knowledge of the cross staff and its use exists in his unpublished notes of lectures that he gave to Raleigh's sea captains (see Shirley, 1983, pp. 90–91). Recognizing the damage to the eye that staring at the noon-day sun could cause, Hariot recommended that navigators and surveyors sight the top edge of the sun so that the instrument blocked out most of the brightness. He rejected the use of smoked glass lenses attached to the sighting end of the cross bar to filter the light because such lenses could refract rays and distort readings. Hariot also suggested that users uniformly press the end of the staff against the edge of their eye socket, as this technique allowed them to take into consideration the eccentricity of the individual eye. For Raleigh's 1595 Guiana voyage, Hariot worked out the individual figures for Raleigh, Captain Jacob Whidden, and John Douglas, the shipmaster, so that each could figure corrections due to facial structure into their final calculations.

Although the cross staff was one of the most common Renaissance angle-measuring instruments, several others existed that Hariot and White might have used. These included the sextant, the astronomical astrolabe, and the simplified mariner's astrolabe, all of which were used to determine latitude.

Establishing Longitude

While methods for accurately determining latitude had been worked out by the 1580s, those for determining longitude had not. While calculating latitude required surveyors and navigators to establish their north-south position in relation to the equator, establishing a longitude required them to calculate their distance east or west of an arbitrary meridian, which for Hariot and White probably would have been near London. Fixing the longitude of the Outer Banks would require White and Hariot to determine how many degrees they were west of London, and they could accomplish this only by knowing what time it was in London when the sun was at high noon in

Virginia. Once they had these two times, they could easily calculate their longitude because each four minutes of time equaled one degree of longitude. That Hariot and White had timepieces with them suggests that they were to attempt a calculation of longitude, but their primitive wind-up watches could not keep London time accurately enough for the task. Although the theory of determining longitude was advanced by Gemma Frisius in 1530, it was not until the 18th century that John Harrison invented a chronometer accurate enough to determine longitude.

Orienting the Map to the North

In addition to establishing the position of the Outer Banks on the earth's surface, Hariot and White had to orient their map to the four directions. Two instruments surveyors and navigators used to achieve this orientation were the magnetic compass and the universal dial, often called the universal ring dial, both of which Bavin's instructions mention. These two were sometimes combined into a single instrument, and both instruments oriented the user to the north.

By the late 15th century, the compass had become a standard instrument in English surveying. As early as 1523, Master Fitzherbert, who wrote the first book on English surveying, recommended that the compass be used to establish the alignments of land in estates (Richeson, 1966, p. 34). In fact, basic surveys could be conducted using the compass alone, and these are now known as compass traverse surveys (Greenhood, 1964). Renaissance surveyors and navigators would first determine north, set their compass, and then walk or sail along a line (such as a coast) or around an area they wished to survey, using the compass to take bearings at every turn made. This method would provide a rough outline of a coast or an area of land, and Hariot and White might well have used this method to construct their rough sketch map of Virginia (for a reproduction, see Cumming, 1988, p. 131). Traversing, however, allowed errors to creep into a map. Because surveyors had to keep track of distance by pacing or sailing, the various legs of the traverse would by necessity be approximations. Also, because the compass was subject to outside influences, such as iron deposits, the bearings and angles based on compass readings were not always accurate. Hariot and White probably used the more sophisticated method of triangulation with a plain table or a theodolite to conduct part of the surveys of their finished maps, which are highly accurate.

The Renaissance commonly used two kinds of compasses, marine and land. The marine compass consisted of a box with a mounted needle upon which the compass card, or rose, which showed direction, was attached. The card revolved with the needle, and the box, mounted in a ship, had a fixed mark pointing to the bow of the ship to show direction. The land compass, similar to modern varieties, had a needle that revolved above the compass card upon which the directions were marked. Early compasses had needles of soft iron that held their magnetism for a short time. Sailors and other users carried a loadstone, which was a piece of magnetite. This naturally magnetic ore, which was usually carried in a brass or silver box, was used to "touch" the needle to remagnetize it. Hariot mentioned in his *Report* that he possessed a compass and a loadstone.

One problem with the compass was that it showed magnetic, not true, north. The difference between these two came to be called the declination, or variation, of the compass, a phenomenon that Columbus discovered on his first transatlantic voyage and that the Englishman William Borough explained thoroughly in his 1581 book titled *Discourse on the Variation of the Compass or Magneticall Needle*, a work which Hariot probably knew. European compass makers often took local variation into consideration in marine compasses by attaching the compass rose to the needle so that the compass read true north for that region. This practice, of course, made such compasses useless for transatlantic voyages. One hope was that the declination of the compass could be used to determine longitude, but this proved to be false because the compass variation did not change uniformly with the change in longitude, and the change in variation occurred so slowly that precise measurements were impractical.

To determine true north, Renaissance instrument makers devised the universal ring dial, an instrument that could either be used alone or attached to a land compass. Introduced to England in 1542 by Jean Rotz, the French mapmaker, this was a valuable instrument which, in addition to identifying true north, could also tell local time. The instrument accomplished this both by measuring the altitude and direction of the sun or a bright star. To use it, the surveyor would adjust the dial to the proper latitude, which he could find by using his cross staff and tables. Next, he would set the pinhole slide to the proper date. Then, at noon, he would turn the dial until a ray of light, shining through the pinhole, fell on the hour circle, which could only occur when the dial was aligned exactly to true north. When mounted over a compass, as Bavin's apparently was, the surveyor could measure the compass variation directly. That Bavin was to use a dial is clear from his instructions: "by your universall dyal you may alweis finde the variacion of the Compasse att noonetyde by observing how fare your Compasse dothe differ from your just meridean" (Anonymous, 1979, p. 242). The instructions go on to emphasize the care in which Bavin was to take in making and then checking his directions. He was to mark both magnetic and true north on his draft map and then check true north again by taking a reading of the pole star from the same location.

Determining true north was essential for accurate mapmaking because magnetic north could change as the surveyor moved from place to place. The magnetic needle was also subject to local influences, especially iron deposits. By determining true north, mapmakers such as Bavin, Hariot, and White could accurately establish the relative locations of different parts of their survey, and they could align their surveying instruments in the same direction each time they set them up.

Use of Triangulation

Because of the high degree of accuracy of their finished maps, Hariot and White probably used methods other than the compass traverse method to survey the Outer

Banks. By the late 16th century, triangulation had become well established in England. William Cunningham had first explained to an English readership its principles in his *Cosmographicall Glasse* in as early as 1559, and estate surveyors had rapidly adopted the method because it allowed them to survey 300 to 400 acres per day, using an angle measuring instrument such as the plane table or a theodolite (see Richeson, 1966, p. 80, for a picture of a plain table and its instruments, and Figure 8.3 for a theodolite). By 1585, triangulation was a standard method of English surveying and cartography, and White and Hariot, as the Bavin instructions make clear, probably used it in at least part of their survey. Figure 8.2 shows the method of triangulation that Arthur Hopton recommended in *Speculum Topographicum, or the Topographical Glasse* (1611) for surveying relatively large areas, in this case the county of Sallop. As the term *triangulation* suggests, the surveyor made observations in the shape of triangles to determine the area of a piece of land and either recorded these angles in a notebook and used them to sketch out a map later, or sketched these triangles in the field directly on a sheet of paper attached to a plane table (see Richeson, p. 80).

Triangulation, based on the geometry of the triangle, remains today a standard method of mapmaking. To use it, the Renaissance surveyor established two stations of observation. In Hopton's example in Figure 8.2, these two points were "Cordocke Hill," labeled F, and "Wrenkin Hill," labeled G, both of which were elevated so that the surveyor could see all the surrounding territory. Hopton then recommended that the surveyor set up his angle-measuring instrument (such as a

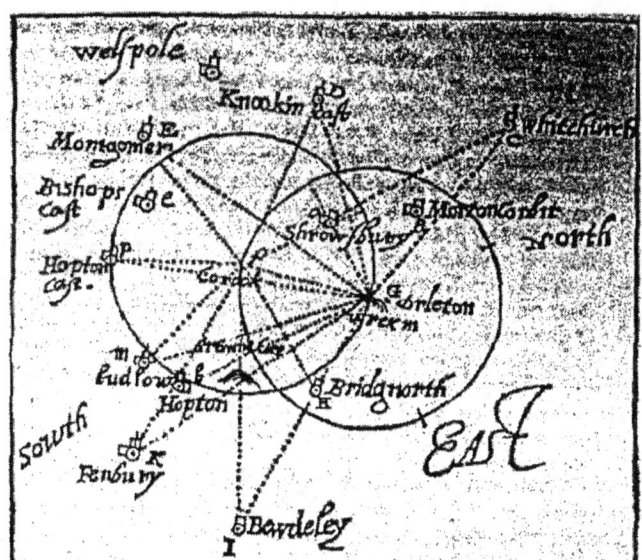

FIGURE 8.2. Method of triangulation (from Hopton, 1611). Photo courtesy of the Newberry Library, Chicago.

plain table or a theodolite) on point F and sight the prominent features of the landscape, such as Hopton Castle, Montgomery, and other places in the region. The surveyor would then record the angles between his sight lines and his base line (F to G), moving systematically around the circle. He then would transport his instrument from F to G, use a compass to orient the instrument in the exact direction it was at F, and sight the same prominent features. By taking two sightings of the same objects, the surveyor created a network of triangles. Once the surveyor determined the length of any one line in the network (Hopton mentioned that nine miles separated Cordocke Hill and Wrenkin Hill), he could easily determine, using a two-legged compass, the proportional distance of all lines in the skeleton map he had sketched. To create the map, the cartographer would flesh out his skeleton to create a finished product. In Hariot and White's collaboration, Hariot most likely, with White's assistance, conducted most of the actual survey and carried out the necessary mathematical calculations. The artist probably helped in the field and then drew the finished maps from the mathematician's calculations.

One of the primary instruments of triangulation that both Frobisher and Bavin had was the plain table, which had come into use in England by the beginning of Elizabeth's reign. With the introduction of this instrument into English surveying, the "landmeter," carrying a perch rod of knotted cord to physically measure each parcel of land, was replaced by the professional surveyor, with his plain table, using geometric calculations to plot surveys and construct maps without having to walk off the entire territory (Greenhood, 1964; Richeson, 1966). Hariot and White undoubtedly had such a table, and if mathematical calculations were beyond White's abilities, they were well within Hariot's.

As Richeson's (1966) reproduction on page 80 (from Lucar, 1590) shows, the plain table consisted of several parts. Its main component was the table itself, which was often made of three slats of wood that could be fitted together by means of grooves to make a flat, rectangular surface. Upon this surface, the surveyor would fit a piece of paper upon which to draw the map in the field using angular readings. The paper was held in place by a frame, which fit snugly around the four edges of the table. The four sides of the frame were often indexed or ruled off so that the surveyor could draw a grid of lines intersecting at right angles. The table was mounted on a stand that attached to the back of the table to hold it level with the horizon. Upon the table, on top of the paper, the surveyor placed a ruler, which, if it had a sight at each end, was called an alidade. The surveyor used this rule to sight distant objects along its length and to draw proportional lines on the paper to mark accurately the direction and distance of these objects using the method of triangulation. Lucar pictures other traditional instruments used with the plain table, such as the two-legged compass to measure distance on the map and to draw accurate circles from the point of location, and the right-angled square, used to draw 90-degree angles. Also pictured is the "wyer line," which in the 1580s would have been a waxed and knotted cord used to measure physically the length of the parts of the land being surveyed. The drum and stool in the illustration imply that the plain table developed from more informal instruments, the drum in particular pointing to the military influence on the development of surveying.

Bavin's instructions make clear the extent that he was to rely on his plain table to make his maps. He was told to draft his maps on sheets of "paper Royall," sized according to "the biggness of the [plain] Table" (Anonymous, 1979, p. 243) he was to use. He was then to begin by mapping the coast, sheet by sheet, marking each sheet with a letter from the alphabet (A, B, C, etc.) until the coast had been surveyed. Bavin was to note in particular the natural boundaries between parts of the coast, such as rivers and headlands. After mapping the coast, he was to move inland, marking his sheets using double letters (AA, BB, CC, etc.) so that he could determine how to join the sheets after the survey. The writer of the instructions, who apparently had experience as a field surveyor, recommended using letters to identify sheets so as to avoid confusing these letters with degrees of latitude, which Bavin was also to mark on the sheets. Bavin was also warned to keep a single scale throughout the mapping process, for "yf you alter your skele you shall never make your plottes [finished maps]" (p. 244). Also on the sheets, Bavin was to set down the distances of all prominent features, such as hills, headlands, bays, lakes, oyster or mussel beds (for pearls), and woods, that would suggest the value of the area to the English. The instructions provided Bavin with symbols to use for various natural phenomena, but he was also told to write the name of each thing beneath the symbol to avoid confusion. He was told, for instance, to mark first the location of a wood with oblique hatch marks and then to write the names of the particular trees in the woods under the marks. These instructions make clear that Bavin was to conduct his survey and create the first draft of his map using the plain table and, hence, the method of triangulation. Since both the instrument and the method had become standard in England, Hariot and White probably used the same equipment and similar methods to map the Outer Banks.

Another surveying instrument that was well established by the time of Hariot and White's voyage was the theodolite, which is pictured in Figure 8.3 (from the title page of Rathborne, 1616). In England, Leonard Digges first described this tool, which he named the "topographical instrument," in his posthumous *A Geographical Practice, Named Pantometria* (1571), a book that Hariot might well have known. Surveyors immediately adopted this instrument, which Humphrey Coles, one of the best English instrument makers of his day, constructed and sold. The theodolite allowed the surveyor to take readings simultaneously on the horizontal and vertical planes. Consequently, the surveyor could perform triangulation measurements to establish relationships between objects on the ground, while at the same time measuring the height of vertical objects such as hills or mountains. The theodolite was often used in conjunction with the plain table and was sometimes mounted on it. Since Bavin probably had such a combined instrument (his instructions mention "the Table of your Instrument" [Anonymous, 1979, p. 243], suggesting that he had a plain table with a more complex instrument attached), Hariot and White might have used one too (Taylor, 1951).

As Figure 8.3 suggests, the 16th-century theodolite was a fairly complex instrument. Rathborne's version consisted of a horizontal circle marked in degrees set upon a tripod. In the center of the circle sits a land compass to give direction, and this suggests that the instrument was made of brass, a metal that would not interfere with the magnetic needle. The compass allowed the surveyor to orient the instrument in the same direction every time it was set up. Above the horizontal circle was a vertical half circle that made a sight or alidade along its diameter. This half circle rotated, pulling along a pointer to mark degrees of arc on the horizontal circle and to determine the angles between distant points during the process of triangulation. The half circle also swiveled on the vertical to measure the angle of the height above the horizon of objects such as buildings and mountains.

The advantage of the theodolite over the plain table was largely a matter of expediency. As Figure 8.3 suggests, when using the theodolite, the surveyor measured angles and noted them down in a tablet, without having to pencil them directly on paper on a plain table. The fact that the plain table is lying discarded on its top in the background suggests that Rathborne rejected it in favor of the theodolite, which allowed rapid reading and recording of angles (as the surveyor in the figure is doing in his notebook), which could later be used to construct a map or estate survey. The theodolite existed in various sizes ranging from this one, which was large and freestanding, to much smaller ones that were attached to a plain table, as was Bavin's.

FIGURE 8.3. Theodolite (from Rathborne, 1616). Photo courtesy of the Newberry Library, Chicago.

Such are the surveying instruments and techniques that Hariot and White probably used in Virginia. I will now turn to the most important result of their survey, the Hariot-White-de Bry map.

THE HARIOT-WHITE-DE BRY MAP AND ITS DERIVATIVES

Hariot and White's survey formed the basis for a number of important maps, but three of them—the preliminary sketch map (see Cumming, 1988, p. 131), the map of the Outer Banks, and the composite map of the East Coast from Florida to Chesapeake Bay (see Hulton, 1984, pp. 85–86)—remained in manuscript and were consequently not widely circulated. The map that received the widest audience is the one reproduced in Figure 8.4 from Theodor de Bry's French edition of Hariot's *A Briefe and True Report*, published in 1590, along with etchings of many of White's illustrations of Native Americans, their cultural activities, ancient Picts, and one other map of the Outer Banks, focused more tightly on Roanoke Island, entitled "The Arrival of the Englishmen in Virginia." De Bry received encouragement on the project from the younger Richard Hakluyt, one of England's foremost advocates of the colonization movement and one of Raleigh's close associates. Hakluyt arranged for de Bry to obtain White's manuscript drawings from which to create his

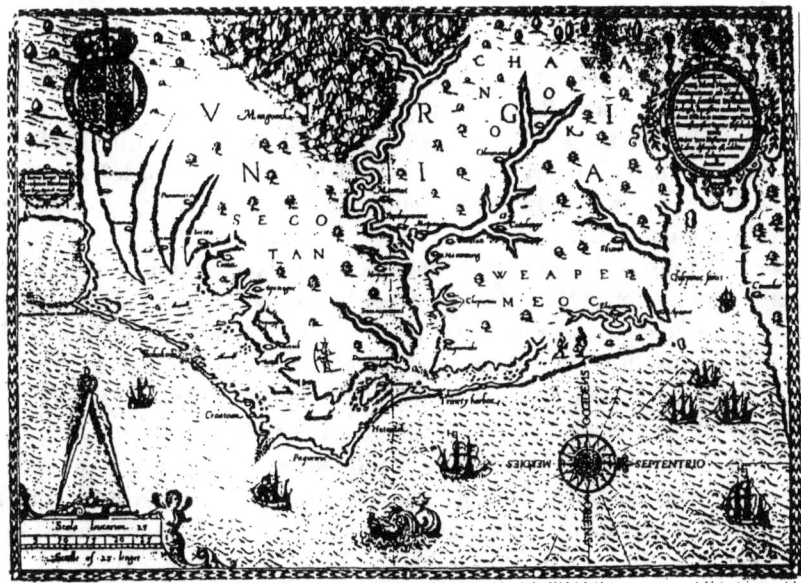

FIGURE 8.4 Hariot-White-de Bry map of the Outer Banks (from de Bry's French edition of Hariot's *A Briefe and True Report of the New Found Land of Virginia*, 1590). Photo courtesy of the Newberry Library, Chicago.

etchings. De Bry based his etched map on a now-lost version of White's manuscript.

The relationship between Hariot's report and the map supports Williams's and Bertin's distinction between the different functions of verbal and visual signs. Hariot's report falls into three parts, each of which analyzes a specific element of Virginia's natural resources and native peoples. The first part lists, describes, and analyzes "marchantable commodities" (1972, p. 7), mostly plants and minerals, which colonists could harvest or mine and which merchants could import to England. The second part discusses the food sources on which an English colony could live, and the third part discusses both the natives of the region and the raw materials available to the English for building dwellings. While these three sections provide convincing details about the suitability of Virginia as the location for an English colony, they are highly analytical, present information sequentially, and include no meaningful overview of the region. The map, on the other hand, while it lacks the specificity of the report, provides the necessary overview, allowing viewers to picture the Outer Banks and conceptualize the relationships among the map's systems of representation. For instance, while the map cannot give the kind of specific information that the report does about the trees of the region, it can suggest their existence and location by picturing them. In this sense, Hariot's *Report* and the map serve as an early model of the effective relationship, Killingsworth and Gilbertson's (1992) principle of complementarity, that should exist between the two major sign systems used in technical communication.

Furthermore, as Shriver's and Read's arguments suggest, the map functions rhetorically. Like many effective visuals, it achieves a complex rhetorical goal in that it appeals to a variety of audiences to achieve different aims. Like Hariot's *Report* itself, the map's primary purpose is both to inform and persuade by addressing several audiences involved in colonization, including navigators, potential colonists, and merchant/investors.

The map must be read within the rhetorical situation of Hariot's *Report*, which Raleigh commissioned Hariot to write when he returned to England from the 1585 colonization attempt under the governorship of Ralph Lane. The 1585 colony returned in confusion after the Roanoke Indians, realizing that the colony intended to stay on their land permanently, turned on the English, refused to feed them, and planned a surprise attack. Learning of this plan, Lane, drawing on his military experience, led a preemptive strike against the Roanokes and decapitated their leader, Wingina. Upon returning to England, Lane and other colonists told of their unhappy experiences, jeopardizing public support for Raleigh's 1587 colony, which became the famous Lost Colony. To address this problem of waning support, Raleigh commissioned Hariot to write a report demonstrating the value of Virginia as an English colony. This report combined informative and persuasive aims in order to convince a complex audience of potential colonists and investors to support the 1587 effort. The report originally circulated in manuscript form before being published, without illustrations, in England in 1588. To appeal to an even wider audience, Raleigh, through Hakluyt, arranged for de Bry to publish the report two years later in Frankfurt, Germany, with etchings based on White's illustrations and the maps based on the White-Hariot survey.

One example of a purely informative purpose of the map is the hydrographical information that it contains. This information would be useful to pilots and navigators sailing from the Atlantic into the far more dangerous waters of the barrier islands and the sounds beyond them. To guide pilots, de Bry carefully marked shoals, with dots indicating shallow waters in the inlets, information without which even the smallest landing vessels could not safely enter the sounds. De Bry marked a particularly hazardous system of shoal water which he pictures jutting out and up at the far left of the map with the Latin phrase *Promontorum tremendum*. Since Roanoke Island was the site of the colony and required the docking of vessels, de Bry indicates the shallows around the island and at the mouth of the Roanoke River that landing vessels would have to avoid. In the more focused map entitled "The Arrival of the Englishmen in Virginia," de Bry heightens his warning by placing sunken ships along the Atlantic side of the Outer Banks to indicate shallow water (see Hariot, 1972, p. 45).

On the depiction of the land mass itself, de Bry marked various features to persuade English viewers to settle the region and exploit it commercially. To create this kind of desire, de Bry pictured mountains, trees, rivers, and natives, all of which suggested the potential value of the land. Although de Bry's map covers the same general coastal area as White's extant manuscript map, it extends a little less than 100 miles further inland, thereby showing topographical features not on the manuscript map and suggesting the potential value of the land to investors. This extension allows the de Bry map to show the upper ends of the Chowan, Roanoke, and Neuse rivers, this portrayal implying that colonists would have easy access by boat to the interior. That the Neuse River ends in a series of dots implies that Hariot and White, or perhaps other members of the colony, actually surveyed, or at least explored, the headwaters of the other rivers. The map also contains symbols for mountains, using a bird's-eye view that was typical of many Renaissance maps. These mountains would connote to the Renaissance viewer the possible existence of gold or other valuable metals commonly found there. De Bry's etching provides other topographical features, indicating flat lands with straight horizontal lines which would suggest areas suitable for cultivation. The de Bry map also has scattered across it miniature symbolic trees to identify forests and woods. The map clearly attempts to distinguish among the types of trees, with some looking like hardwoods near the coast and conifers further inland. Both kinds of trees would represent value to the Renaissance viewer because the conifers could provide pitch and the hardwoods masts for ships, two vital resources that the growing English navy desperately needed.

In addition to portraying topography and suggesting the potential value of the land, de Bry's map also pictures the contact between the Native Americans and the English. This contact would have appealed especially to merchants, who needed a willing commercial partner interested in trading with them. Careful to show the Indians as a peaceful people, de Bry sketched oversized figures clearly based on some of White's ethnographic Indian portraits. The most noticeable figure is the Indian warrior, standing in a stock mannerist pose holding his bow and arrow. It is important to note that the Indian looks peacefully at the ships approaching the shore and that his primitive weapons offer little threat to the English with their firearms and cannon. The other picture, that of the Indian woman and her child, is a reversed reproduction based on another White portrait. It is significant that the child is playing with a rattle and a doll, both European toys; these details suggest to merchants the potential value of the natives as trading partners who desired English goods. Scattered throughout the area are Indian villages marked by symbolic palisades, which show that the land is at least sparsely populated with Indian trading partners. The map also names what appear to be tribal regions, including those of the Mongoake, the Secotan, the Chawnok, and the Weapemeoc. Ten small Indians paddle their frail canoes along the sounds, which the English viewer must have compared with the huge English ships that sail along the Atlantic coast, the comparison suggesting the strength of the English and the vulnerability of the Indians—a distinction that would have convinced potential colonists that the region was safe to settle. These symbolic ships probe the coast, with two of them moving toward inlets that will admit them into the sound, one heading north toward the Chesapeake Bay, four of them sailing across the Bay's mouth, and a single, smaller craft apparently in the process of landing at the Indian village of Comokee on the Bay's northern coast. These busy little ships represent the potential for English trade in the region and the naval power that the English could exert throughout Virginia.

The map also contains many conventional symbols of power. In the upper left appears the royal arms of Elizabeth, asserting her claim to this section of the New World. This seal implies that she supports the colony and encourages others to participate as colonists, investors, or traders. In the upper right corner, under the arms of Raleigh, that claim is reasserted in Latin, stating that this part of America, now called Virginia, was discovered in 1585 (it was actually claimed in 1584) by Sir Walter Raleigh for the Queen. In the lower left is a cartouche that contains the scale of the map in leagues, which suggests the vastness of the region.

The orientation of the map is unusual. Unlike White's manuscript map of the Outer Banks, which is oriented to the north, the de Bry etching is oriented with west to the top. Since the invention of the compass and the development of the portolan chart in the late Middle Ages, the convention of orienting maps to the north had been fairly well but not universally established. The western orientation gives the land an inviting feel, as if White and de Bry were trying to draw the European viewer in toward the new land. This feel is certainly consistent with Raleigh and Hakluyt's attempts to persuade investors to support the colonization movement.

CONCLUSION

The Hariot-White-de Bry map of the Outer Banks is one of the most important maps created of North America in the 16th century. As William P. Cumming, historian of American maps, comments, this map "became a prototype not only for maps of the Southeast but also, by the middle of the seventeenth century, for western hemispheric and world maps" (1988, p. 137). Its influence extended beyond its function as a map of a particular region largely because it set new standards for mapmaking in America. These high standards resulted from the careful survey that Hariot and White performed, using the best instruments and techniques available at the time to expedition surveyors. These instruments and techniques included the cross staff, the compass and universal ring dial, the plain table, and perhaps the theodolite and the use of triangulation, all of which allowed the Renaissance surveyor to map out large tracts of land quickly and accurately. The accuracy of the survey and the care with which the map was integrated into Hariot's *Report* make the map an important example of effective Renaissance technical communication.

REFERENCES

Anonymous. (1979). Instructions for a voyage of reconnaissance to North America in 1582 or 1583. In D. B. Quinn (Ed.), *The new American world: A documentary history of North America to 1612* (vol. 3, pp. 239–245). New York: Arno Press. (Original work n.d.)

Barton, B. F., & Barton, M.S. (1985). Toward a rhetoric of the computer era. *The Technical Writing Teacher, 12*(2), 126–145.

Bertin, J. (1983). *Semiology of graphics: Diagrams, networks maps* (W. J. Berg, Trans.). Madison, WI: University of Wisconsin Press.

Bourne, W. (1578). *The arte of shooting in great ordanaunce*. London: Thomas Woodcocke.

Bourne, W. (1581). *Discourse on the variation of the compass or magnetic needle*. London: Dawson & Gardyner.

Brown, L. A. (1977). *The story of maps*. New York: Dover.

Cumming, W. P. (1988). *Mapping the North Carolina coast: Sixteenth-century cartography and the Roanoke voyages*. Raleigh, NC: Division of Archives and History, North Carolina Department of Cultural Resources.

Cunningham, W. (1559). *The cosmographicall glasse*. London: John Day.

Digges, L. (1571). *A geometrical practice, named pantometria*. London: Henrie Bynneman.

Digges, L. (1637). *A booke named tectonicon*. London: Felix Kingston.

Greenhood, D. (1964). *Mapping*. Chicago: University of Chicago Press.

Harriot, T. (1590). *Merveilleux et estrange rapport, toutesfois fidele, des commoditez qui se trouvent en Virginia*. Frankfurt, Germany: T. de Bry. (Original work published 1588)

Harriot, T. (1972). *A briefe and true report of the new found land of Virginia: The complete 1590 Theodor de Bry edition*. New York: Dover. (Original work published 1588)

Hopton, A. (1611). *Speculum topographicum, or the topographical glasse*. London: S. Waterson.
Hulton, P. (1984). *America 1585: The complete drawings of John White*. Chapel Hill, NC: University of North Carolina Press.
Killingsworth, M. J., & Gilbertson, M. K. (1992). *Signs, genres, and communication in technical communication*. Amityville, NY: Baywood Publishing.
Kostelnick, C. (1994). From pen to print: The new visual landscape of professional communication. *Journal of Business and Technical Communication, 8*, 91–117.
Kupperman, K. O. (1984). *Roanoke: The abandoned colony*. Totowa, NJ: Rowman & Allanheld.
Lucar, C. (1590). *A treatise named Lvcarsolace*. London: I. Harrison.
Moran, M. G. (1990). John White: Renaissance England's first important ethnographic illustrator. *Journal of Technical Writing and Communication, 20*(4), 343–56.
Moran, M. G. (1990). *Renaissance survey techniques and the mapping of Raleigh's Virginia*. Chicago: The Hermon Dunlap Smith Center for the History of Cartography, The Newberry Library.
Quinn, D. B. (1985). *Set fair for Roanoke: Voyages and colonies, 1584–1606*. Chapel Hill, NC: University of North Carolina Press.
Rathborne, A. (1616). *The surveyor in foure books*. London: W. Burre.
Read, H. (1972). *Communication: Methods for all media*. Urbana, IL: University of Illinois Press.
Richeson, A. W. (1966). *English land measuring to 1800: Instruments and practices*. Cambridge, MA: MIT Press.
Rivers, W.E. (1994). Studies in the history of business and technical writing: A bibliographical essay. *Journal of Business and Technical Communication, 8*(1), 6–57.
Shirley, J. W. (1983). *Thomas Harriot: A biography*. Oxford, England: Clarendon.
Shriver, K. A. (1997). *Dynamics of document design*. New York: Wiley Computer Publishing.
Stick, D. (1983). *Roanoke Island: The beginnings of English America*. Chapel Hill, NC: University of North Carolina Press.
Taylor, E. G. R. (1951). Instructions to a colonial surveyor in 1582. *The Mariner's Mirror, 37*, 48–62.
Taylor, E. G. R. (1954). *The mathematical practitioners of Tudor and Stuart England*. Cambridge, England: Cambridge University Press.
Tebeaux, E. (1991a). Ramus, visual rhetoric, and the emergence of page design in medical writing of the English Renaissance. *Written Communication, 8*, 441–445.
Tebeaux, E. (1991b). Visual language: The development of format and page design in English Renaissance technical writing. *Journal of Business and Technical Communication, 1*, 246–274.
Tebeaux, E. (1997). *The emergence of a tradition: Technical writing in the English Renaissance, 1475–1640*. Amityville, NY: Baywood Publishing.
Williams, T. R. (1993). What's so different about visuals? *Techncial Communication, 40*(4), 669–676.

part III
Key American Movements in the History of Technical Communication

9

Landmark Essay: The Rise of Technical Writing Instruction in America

Robert J. Connors
University of New Hampshire

THE EARLY YEARS: 1895–1939

For as long as men have used tools and have needed to communicate with each other about them, technical discourse has existed. Scholarship has traced technical writing of a quite familiar sort back to the Sumerians, and we need come no farther forward in history than the Roman Empire to find technical writing as lucid and sophisticated as any that is done today. The tradition of technical writing is ancient, and, as Michael Connaughton's 1981 work shows, can be traced historically. But systematic instruction in the methods of technical writing, though a relatively recent development and thus not difficult to trace, has been the subject of few studies; for all that many technical writing teachers know, their discipline sprang full-blown from the brows of Gordon Mills and John Walter in 1954. Although technical writing has been accepted as an important part of the discipline of English and seems in many ways to have come of age, students deserve to know more about its history and development as part of their college curricula. In this article, therefore, I will trace instruction in technical writing from its beginnings in a few schools of engineering through its lean times, when it was a poor cousin to literary studies in English departments, to its present eminence as a center of vital scholarly and pedagogic activity.

ENGINEERING EDUCATION IN THE 19TH CENTURY

It is only in the last century that technical writing courses have been taught in American colleges. To understand the genesis of these early courses, we must first understand the context from which they grew; the vast changes that took place in all American colleges during the period of 1860 to 1900. Prior to the Civil War, colleges in America had been predominantly based on religion, usually fairly small in size, and reliant upon a classically descended curriculum. However, with the passage of the Morrill Act, the first in 1862 and the second in 1877, the foundations were laid for a revolution in American college study. The Morrill Act founded and promoted the land-grant agricultural and mechanical (A & M) colleges that were to make college education available in the late 19th century to a hugely increased percentage of the population. These colleges were to broaden and specialize the college curriculum in many ways.

In the last 40 years of the 19th century, the traditional study of the classics of Greek and Roman philosophy and literature began to be supplemented by studies in mathematics, modern languages and literatures, liberal arts of all sorts, and by an ever-growing field of technical and applied specialties—chief among them engineering. The Civil War was largely responsible for this change in the status of the technical fields. During that conflict, as never before, field engineers had been important figures, and with the burgeoning Industrial Revolution, the establishment of A & M colleges, and the growing technical needs of postwar America, the creation of schools and colleges of engineering (usually adjunct to the "arts" college in non-A & M schools) was a natural step. It was within these schools of engineering that technical writing courses began.

These specialized upper-level courses in writing, however, hardly existed during the 19th century, and this rarity is explained by the manner in which engineering schools developed in America. Prior to 1870, the canon of established engineering materials was fairly small, and engineering curricula contained a large percentage of humanities-based courses as a result. Some of this coursework was classical, but much was relatively recent—modern English and foreign languages, freshman composition (which was itself a new course, at least in its modern form), and the "philological studies" of the early literary scholars, along with a few science courses, history courses, and varied electives. Before 1870, an engineer graduated with a good bit of knowledge of the "humane subjects." The engineering discipline was rapidly being awakened by the fantastically swift industrial development of the Gilded Age, and new engineering materials were not long in appearing in the curricula. As a result, during the 1870s, courses in the humanities dropped almost completely out of sight in engineering schools. In a retrospective article published in 1931, H. L. Creek and J. H. McKee described what happened:

> The great decline in the amount of time given the humane subjects came between 1870 and 1885, the time at which the largest increase in the number of engineering

colleges occurred, when more than one-third of the time given humane subjects disappeared from subject matter. (p. 819)

Losing ground most seriously during this period were the foreign and classical languages and literary study (Anonymous, 1926).

As we might expect, what were left after this rapid takeover of the engineering curriculum by technically based courses were the "old reliables" of language instruction—freshman English courses. Freshman composition requirements were almost universal, and the tacit assumption in engineering schools between approximately 1880 and 1905 was that these first-year courses were all the introduction to writing that engineers needed. This period was, understandably, a rather dark time in the history of engineering education, a time when, by the schools' own later admissions, they turned out a large number of otherwise competent engineers who were nearly illiterate.

Despite the fact that some freshman composition courses in engineering schools were specialized to the needs of technical students, there seems to have been no courses before 1900 that dealt with the needs of upperclassmen for knowledge of the writing demands of the engineering profession. Although engineering education itself became vastly more sophisticated as the 19th century drew to a close, it almost completely ignored the linguistic needs of its students. The Society for the Promotion of Engineering Education (SPEE), founded in 1894, had no members from English departments until after 1905. In general, engineering schools acted as if their students needed only technical courses.

It took time, but a wave of reaction to this attitude began to build in the early 20th century, and after approximately 1903, an increasingly bitter series of condemnatory articles about the illiteracy of engineering school graduates began to appear in the engineering journals and weeklies. Letters and essays in the most important professional organs decried the inability of the new men to write coherent engineering reports, or even simple business letters. The *Engineering Record* spoke for many practicing engineers when it charged that "It is impossible, without giving offense to college authorities, to express one's self adequately on the English productions of the engineering students.... Most of them can be described only by the word 'wretched'" ("English for Engineers," 1915, p. 763).

The causes of the problem are not hard to trace. As pioneer technical writing teacher J. Martin Telleen explained in 1908, the standard freshman English course came too early in students' careers, it was too general in scope to be very helpful, its orientation was not practical enough, and there was almost no interdepartmental cooperation between English and engineering faculties (p. 68). Even as early as 1900, the familiar "two-culture" split had been established in colleges: English teachers saw engineers as soulless technicians, while engineers saw English teachers as dreaming aesthetes, promoting "refinement and culture" to the exclusion of reality (O'Brien, 1913, p. 715). Clearly, though, some cooperation would have to be achieved if the problem that all admitted exlisted was to be solved, and technical writing courses in their earliest forms were the solution of choice.

SAMUEL EARLE AND EARLY TECHNICAL WRITING THEORY

The period of 1900 to 1910 was the gestation period for technical writing courses. Although the surface of college life seemed quiet, at many schools there was furious activity that would soon come to a climax. Beginning around 1899, a number of engineering schools established separate English departments within themselves in order to serve the special needs of engineering students, the most famous of which that still exists is the Department of Humanities at the University of Michigan. Initially, these in-house English departments taught only a specialized freshman and sophomore sequence, but they also provided a natural climate for upper-level courses in specialized composition.

This activity was not long in producing prototypical technical writing materials. The first notable textbook devoted to technical writing, published in 1908, was T. A. Rickard's *A Guide to Technical Writing*. This was a transitional text that dealt mostly with usage— primarily meant for practicing engineers rather than for college courses—but it sold well and was adopted at a number of schools. Rickard's book, though, was merely a precursor to the first genuine technical writing textbook written for use in college courses. The book was called *The Theory and Practice of Technical Writing,* and it was published in 1911 by Samuel Chandler Earle of Tufts College, a man who, more than any other, deserves the title of "The Father of Technical Writing Instruction."

The Theory and Practice was a genuinely new sort of textbook when it appeared, sharing only a few elements with the general composition texts of the early 20th century. It grew out of courses in "engineering English" that Earle had been teaching at Tufts since 1904, courses that were perhaps the first recognizable technical writing courses. In addition to authoring the first real text, Earle came to be the philosophical voice of the early technical writing movement as well. In an important article in 1911, he ably defended specialized composition for engineers, stating that although technical composition requires great specialized skill, "it has commonly been assumed that for such writing a course in general composition is enough." It was not enough, obviously, and Earle went on to describe the reforms he had instituted for engineering students at Tufts. "We have departed," he wrote, "from common practice mainly in three ways: in shaping the work in English more frankly and more completely for engineers; in giving systematic training in technical writing; and in adapting special means for increasing the efficiency of the work" (p. 33).

Just as important as systematic technical writing training for students, in Earle's eyes, was the problem of the cultural split between English and engineering teachers. He condemned the attitude of English teachers that saw engineers as philistines, to be proselytized to about the superior virtues of culture and literature over engineering. In words that might be carved over the doorway of every technical writing division office, Earle wrote:

> We find, as I believe everyone will who studies the case without prejudice, that for those who have already entered upon what is to be their life work, true culture comes not from turning aside to other interests as higher, but from so conceiving their special work that it will be worthy of a life's devotion. (1911a, p. 35)

Such advice fell, unfortunately, on the mostly deaf ears of English departments.

Earle's text is dissimilar to present-day technical writing texts, due primarily to the fact that he approached his subject from a "modes of discourse" perspective that has since lost popularity. Technical writing for Earle was "narrative, descriptive, expository, or directive"; he did not cover any technical forms *per se*. *The Theory and Practice* included many examples of engineering writing (and it is important to remember that, until the 1950s, technical writing and engineering writing were synonymous), but only advocated the plain style for engineers, and approached questions of audience in a surprisingly sophisticated way. The book was a prototype and not a completely successful text, in that it found a ready audience, but was superseded by other books in the early 1920s.

Samuel C. Earle was a true educational ground breaker. Said the obituary after his tragically early death in 1917 at the age of 47, "To him is largely due the present method of teaching English in engineering schools." Though his passing was a loss to the profession he had helped to found, Earle's work gave direction and impetus to "the decade of great awakening" that followed the publication of his textbook (Earle, 1911a, p. 33). Between 1911 and 1920, the basic elements of technical writing courses as we now teach them were limned out at a number of schools around the country. The early centers of interest in technical writing were established by 1916—at Tufts, the University of Cincinnati, Princeton, the Massachusetts Institute of Technology, the University of Kansas, and Rensselaer Polytechnic Institute, to name the most active.

By 1916, the stream of professional complaints about technical school graduates had become a torrent, and engineering curricula began to change in an attempt to improve the situation. Writing in 1931 (and reflecting the fears and wishes of teachers during that time), J. Raleigh Nelson called the period from 1915 to 1930 a time of "complete reaction" to the nonhumanistic training given to engineers during the period from 1870 to 1910. Nelson saw a "unanimous demand for a more liberal and humanistic scheme of education" arise around 1915 (p. 495). In actuality, the demand was for basic literacy in engineering graduates, but English teachers often put their own interpretation on the dissatisfaction with the older curricula. If engineering schools wanted English instruction, they would have to accept literature along with writing, because the English graduate schools of the time were not producing anything but literary scholars, all of which wanted work. And thus grew another problem of understanding between English and engineering faculties.

The essentially literary nature of nearly all available English teachers throughout the early years of engineering education led to real disagreements, both between engineering and English teachers and among English teachers themselves. On the East Coast grew a movement, led by Frank Aydelotte of the Massachusetts Institute of Technology, whose aim was frankly to "humanize the engineering student's character and his aims in life" through literary study. Aydelotte and his followers (of which did *not* include Samuel Earle) claimed that the demand that engineering graduates "should be better able to write and speak their mother tongue is really a demand that they have better literary education" (1917, p. 300). Aydelotte's 1917 textbook, *English and Engineering,* was a reader of essays meant to "furnish something of the liberal, humanizing, and broadening element which is more and more felt to be a necessary part of an engineering education" (p. xiv).

Opposed to what Aydelotte terms as this "broad view of engineering education" was the "narrow view," which saw the promotion of reading and writing skills alone as the practical and proper goal. This position was most evident in the Midwest and far West, where English courses were taught most often by in-house English departments working closely with engineering faculty members. The A & M schools and schools of mines that grew up during 1900 to 1915 were especially uninterested in literary studies. In general, the more established the "arts college" at a school was, the more disagreement over the content of engineering English courses there would be.

During this period, there were generally three kinds of English courses available for engineering students: the required freshman composition course, a sophomore literature sequence that was sometimes required, and the junior- or senior-level courses in "exposition for engineers" that were the prototypes of today's technical writing courses. This entire three-pronged sequence, however, was plagued from the beginning by certain problems, the most serious of which were the lack of interest in learning to read or write literature on the part of the students, the quality and experience of many English teachers assigned to technical writing courses, and the lack of cooperation between English and engineering faculty members. The lack of student interest in English courses was in part a result of the way in which such courses tended to be taught. They were typically assigned to young and inexperienced faculty members who often looked down on engineering students as mere technicians, patronizing them while preaching a gospel of literary sweetness and light. Engineering professors did not help either, often referring to English courses disparagingly as unrealistic and less worthy than technical courses. As an editorial in the *Engineering Record* stated in 1917, "Students usually regard the [English] courses as necessary evil" ("A New Era," p. 291).

The fact that technical writing courses were seen by English departments as second rate and often staffed with younger faculty members or departmental fringe people meant that there was no glory and no real chance for professional advancement in technical writing. Thus, the quality of technical writing courses was often low in the early days, as departments rotated unwilling and uninterested teachers through them. Because of this second-class status given to engineering English, relations between English and engineering teachers ebbed. Engineering faculties had little patience with the stance of moral superiority assumed by many English teachers, or with the idea that students must be "humanized" through English literature. In fact, when the English Committee of the SPEE conducted a survey in 1918, they found that although English as "training in thinking," "a guarantee against illiteracy," and "a tool for use in technical work," got support from 72% of engineering faculty members, the idea of English courses as "a cultural and recreational escape from the monotonous literalism of vocational study" (the English Committee's wording, not mine) was supported by only 5% (Park, 1918, p. 209). The engineering faculty and the English faculty had clearly different agendas.

THE FORMATION OF A DISCIPLINE

Despite the peripatetic wrangling over literature versus vocationalism, the interest in technical writing grew apace. Prior to 1912, there had only been two English teachers in the SPEE; seven more joined in that year. The English Committee had been formed by 1914 (chaired by Samuel Earle until his death), and by 1918 there were 16 English teachers in the SPEE (Chatburn, 1919). In 1918, the Mann Report on Engineering Curricula recommended more time spent on English, and by 1920, 64% of all engineering schools required some sort of technical writing course for their students (Park, 1920). As J. Raleigh Nelson (1931), whose technical writing course at Michigan began in 1914, suggests, it was during this period, 1915 to 1920, when the engineering-only hardliners threw up their hands and integrated English into the curriculum.

As the 1920s opened, technical writing was beginning to become more of a presence in the curricula. The amount of time devoted to it increased, new courses were proposed on both the freshman and upperclass levels, and new textbooks began to appear that were aimed specifically at the technical writing student. T. A. Rickard published a new textbook in 1920, this one meant specifically for classroom use, but, like his first text in 1908, Rickard's *Technical Writing* was essentially concerned with good usage rather than with technical formats. A much more important step forward came in 1923, with the publication of the first "modern" technical writing textbook. It was called *English for Engineers,* and was the work of a toughminded and professionally determined assistant professor at Ohio State University named Sada A. Harbarger. (The author was referred to in the book only as "S. A. Harbarger," perhaps because the publisher felt that many readers might resent a woman claiming to be able to teach technical writing.)

English for Engineers is the first textbook that is organized according to the "technical forms"—reports and letters—that still remain the basis for most textbook organization today. Chapters include treatments of many sorts of letters, as well as explanations, abstracts, summaries, book reviews, editorials, articles, reports, and papers at meetings. But although this "forms" approach now seems natural to us, it was not immediately recognized as the best. Textbook organization by forms caught on slowly throughout the 1920s, and Harbarger's text was not initially as popular as Rickard's nonforms text (though it outlasted Rickard by a decade, being reprinted last in 1943). Harbarger was extremely active in the SPEE, and her views of the profession, as well as her textbook, were to be influential in shaping technical writing instruction.

The mid-1920s saw two new developments in the profession, one of them practical and the other philosophical. The practical development was the introduction of technical writing texts that were concerned only with the writing of technical reports. Ralph Fitting's *Report Writing* of 1924 and the immensely popular and influential *Preparation of Scientific and Technical Papers*, written by Sam F. Trelease and Emma S. Yule in 1937, found immediate audiences in technical writing classrooms and their narrowly focused, formal approach was to influence a whole generation of technical writers. Texts following these two works treated many different sorts of reports—preliminary, investigative, field work, recommendation, etc.—but seldom dealt with other technical writing tasks. They might be considered the apotheosis of the technical-forms approach to textbooks.

The philosophical development of the mid-1920s involved the rise of a younger group of technical writing teachers who defined themselves primarily as teachers of technical writing rather than as teachers of literature. The number of writing teachers grew slowly, of course—in 1926, J. Raleigh Nelson complained that "the little company of enthusiasts who have pioneered in this field the past 20 years do not see their ranks filling with recruits as rapidly as they might wish" (p. 813)—but some of the younger technical writing teachers, seeing the doors of conventional literary departments closing to them, began to downplay the call for more literature for engineering students that had been part and parcel of the English lament for years. In 1924, Bradley and Merwin Roe Stoughton made the shocking statement that "the habit of creative literary imagination is a detriment to an engineer.... Literary activity...is not desirable training for engineering students and does not help them present engineering data in brief and attractive form..." (pp. 144–147).

This was still a minority position in the 1920s, though; at that time, most English teachers were fighting together to accomplish goals on a broad front. There was also evidence that these goals were being accomplished. A 1924 SPEE survey (the organization was fond of surveys) found that it was no longer necessary to urge the importance of English for engineers; the uproar over illiteracy since 1910 had done its work well. The survey also found that English requirements at engineering schools had doubled since 1914 and that more colleges were instituting technical writing courses each year (Nelson, 1924). There was no question that, by 1924, English was an important part of engineering education once again.

EXPANSION AND DEPRESSION

Changes continued within the discipline throughout the late 1920s as new textbooks appeared, both traditional "usage" texts like Clyde Park's *English Applied in Technical Writing*, which enjoyed modest sales until the mid-1940s, and the increasingly popular "technical forms" texts, the best known of which was *Report Writing* by Carl Gaum, Harold Graves, and Lyne Hoffman, lasting well into the 1950s. Technical writing courses of the period were gradually refining themselves, taking on the characteristics and beginning to teach some of the forms that we still use today. Most technical writing courses of the late 1920s stuck to a few relatively rigid forms, though, and a contemporary description of an average course called it an "intensive study of the logical organization and effective presentation requirements of technical articles, reports, and business letters" (Hall, 1931, pp. 419–420).

This was a time of experimentation, as J. Raleigh Nelson was to say later, but the experimentation was conducted in an ever more secure atmosphere as English teachers realized how much they were needed. Another SPEE survey, in 1930, showed that of 1,300 engineers and teachers, 95% approved requirements in English composition, 75% approved speech requirements, and 45% approved literature requirements (Nelson, 1931). (This interest in speech, incidentally, was largely due to the vast influence of Dale Carnegie's books on *Public Speaking and Influencing Business Men*, published in the late 1920s. This first entrepreneur of self-improvement, on a grand scale, created a huge demand for speech courses in all technical fields, but that is another essay [Parker, 1931]).

Despite this acceptance, the early 1930s were not a happy time for engineering English teachers. The Depression had hit engineering schools hard, and the professional publications of the time reflected a pervasive discontent. Despite the demand for technical writing, most English teachers who made a specialty of it were still underpaid and little recognized in their own departments. Interest in composition teaching caused teachers to "lose caste" among their departmental peers and were seen as "professional suicide" by younger teachers (Birk, 1939, p. 426). Engineering teachers still did not give English teachers the cooperation they felt was necessary, and engineering students often seemed to have little respect for the sorts of teachers being turned out by graduate schools in English. It was said in the 1930s that many English teachers "appear to their critics as not of a sufficiently masculine type or of enough experience in the world outside their books to command the respect of engineering students" and they were called "effeminate" by some students. (In 1938, one student was quoted as calling his teacher "a budding pinko" [Creek, 1939, p. 301].)

In spite of these problems and discontents, and in spite of the Depression, technical writing courses continued to fill in attendance. More sections were taught each year and new textbooks began to pour off presses at an ever-faster rate. (Would-be authors soon realized that the success ratio for technical writing texts was the highest for any type of composition text.) The most popular texts throughout the 1930s were W. O. Sypherd, Sharon Brown, and A. M. Fountain's *The Engineer's Manual of English* (1943), primarily a technical forms text, and Thomas Agg and Walter Foster's *The Preparation of Engineering Reports* (1935), which was a narrowly formal approach to report writing that practically led the reader step-by-step through writing a report.

In 1938, a study appeared which showed the degree to which technical writing had come of age: a dissertation, later turned into a published report, called *A Study of Courses in Technical Writing,* by Alvin M. Fountain. Fountain's exhaustive survey-and-interview study showed that of 117 engineering schools in America, 76 schools offered 93 different technical courses in 1937 (p. 82). Fountain's study is an important diachronic slice of history, the only one extant; it covers the content of technical writing courses during the mid-1930s, the textbooks that were most popular, and the methods used by teachers of technical writing. The most important information in Fountain's study—and this is corroborated by the textbooks of the period—is that he shows how a technical-forms approach of a rigid and mechanical sort had become all but absolute by the late 1930s. Fountain's study also indicated the range of forms taught at the time. Essentially, every technical writing course Fountain examined covered the report form; it seems to have been a *sine qua non* in such courses after 1935. Thirty of the 93 courses studied used "Reports" in their course titles, and more than one third of the courses devoted the majority of the course time to report writing. Fifty-one of the 93 courses also covered business letters of various sorts, usually of a technical nature, and only 37 reviewed fundamentals of usage—the hallmark of the older, fading form of the upper-level technical English course. In addition, 33 of the courses involved technical articles, and oral presentations were important in many as well (pp. 83–98).

Fountain's report shows clearly that technical writing was a thriving industry in 1938, having produced its own authors, experts, and directors. The courses were more advanced and taught more forms. The study also showed, however, that little progress had been made on the professional front for teachers of technical writing. Conditions were still poor for many; there was still little chance for advancement, and the majority of technical writing courses was still staffed by instructors and assistant professors. At the same time, the understanding gap between engineering and English teachers was widened, and, by 1939, an important bastion of interdepartmental understanding was on the way out: A survey showed that of the more than two dozen departments of English that had once existed within engineering schools, only five remained (Birk, 1939, p. 426).

The dissatisfaction in the journal articles grew more shrill. Graduate schools still turned out nothing but literary scholars, and only the less talented of them gravitated to engineering English. After all the fruitless complaining of the past, little seemed likely to be done. In fact, technical writing pioneer W. O. Sypherd noted in a retrospective article in 1939 that "the prevailing notes are of uncertainty, discontent, and vague longing." Sypherd complained bitterly that literature courses were too few to matter, that freshman courses were ineffective, and that lack of writing in other engineering courses, bad student attitudes, and no interdepartmental cooperation had brought engineering English to a critical point. "I see little hope for any marked improvement," Sypherd concluded, "unless some radical upheaval should come to pass" (pp. 161–164).

A "radical upheaval" was certainly on the way. The first five years of the 1940s brought activities that would result in a complete restructuring of engineering education, and the beginnings of the final transformation of technical writing courses into those still taught today. World War II, of course, was the greatest single factor in these changes, creating a new technological imperative that swept all before it. Back in the United States, however, other forces were at work that would also transform the postwar engineering scene.

A DISCIPLINE COMES OF AGE: 1940–1980

Developments During World War II

On the surface, World War II brought the engineering English industry, at least as it appeared in journal articles, to an almost dead stop. In my research for this article, I found a huge hiatus in the production of articles of technical writing during the period from 1940 to 1946—almost as if the concept of "engineering English" had dropped off the face of the earth. This journal silence, though, is not to be construed as meaning that the teaching of technical writing slowed or stopped during the war—it did not. Business in both technical schools and arts colleges went on much as usual, despite lower enrollments. Technical writing courses continued to fill and new textbooks appeared, as did revisions of older books. Sada Harbarger passed away in 1942 and was duly eulogized by the English Committee over which she had tyrannized for so long. In most ways, technical writing continued along by a now well-trodden path.

Despite the lack of journal articles, English teachers were not unoccupied during this period. They were responsible, in fact, for two wartime SPEE reports that, although they had no immediate effect, were to change the course of postwar engineering education in America. These were the reports of the SPEE Committee on the Aims and Scope of Engineering Curricula, produced in 1940 and 1944. The committee was chaired and directed by H. P. Hammond, and the works of his committee came to be known as the Hammond Reports. Both reports dealt with the same questions, and, taken together, had an important impact.

The Hammond Report of 1940 brought together many of the fears and complaints that English teachers had been voicing throughout the previous decade in a new and powerful form. It condemned the narrow vocationalism of the engineering curriculum and put stress on a proposed platform of "science, of humanities, and of social relationships rather than on the practical techniques of particular occupations or industries" (Fatout, 1948, p. 717). To this obviously Dewey-influenced pronouncement, the Report added a charge recommending "the parallel development of the scientific-technological and the humanistic-social sequences." These two "stems," as they became known, were at the heart of the Hammond Reports. There were, of course, already extant humanities requirements at most schools, but the Hammond Committee wanted more, and a second Hammond Report was issued in 1944, this one suggesting a complete four-year program that required 20% or more of the student's time to be devoted to humanistic-stem courses—mostly literature, economics, history, and social studies. There was controversy about these reports during the late 1940s, but during the experimental period in education after the war the conception of the humanistic stem gradually won out, and by the early 1950s, the arm-twisting propaganda of the humanistic-stem proponents had achieved final victory.

What is interesting about this minor struggle in the history of engineering education is the fact that neither freshman composition nor technical writing courses were claimed or championed by either side. The engineering professor who saw no pressing need for curricular changes viewed composition courses as service adjuncts to his activities—not important to fight for—and the humanistic-stem supporters did not see writing courses as humanistic enough to be included under their rubric. As Paul Fatout said in a 1948 article on the growth of the humanistic stem, "...composition is not considered a legitimate offshoot of the humanistic stem" (pp. 715–716). There seemed to be no niche for technical writing in the controversy.

The Postwar Technical Writing Boom

In spite of the lack of champions, technical writing courses continued and even expanded, particularly after the end of the war. Part of this expansion was due, of course, to the thousands of new students attending college on the G.I. Bill, but the striking growth of technical writing was also in part a result of the nature of World War II, the first truly technological war. During the duration of the war, from 1939 to 1945, necessity had mothered thousands of frightful and complex machines, and the need for technical communication had never been greater. (In 1939, British officers were ordered to prepare for the war by sharpening their swords—an eloquent example of how much technology had changed the world by the time of Nagasaki six years later.) Technical writers were in great demand during this time, for each new airplane, gun, bomb, and machine needed a manual written, and the centrality of the lucid explicator of technology was obvious as never before. As Jay R. Gould wrote later:

> World War II is an important date for the technical writing profession. Reports had to be written for the men and the women who were inventing the machines and the electronic systems...much more importance was given to the technical writer, a man or woman who spent all of his time in communicating.... The need was so urgent that technical writers entered the profession from many sources. (1964, pp. 12–13)

For the first time, technical writing was more than an adjunct function of some other activity—it was a job in itself.

After the end of the war, technical writing finally became a genuine profession as wartime technologies were translated into peacetime uses. The giant technological corporations—General Electric, Westinghouse, General Motors—opened separate departments of technical writing after finding that it was no longer cost effective to pay engineers to both design and write. The technical writing and editing profession became more aware of itself during these years, but in spite of these changes, few colleges offered technical writing majors or structural changes in technical writing courses. Schools seemed to ignore the changing conditions of the field, and when the journals began to print technical writing articles again after war's end, the articles had subtly changed focus. Now they dealt with tasks and techniques within the teaching of technical writing, rather than being concerned with the status and conditions of the teaching. It was during this time that the first "modern" technical writing articles were written and published, but what the profession gained in techniques it lost in self-awareness.

The postwar era was a demanding time for teachers of technical writing; after 1945, the demand for their courses rose dramatically as the colleges were deluged with returning veterans. This was, as Alvin Fountain put it in a retrospective article, "the frantic era of the G.I. Bill, the quonset hut, the barracks classroom, and the tar paper apartment, infested by returning veterans armed with wives and children, a bunch of common sense, and a serious purpose" (1959, p. 47). Teachers tried to cope as best they could with the population explosion in their classrooms.

This period brought more complexity, as well as more students, to technical writing, and the late 1940s saw further expansion of the number of forms taught in typical courses. Initially, of course, the report had been central to courses in technical writing; only gradually had business letters been added, and before the war the technical article became a fairly common form as well. By 1951, however, these simple and basic forms had been heavily supplemented and diversified. A report of common forms taught in that year indicated that at least six different report forms were widely taught, and correspondence forms often proliferated until more than 10 letter types were taught (Rider, 1951). As might be expected, manual-writing also became a popular skill to learn in postwar writing courses. This was partially a result of the military influence of the war, but it was also due to the increasing number of technically-based consumer products that the United States was turning out.

A New Professionalism

The decade of the 1950s saw technical writing "grow up," assuming the essential form we know of today. The profession of technical and scientific writing grew and matured during this period with the establishment of the Society of Technical Writers and, in 1958, the influential work entitled *Transactions on Engineering Writing and Speech* (now titled *Transactions on Professional Communication*) of the Institute of Electrical and Electronics Engineers (see MacNamara, 1958). During the 1950s, the importance of the profession of technical writing became apparent to industry, and colleges gave more serious consideration to turning out trained technical writers. The programs at the Massachusetts Institute of Technology, the University of Michigan, and at Rensselaer Polytechnic Institute assumed during this decade the leading place still held today, and, in 1958, RPI established the first Master's degree in technical and scientific writing in the United States.

On most campuses, the problems that had always plagued technical writing programs continued as usual, but in spite of them there was a continuing refinement and sophistication to the courses being offered (Rider, 1951). Around the mid-1950s, the humanistic-stem requirements that had been so heavily accentuated during the immediate postwar era began to be replaced with technical writing requirements at some schools; this was largely the result of pressure from the engineering faculty and the continuing complaint of industry that new graduates still could not write well (Kobe, 1956; Pierce, 1958). By 1957, nearly all colleges offered a technical writing course, and 64% of engineering schools made such a course a requirement during the junior or senior year (Wellborn, 1960). The courses that were being required were often more carefully planned than technical writing courses had been in the past, and experimental methods of teaching became much more common during this period. The most successful experiments of the 1950s were probably the cooperative courses that were team-taught by English and engineering teachers (Rathbone, 1958).

Textbooks throughout the decade were still largely derivative of one another, but several stand out as being particularly popular and important. Joseph N. Ulman, Jr., and Jay R. Gould's *Technical Reporting* (1952) was a conservative textbook that concentrated on the report, but its completeness has made it popular for over 20 years. Ulman and Gould presented a clear bridge to the traditional textbooks of the 1930s, and many of the texts of the 1950s followed its conservative lead. In 1954, however, there appeared a textbook which was not only extremely popular in its own time, but which is arguably the single most important postwar technical writing text: Gordon Mills and John Walter's *Technical Writing*. Mills and Walter had begun working together in the late 1940s, and they were determined to try to reinvigorate technical writing instruction by bringing it closer to the businesses and industries that actually used the forms that were taught. Mills and Walter conducted a survey of over 300 actual technical writing situations in industry, and from this came a number of changes in the approach that informed their textbook.

As Walter explained in 1973, two of the most important assumptions that he and Mills had gleaned from their survey had been these:

> 1. A rhetorical approach rather than the rigid "types of reports" approach that most texts used was best. Most reports are made up of several common processes: definition, description, explanation of process, etc.

2. The only good criterion for technical writing is "does it work?" This indicates that in technical writing as well as in other rhetorical forms, the writer-reader relationship is most important. (pp. 5–6)

Technical Writing reflected these assumptions and went on to be the most popular and paradigmatic text of the 1950s, pointing the way to a new rhetorical approach to technical writing that was to revivify what had been in danger of becoming a sterile and mechanical course (Freedman, 1963).[1]

Mills and Walter were not alone in their concern with creating a sense of reader-writer relationship in technical writing instruction. A growing awareness that audience considerations had long been scanted in technical writing was one of the important developments of the 1950s. In a prescient article originally appearing in the *Journal of Chemical Education* in 1951, then published again in 1963, James W. Souther mentioned this new awareness:

...more and more, writers in industry are becoming aware of their readers' interests. They are placing conclusions, summaries, and recommendations at the beginning of the report because the administrators are most interested in such material. The more widespread use of such devices as statements of purpose and background...is ample proof of the writer's growing awareness of the reader. (p. 225)

In 1955, J. H. Wilson blasted college technical writing courses for their traditional dismissal of audience considerations in the much-discussed article, "Our Colleges Can Teach Writing—If They Are Made To."[2] And, in 1959, Joseph Racker presented his influential concept of writing to audiences with different levels of technical expertise in his essay, "Selecting and Writing to the Proper Level."

Another important change that the 1950s and the early 1960s saw was the expansion of technical writing into fields other than engineering. Other applied sciences had long existed at many colleges, but only during this period did departments of agriculture, architecture, chemistry, pharmacy, and even home economics begin to send their students to technical writing. The course began to gain campus-wide recognition as a useful, no-nonsense addition to the curriculum of any serious student. Textbooks soon reflected this broadening of audience; for example, Theodore Sherman's *Modern Technical Writing* (1955) claimed that it was "appropriate to a wide range of subjects so that any technical writer, regardless of his field of specialization, will find the help that he needs" (p. v).[3] At this time, too, we find the first published mention of technical writing courses built around a single, long project, with a series of check-in assignments preceding the long report, a course arrangement that was to become widely popular in the 1960s and 1970s (Thomas, 1955). Courses began to consider graphic presentations as well as verbal ones, due mainly to the effect of a Iowa State University technical writing course, which provided a successful model (Sweigert, 1956).

Breakthroughs and Problems

By 1959, new textbooks were appearing in such profusion that even to list them would take too much space. Many technical writing texts from the late 1950s are still in print (since the mortality rate for technical writing textbooks is still much lower than that for any other sort of composition text), and as the decade drew to a close, more and more texts began to copy the rhetorical approach of Mills and Walter, as well as the general-coverage approach of Sherman. But in spite of these successful new texts and the experimental advances, technical writing still had its problems. Although technical writing had by this time a long and honorable history and was obviously in English departments to stay, it got as little welcome from literary departments in 1959 as it had in 1929. Still considered a low-level service course, technical writing was still being assigned to graduate students and instructors.

After *Sputnik* was launched into space in 1957, an alarmed United States began a war of technology with Russia that was to last into the early 1970s, and as the 1960s opened, there was a serious shortage of technical writers in industry. Most English majors still saw technical writing as hack work and most engineers could do better if they remained specialized; as a result, industries engaged in bidding wars for those few technical writers were not lost. The Society of Technical Writers grew quickly and went through several name changes, emerging as today's thriving and influential Society for Technical Communication. As a group, technical writers advanced greatly, in both pay and prestige, during this period (Smith, 1963).

On college campuses, however, things were not so smooth. The 1960s were a time of disturbance and change for technical writing instruction, as for so many other elements in American culture. Technical writing courses were struggling to define themselves, its teachers were wrangling over what their jobs should entail, and its students were getting objectively more intelligent, but were fewer and fewer in number as the decade proceeded. It was a confused time for American colleges and for technical writing instruction, but it was a period that prepared the ground for the great leap forward that was to come in the 1970s.

A sort of critical self-examination and desire to define technical writing itself was an important element of the intellectual effort of technical writing teachers during the early and mid-1960s. As early as 1954, Mills and Walter had stated in the preface of their text that "nobody had ever seriously explored the concept of technical writing with the purpose of trying to say exactly what technical writing is" (p. vii). In the 1960s, that investigation was taken up by a number of teachers. In 1963, Robert Hays investigated the linguistic nature of technical writing in a widely reprinted essay entitled "What Is Technical Writing?" W. Earl Britton, in an article in *College Composition and Communication* in 1965, wrote what is probably the most comprehensive early definition of technical writing, defining it by subject matter, linguistic nature, thought processes involved, and purpose. Britton's conclusion was that technical writing is defined more than anything else by "the effort of the author to convey one meaning and only one meaning in what he says" (pp. 113–116).

THE RISE OF TECHNICAL WRITING INSTRUCTION IN AMERICA   189

This interest in the ultimate nature of technical writing was matched in the 1960s by an awakening interest in the process of teaching it and the methods available for doing so. The first steps in the direction of empirical research into technical writing and the teaching of it were made during this period; for the first time, researchers were gathering and analyzing facts about technical writing in a scientific manner, obtaining results and conclusions that could not be dismissed as mere opinion. Several important early experiments were Harry E. Hand's 1964 study of the relative seriousness of different sorts of errors in technical writing papers and Richard M. Davis's (1967) massive investigation of the efforts of variables in the writing of a technical description.

In the midst of this growing professionalism and increasing self-consciousness, however, technical writing courses were still beset by the same old problems that had always plagued them. The ascent of literary studies throughout the 1960s meant that the age-old battle raged on between those who wished to teach technical students to write and those who wished to teach them to read and appreciate great literature. Throughout the 1960s, many technical writing teachers continued to be graduate students and lower-level faculty members who had been dragooned into technical writing and whose primary interests remained literary studies. This split between interests and assignments came out strongly at a Conference on College Composition and Communication (CCCC) workshop on technical writing instruction in the early 1960s, which was stated, rather ironically, as:

> That this [technical writing course] is frequently thought of as a "service course" was recognized. Several expressed strong disapproval of the attitude, and stronger disapproval of admitting it. A few spoke in defense of the designation. Most confessed that the fact was inescapable, though the name was nauseous....
>
> Some piously professed to see an encouraging resurgence of interest in and demand for the humanities. These strains, sweet to CCCC professional ears, played softly for the duration of the Workshop.... The Workshop believed, in the main, that recommended reading should be largely literary rather than scientific and technical.

The co-secretary of the workshop summed up her colleagues' responses this way:

> I sense that the Workshop members believe that technical writing must be about scientific matters in which they have no training and less interest. They see themselves doing grease monkey work on physics papers for spelling errors and would die first.... My minority opinion is that there is such a thing as technical presentation and reading and writing about literature doesn't teach it. (Power, 1961, pp. 163–164)

That was indeed a minority opinion in 1961, and although it came to be more widely held over the following 10 years, literary studies continued to be the main interest of most technical writing teachers well into the 1970s.

In terms of course content, the 1960s were not a time of major change. The content and form of the reports and other forms traditionally taught continued to evolve as they always had, but only one new and important form emerged as a result of the expansion of the field of technical writing that the 1950s had brought: the proposal. During the late 1950s and early 1960s, it was estimated that industry spent in excess of one billion dollars per year on the writing of proposals, and the importance of this new form soon became obvious to the writers of technical writing texts (Kendall, 1963). The first textbook to seriously treat this form, Seigfried Mandel and David L. Caldwell's *Proposal and Inquiry Writing*, was published in 1962. The book was popular and influential, prompting a reviewer to note that "At long last another relatively new and lustily growing American industry is beginning to have its folklore committed to writing...the 'technical-proposal generation' industry" (Collins, 1962, p. 31). The proposal, as a technical form, quickly spread to all texts; partially because of the influence of Mandel and Caldwell, but also due to the influence of the second edition of Mills and Walter, which featured the proposal.

Retrenchment and A New Sense of Identity

At the end of the 1960s, and well into the early 1970s, there was a serious drop in the number of undergraduate students enrolled in engineering programs. In the Fall of 1968, there were over 239,000 undergraduate engineering students, but by the Fall of 1973 this number had fallen to fewer than 187,000—and this in a time of skyrocketing general enrollments (Sheridan, 1979). Enrollments in technical writing classes shrank accordingly, with the result that fewer unwilling conscripts were forced to teach the course. Still, though, many technical writing teachers were merely time-servers as the 1970s opened. In a well-known experiment in 1970, Juanita Williams Dudley complained about the character of many technical writing classes in tones that are by now familiar: "Frequently the technical writing conscript regards his assignment as a humiliating, dehumanizing hairshirt that must be endured until advanced degrees and seniority confer upon him enough power to bargain for courses in literary criticism and creative writing" (1971, p. 42).

However, a new day was dawning for technical writing instruction. Due in part to declining enrollments in courses, a solid core of committed technical writing professionals was forming by the late 1960s, a growing number of teachers who considered technical communication their primary area of interest and expertise. In 1970, the *Journal of Technical Writing and Communication* was started, a journal which quickly became the most respected organ in the field of technical writing instruction. The journal reflected an ever-increasing sense of pride and self-consciousness on the parts of many experienced technical writing teachers who had served faithfully through the "lean years." Tools became more sophisticated; in 1971, Stello Jordan edited the two-volume *Handbook of Technical Writing Practices* that was called "the most complete and sophisticated technical writing guide ever published...a true and important picture of the many-sided profession of technical writing, and an impressive, diversified explanation of why and how technical information is communicated" (Schlesinger, 1973, p. 45). In 1973, the Association of Teachers of Technical Writing was formed, and their journal, *The Technical Writing Teacher*, began publication. Though early issues were somewhat crude, the journal underwent marked improvement throughout the decade and now ranks only behind *JTWC* in the opinions of many technical writing teachers.

Finally, in 1974, technical enrollments began once again to rise, and by the late 1970s, they were going up at a rate of more than 10% per year at a time when general college enrollments were static. As the demand for courses in technical communication grew during the decade, the demand for literature courses fell. Soon many chairpersons of English departments became uncomfortably aware that the only thing supporting their sparsely populated Milton courses was the credit generated by the quondam poor relation, the technical writing division. The Modern Language Association, which had for over 50 years refused to recognize technical writing as a legitimate function of English scholars, caved in during the mid-1970s and gave technical writing belated recognition in 1976, when the first technical writing panel was presented at an MLA convention.

This demand was partially due to one thing that had not changed: the need for technical communications specialists in industry. A survey during the late 1970s showed that over 50% of an engineer's time was spent dealing with writing, and over 85% of professional engineers polled said that a technical writing course should be required of all technical students (Davis, 1977). More and more departments, some of them only quasi-technical, began to require a technical writing course for their students, as the good reputation of these courses became more widely known. As John Walter put it in 1977,

> The widespread emphasis on technical writing (coming primarily from students, I think) has led to considerable growth in the number of schools offering courses in technical writing and to increased enrollment in those schools which have offered the course for years.... We've come a long way, and more and more departments are compelled to recognize that technical writing is a legitimate concern of conscientious teachers, and one which must be rewarded when teachers do a good job. (p. 2)

Technical writing teachers were not always rewarded by their departments, but many found freedom and credit in the 1970s which previously had only been dreamt of. Their courses were crowded, and students had never been so eager to learn. Teachers of technical communication began to be tenured and promoted on the basis of their skills and publications within the field, a situation that had been rare prior to the 1970s. Many found lucrative sidelines in consulting for industry, and such consulting nearly always rebounded to enrich the technical writing classroom with new insights into the contemporary world of industry. Each *MLA Job List* brought news of more and more tenure-track positions specializing in technical writing. Professionally, it was a satisfying decade.

Textbooks during the 1970s grew ever more sophisticated and began to appear in versions aimed at two-year as well as four-year colleges. Old favorites such as Mills and Walter (1954) continued to sell well, but it was supplemented by a new, more rhetorically based text, exemplified by Lannon's *Technical Writing* (1979). Perhaps the most influential—though not the most popular—text of the decade was Mathes and Stevenson's *Designing Technical Reports* (1976), with its elegant audience-analysis procedure and its determined investigation into the purposes behind technical writing. It can truly be said that the decade brought technical writing instruction to a state of efficiency and productive professionalism it had never known before.[4]

In the 1980s and presently, in the 1990s, technical writing is not without problems, but its prospects have never been brighter. There are still arguments being made that the technical writing course should be taken out of the hands of English teachers, but these arguments are as old as technical writing instruction itself and will likely prove no more effectual now than they were in 1920 (Mathes, Stevenson, & Klaver, 1979).[5] Technical writing scholarship is thriving, and there is a healthy tone of innovation and skepticism in the essays found in today's technical writing journals; the received wisdom is being tested against new situations and needs as never before, and the field is more vital than ever because of it. It now seems likely that technical communication is an acceptable field of study for English graduate degrees in many schools. The field has generated its own patriarchs and scholars, and some English departments have already begun to trade heavily on their technical writing fame. There is finally evidence that many colleges see and appreciate the dedication of their technical writing staffs, and the technical writing division is no longer the repository of callow youths and second-raters that it once tended to be. In general, the prospect is excellent for both teachers and students of technical writing. We have come a long way from 1939, when teaching technical writing was called "professional suicide," and, we can say with pride, an even longer way from 1915, when technical students' papers could be "described only by the word 'wretched.'" It has been a long road, but one well worth the traveling.

NOTES

[1] See, for instance, the complaints in Morris Freedman's "Technical Writing, Anyone?" (1963, pp. 4–7).

[2] The problem was also addressed in John I. Mattill's "Writing as Communication: The Engineer Must Learn How to Reach His Constituents" (1954, pp. 476–479).

[3] Much earlier, Sada Harbarger had tried with some Ohio State colleagues to expand the potential audience of technical writing texts with a 1938 textbook, *English for Students in Applied Sciences*. The profession was clearly not ready for it at the time, and so it failed.

[4] Special thanks to Fabian Gudas and Barbara Sims of Louisiana State University for sharing with me their experiences of teaching technical writing during the period of 1945 to 1980.

[5] This argument was most recently resurrected in Mathes, Stevenson, and Klaver, "Technical Writing: The Engineering Educator's Responsibility" (1979, pp. 331–334). It brought, predictably, a rash of responses from English teachers and no action at all from engineering teachers.

REFERENCES

Agg, T. R., & Foster, W. (1935). *The preparation of engineering reports*. New York: McGraw-Hill.
Anonymous. (1926). A study of evolutionary trends in engineering curricula. *Proceedings of the Society for the Promotion of Engineering Education, 34*, 551–585.
Aydelotte, F. (1917). *English and engineering*. New York: McGraw-Hill.
Aydelotte, F. (1917, February 24). Training in thought is the aim of elementary English course as taught at M.I.T. *Engineering Record*, pp. 300–302.
Birk, W. O. (1939). Organization and conditions. *Proceedings of the Society for the Promotion of Engineering Education, 47*, 408–434.
Britton, W. E. (1965). What is technical writing? *College Composition and Communication, 16*, 113–116.
Chatburn, G. R. (1919). The SPEE: A survey of its past and a reconnaissance of its future. *Proceedings of the Society for the Promotion of Engineering Education, 27*, 180–223.
Collins, W. E. (1962). Review of *Proposal and inquiry writing* [book]. *Institute of Radio Engineers Transactions on Engineering Writing and Speech, 6*, 31–32.
Connaughton, M. E. (1981). Technical writing in America: A historical perspective. In J. C. Mathes & T. E. Pinelli (Eds.), *Technical writings past, present, and future* (pp. 31–42). Hampton, VA: National Aeronautics and Space Administration.
Creek, H. L., & McKee, J. H. (1931). English in colleges of engineering. *English Journal, 21*, 818–828.
Creek, H. L. (1939). Teachers of English in engineering colleges: Selection and training. *Proceedings of the Society for the Promotion of Engineering Education, 47*, 300–313.
Davis, R. M. (1967). Experimental research in the effectiveness of technical writing. *IEEE Transactions on Engineering Writing and Speech, 10*, 33–38.
Davis, R. M. (1977). How important is technical writing?—a survey of the opinions of successful engineers. *The Technical Writing Teacher, 4*, 83–88.
Dudley, J. W. (1971). Writing skills of engineering and scientific students. *IEEE Transactions on Engineering Writing and Speech, 14*, 42–46.
Earle, S. C. (1911a). English in the engineering school at Tufts College. *Proceedings of the Society for the Promotion of Engineering Education, 19*, 33–47.
Earle, S. C. (1911b). *The theory and practice of technical writing*. New York: Macmillan.
English for engineers. (1915, June 19). *Engineering Record*, p. 763.
Fatout, P. (1948). Growth of the humanistic stem. *Proceedings of the American Society of Engineering Education, 55*, 715–720.
Fitting, R. U. (1924). *Report writing*. New York: Ronald Press.
Fountain, A. M. (1938). *A study of courses in technical writing*. Nashville, TN: George Peabody College.
Fountain, A. M. (1959). Working with electrical engineers in seminar. *Institute of Radio Engineers Transactions on Engineering Writing and Speech, 2*, 46–48.

Freedman, M. (1963). Technical writing, anyone? In H. A. Estrin (Ed.), *Technical and professional writing* (pp. 4–7). New York: Harcourt, Brace, and World.
Gaum, C., Graves, H., & Hoffman, L. S. S. (1929). *Report writing*. New York: Prentice Hall
Gould, J. R. (1964). *Opportunities in technical writing*. New York: Universal Publishing.
Hall, A. V. (1931). English as an essential part of the engineering curriculum. *Proceedings of the Society for the Promotion of Engineering Education, 39*, 416–22.
Hand, H. E. (1964). An attempt to measure success in technical writing. *Proceedings of the American Society of Engineering Education, 72*, 70–72.
Harbarger, S. A. (1923). *English for engineers*. New York: McGraw-Hill.
Hays, R. (1963). What is technical writing? In H. A. Estrin (Ed.), *Technical and professional writing* (pp. 64–69). New York: Harcourt, Brace, and World.
Jordan, S. (Ed.). (1971). *Handbook of technical writing practices* (Vols. 1-2). New York: Wiley Interscience.
Kendall, R. (1963). The proposal digest. *IEEE Transactions on Engineering Writing and Speech, 6*, 79–81.
Kobe, K. A. (1956). What colleges are doing to train chemists and chemical engineers in technical writing. *Journal of Chemical Education, 33*, 55–57.
Lannon, J. M. (1979). *Technical writing*. Boston: Little, Brown.
MacNamara, D. J. (1958, March). Welcome. *IRE Transactions on Engineering Writing and Speech, 1*, 1.
Mandel, S., & Caldwell, D. L. (1962). *Proposal and inquiry writing*. New York: Macmillan.
Mathes, J. C., & Stevenson, D. W. (1976). *Designing technical reports*. New York: Macmillan.
Mathes, J. C., Stevenson, D. W., & Klaver, P. (1979). Technical writing: The engineering educator's responsibility. *Engineering Education, 69*, 331–334.
Mattill, J. I. (1954). Writing as communication: The engineer must learn how to reach his constituents. *Proceedings of the American Society of Engineering Education, 61*, 476–479.
Mills, G. H., & Walter, J. A. (1954). *Technical writing*. New York: Holt, Rinehart, and Winston.
Nelson, J. R. (1924). The department of English. *Proceedings of the Society for the Promotion of Engineering Education, 32*, 560.
Nelson, J. R. (1926). The English department. *Proceedings of the Society for the Promotion of Engineering Education, 34*, 812–814.
Nelson, J. R. (1931). English, engineering, and technical schools. *English Journal, 20*, 494–502.
A new era in teaching English to engineers. (1917, February 24). *Engineering Record*, p. 291.
Obituary of Samuel Chandler Earle. *Proceedings of the Society for the Promotion of Engineering Education, 25*, 246.
O'Brien, H. R. (1913, November 6). Engineering English. *Engineering News*, 914–915.
Park, C. W. (1918). Report of committee #12, English. *Proceedings of the Society for the Promotion of Engineering Education, 26*, 205–217.
Park, C. W. (1920). Report of committee #12, English. *Proceedings of the Society for the Promotion of Engineering Education, 28*, 294–298.
Park, C. W. (1926). *English applied in technical writing*. New York: F. S. Crofts.
Parker, J. W. (1931). The need for speech training for engineers. *Proceedings of the Society for the Promotion of Engineering Education, 39*, 226–228.

Pierce, J. R. (1958). The challenging field of engineering writing and speech. *Institute of Radio Engineers Transactions on Engineering Writing and Speech, 1,* 12–13.
Power, K. (1961). Special problems of the C/C course in technical schools. *College Composition and Communication, 12,* 163–165.
Racker, J. (1963). Selecting and writing to the proper level. In H. A. Estrin (Ed.), *Technical and professional writing* (pp. 236–246). New York: Harcourt, Brace, and World.
Rathbone, R. R. (1958). Cooperative teaching of technical writing in engineering courses. *Proceedings of the American Society of Engineering Education, 66,* 126–130.
Rickard, T. A. (1920). *A guide to technical writing.* San Francisco: Mining and Scientific Press. (Original work published 1908)
Rider, M. L. (1951). Some current practices in teaching advanced composition for engineers. *Proceedings of the American Society of Engineering Education, 58,* 176–178.
Schlesinger, E. K. (1973). Review of *Handbook of technical writing practices* [book]. *IEEE Transactions on Professional Communication, 16,* 44–45.
Sheridan, P. J. (1979). Engineering and engineering technology enrollments, fall 1978. *Engineering Education, 70,* 58–67.
Sherman, T. A. (1955). *Modern technical writing.* Englewood Cliffs, NJ: Prentice Hall.
Smith, R. W. (1963). *Technical writing.* New York: Barnes and Noble.
Souther, J. W. (1963). Design that report! In H. A. Estrin (Ed.), *Technical and professional writing* (pp. 225–229). New York: Harcourt, Brace, and World.
Stoughton, B., & Stoughton, M. R. (1924). Education in English for engineering students. *Proceedings of the Society for the Promotion of Engineering Education, 32,* 144–47.
Sweigert, R., Jr. (1956). A technical writing course that works. *Proceedings of the American Society of Engineering Education, 62,* 262–266.
Sypherd, W. O. (1939). Thirty years of teaching English to engineers. *Proceedings of the Society for the Promotion of Engineering Education, 47,* 161–165.
Sypherd, W. O., Brown, S., & Fountain, A. M. (1943). *The engineer's manual of English* (Rev. ed.). Chicago: Scott Foresman. (Original work published 1933)
Telleen, J. M. (1908). The courses in English in our technical schools. *Proceedings of the Society for the Promotion of Engineering Education, 16,* 61–73.
Thomas, J. D. (1955). Thwarting the two-day term report. *Proceedings of the American Society of Engineering Education, 62,* 138–160.
Trelease, S. F., & Yule, E. S. (1937). *The preparation of scientific and technical papers.* Baltimore: Williams and Wilkins.
Ulman, J. N., Jr., & Gould, J. R. (1952). *Technical reporting.* New York: Dryden Press.
Walter, J. A. (1973). Confessions of a teacher of technical writing. *The Technical Writing Teacher, 1,* 3–9.
Walter, J. A. (1977). Message from the president. *The Technical Writing Teacher, 5,* 2–3.
Wellborn, G. P. (1960). Is the technical student short-changed in college? *College English, 21,* 394–396.
Wilson, J. H., Jr. (1955). Our colleges can teach writing—if they are made to. *Proceedings of the American Society of Engineering Education, 62,* 431–435.

10

Interfacing: Multiple Visions of Computer Use in Technical Communication*

Johndan Johnson-Eilola
Purdue University

Stuart A. Selber
Pennsylvania State University

Cynthia L. Selfe
Michigan Technological University

INTRODUCTION

There is a tendency to think about the history of computers as a series of technological inventions separate from social, cultural, and political conditions. Consider Zakon's (1993) much-cited Internet time line. Tracing a series of events between 1957 and 1996 that led to some staggering Internet growth statistics (in July of 1996, for example, there were 12,881,000 hosts, 134,365 networks, and 488,000 domains on the Internet), Zakon, in part, privileges the following technological inventions: the launching of *Sputnik* by the USSR in 1957; the commissioning of ARPANET by the U.S. Department of Defense in 1969; in 1973, the outlining of the mechanics of Ethernet; the establishment of BITNET (Because It's Time NETwork) in 1981; the introduction of domain name servers in 1984; and, in 1991,

---

* Author's names are ordered alphabetically; all authors contributed equally to this project.

the releasing of Gopher at the University of Minnesota and the World Wide Web at CERN.

Although accurate in chronological terms, such a time line falls short of representing accurately or robustly the environmental contexts within which Internet development first occurred—the Cold War—and continues to occur—the post-Industrial workplace. By reducing the history of Internet development and use to a series of isolated technological inventions, very real social, cultural, economic, and political influences on computing are all but erased from consciousness. In discussing the history of hypertext, Johnson (1995) makes a similar argument about other one-dimensional perspectives on technological developments: "Nearly every discussion of hypertext points to the 'great men' who have invented hypertext, either conceptually or physically—Bush, Englebart, Nelson, Nielsen, Bolter, and so on. Instead of focusing on the 'inventors' of hypertext as discrete individual geniuses, it would behoove us to study the emergence of hypertext from a more culturally constructed perspective" (p. 19). What Johnson encourages us to do here is study the development and use of computer technologies historiographically.

In this chapter, along these same lines, we examine a variety of methods for thinking about the relations between computers and technical communication. Rather than attempting to locate or construct a privileged method for understanding, and tracing, the role of computers in technical communication, we assert the need for a continuing practice of meta-analysis. In other words, we think it is crucial that technical communicators—teachers, researchers, students, and practitioners—gain the ability to think critically, multiply, historically, and contextually about the ways computer technologies are developed and used within our culture and how such use, in turn, intersects with the practice of technical communication in the workplace and the teaching of technical communication in the classroom.

In addition to our interest in illuminating social, economic, historical, political, and cultural forces, our decision to focus on multiple perspectives is driven by a number of factors, including the rapid rate of technological change, the relatively low status of technical communicators in industry, and the failure of computer technologies to increase efficiency or quality of life in any but isolated circumstances.

These factors, we believe, are intimately related. In an age of rapid change, many computer users have limited time to think carefully about how they learn new computer-based activities—the easiest method of adoption is typically the quickest, the most intuitive, and, therefore, the least transformative. Thus, paradoxically, in an industry where technical communication is still frequently considered a secondary consideration—by technologists, managers, and users—technical communicators that work with technology and on technology-related projects now have increasing opportunities to affect change when they can think in critically informed ways about technology and its uses. The potential for change, we have begun to realize, is enacted only by technology designers and users, and does not reside in machines.

INTERFACING: MULTIPLE VISIONS OF COMPUTER USE   199

In a movement articulated with this recognition, the value that many individuals and corporations have come to associate with computer technology has begun to shift away from a focus on traditional "hard" technological products—microchips, hard drives, displays, cables—and toward "soft" technological products and the uses of these products within situated contexts: the ways in which, and the purposes for which, data packets are moved through networks and among individuals; the ways in which users read and employ email; and the ways in which corporations publish information to individuals in dispersed locations. In other words, we have begun to understand that the values associated with computers depend at least as much, if not more, on the people who design and use the machines—their various histories, purposes, relationships, and contexts—as on the technical specifications of the machines themselves.

Unfortunately, the more things change, the more they remain the same. Because so many corporations are scrambling to integrate new technologies into the work and lives of employees, and because technological change continues at such a rapid pace, few technical communicators—indeed, few individuals in any position—are encouraged, or educated, to consider in critically informed ways the complex relationships between people and machines, or the relationships between the machines and the social contexts in which they are used. And yet, these particular habits of mind are essential to the task of rethinking the relationships we have constructed with technology and realizing the possibilities of technological change within particular social, cultural, economic, and political contexts.

Therefore, we think it is critical that technical communicators, and those who educate such individuals, cultivate these habits of mind. We need to learn to think actively, critically, and continually about the relations between technical communication, computers, and human users; and we need to teach students to see, think about, and understand these complex relationships. With an eye on this goal, in this chapter we offer four different perspectives on the relationships between technical communication and computer technologies, situating these perspectives in a wide range of historical, educational, rhetorical, and theoretical contexts (see Table 10.1).

These perspectives are not meant to construct universal Truths about technology or technical communication, nor are they meant to exhaust the possible contexts for understanding the various relationships mentioned here. Instead, they attempt to illustrate a range of ways to think about—and come to understand—technical communication and its relation to, and use of, technology. Some of these perspectives are adapted from our own work, but others have been identified by scholars who have not seen technical communication as their major area of focus. What we consider important, then, is not only that readers apprehend the content of the individual analyses—indeed, given the limitations of space, these analyses are really only sketches—but rather that they develop the habit of moving across them, of constructing these important relationships from multiple and related perspectives.

TABLE 10.1
Possible Interfaces between Technical Communication and Computer Technologies

Framework	Variations		Major Theorists	Examples
Pedagogical	Product	Making products	Selber	Make Web pages
	Computer literacy	Understanding textuality		Understand how people read/use Web pages
	Historical / Political / Social context	Situating uses		Examine issues of access to the Web, such as who controls it
Rhetorical	Objective	Positivist	Berlin	Web increases transmission of accurate information
	Subjective	Expressivist		Email increases viewpoints
	Transactional	Social constructionist		Email encourages participation
Spatial Dynamic	Automating/ Contracting	Limiting communication to increase efficiency	Zuboff/ Johnson-Eilola and Selber	Web site sends individual pages to browsers on request
	Informating/ Expanding	Increasing communication to increase quality		Web site tracks users over time to develop user profiles
Critical	Instrumental	Simple tool	Feenberg	Word processor as typewriter
	Substantive	All-powerful social force		Word processor as deskilling/ dehumanizing office automation
	Critical	Ambivalent, use depends on context and mode of production		Word processor integrates into (affects and is affected by) context, task, user, and community

HISTORICAL, PEDAGOGICAL, AND PROGRAMMATIC RELATIONSHIPS BETWEEN TECHNICAL COMMUNICATION AND TECHNOLOGY

With its professional practices at least partly rooted in, and influenced by, the private and government sectors, including the branches of the military (Connors, 1982; Kynell, 1996; Souther, 1989), technical communication has always been, at least in historical terms, closely aligned with advancements in science, engineering, and technology. It is not surprising, then, that educators in programs that prepare students for the tasks associated with nonacademic writing and communication have considered—and often debated—different approaches to integrating a wide range of technical skills and experiences in classroom settings, and in degree-related requirements such as internships and professional projects. These discussions and debates have often, though not always, taken place at the annual meeting of the Council for Programs in Technical and Scientific Communication (CPTSC), the only organization associated with English studies that specifically focuses on programmatic issues in technical and scientific communication. In fact, in a 1973 letter sent by Thomas Pearsall to the directors of 20 technical communication programs encouraging the founding and first meeting of the CPTSC, Pearsall noted, among other critical questions, that the profession needed to examine the issue, "How much science and technology does a technical communicator need and what kind?" (Pearsall & Warren, 1996, p. 140). In the first meeting of the CPTSC in the spring of 1974, there was, according to Pearsall and Warren, general agreement at the undergraduate level that "All the programs require 25–40 percent of students' programs to be in science and technology. In most cases only about 30 percent of the program is taken up with the communication core" (p. 141).

From 1974 to 1984, Pearsall and Warren note, this ratio remained consistent in the development of new course and program content: "Although the precise mix and content of courses depended on the environment of the home school, there was general agreement, at the bachelor's level, on the three-way split among communication courses, general electives, and technical electives" (1996, p. 143). In connection with the development of new course and program content during this decade, program directors were encouraged to seek help from corporate advisory boards, though at times members were "warned against the tendency of some advisory committees to ask for too vocational an approach" to the study and practice of technical communication in technological contexts (p. 144).

The interest in the curricular integration of technical skills and experiences continued from 1985 to 1995, and perhaps even expanded at the annual meeting of the CPTSC, which is somewhat predictable given the advances in this same time period associated with personal computers, local area networks, wide area networks, the Internet, and a wide range of electronic publishing and production applications. A glance at the last five years of available CPTSC proceedings (1991–1995) reveals papers on how to keep up with computer technologies, design capstone courses in the theory and practice of electronic texts, resist technological inertia in technical communication programs, avoid computers taking over the classroom, teach courses in the design of information systems for multinational companies, and develop Internet-based approaches to the service course, among many others. In 1995, the meeting's keynote address, by Cynthia Selfe, focused on the values associated with the images of computer technologies in both professional and popular discourse. In addition, the 1996 program of the CPTSC meeting listed six papers focusing on computing, though many more included computing as one programmatic issue among many others.

Although, through the CPTSC, we can trace over two decades of general interest in the curricular integration of technical skills and experiences in technical communication programs, in recent years we have taken a more specific look at our pedagogical and programmatic practices as a profession in the area of computing. This interest has been encouraged by a wide range of factors, but from a programmatic perspective, and a practical level, we realize that nowadays our students often end up working in computer-related industries: According to the Society for Technical Communication's 1996 Technical Communicator Salary Survey, over half of STC's 20,000 members work for hardware and software companies. In addition, however, as George Hayhoe, the current editor of *Technical Communication*, argues in a recent editorial, "Far from being the great divider in our profession, computers are increasingly our common ground. No, we don't all write about them, but virtually all of us write on them nowadays. And increasingly, we use them to conduct much of our everyday business and to publish the text, art, and other media we produce" (1996, p. 327).

Selber's 1994 survey of technical communication programs provides one attempt to understand how the profession is preparing students for the kind of work Hayhoe describes, and, in connection with this preparation, what kinds of factors encourage our constructions and uses of computers in both pedagogical and programmatic terms. Selber's project consisted of two main questions: How are we incorporating computers in technical communication curricula? And why are we incorporating computers in technical communication curricula? To explore the first question, Selber collected over 100 computer-related course descriptions from 39 technical communication teachers working in various programs. To explore the second question, Selber collected, from these same teachers, official and unofficial rationales for using computers in technical communication classes and programs.

In looking at how we incorporate computers in technical communication curricula, Selber provides a classification based on an examination of the primary instructional goals of the courses he surveyed. In looking at these goals, most of the courses broadly fell within the following three categories: production, computer literacy, and situating technology in historical, political, and social contexts. As with Berlin's taxonometric overview (offered later in the chapter), Selber's classification is not meant to be exclusive or monolithic. Rather, it "represent[s] a broad sketch" (1994, p. 369), one that identifies technical communication programmatic tendencies as those that are connected with the development and use of computer technologies. In line with this view, the classification recognizes that "many of the courses located within any one section may include discussion about issues pertinent to the others. In addition, many institutions offer several courses in more than one category, and/or single courses that discuss the entire complex range of issues surrounding computer use" (p. 369).

The first category identified by Selber includes production courses, which, in general, "introduce students to the skills and processes involved in using computers to support [the] day-to-day work" of technical communication (1994, p. 370). In terms of computing, this primary instructional goal is common in a wide range of courses at the undergraduate level and, to a lesser extent, at the graduate level, including those in computer applications, document and graphic design, publications management, specialized writing courses such as computer documentation and World Wide Web design, and introductory writing courses in technical communication. Courses in this last category recognize that in order for students to be successful communicators in complex technological contexts, they need a hands-on understanding, on some level, of the workings of computer hardware and software. This claim is difficult to disagree with. Consider the relatively straightforward task of developing a home page for the World Wide Web, a task increasingly common to technical communication in workplace settings (Silker & Gurak, 1996). In December's (1996) methodology for creating World Wide Web pages, all six development areas—planning, analysis, design, implementation, promotion, and innovation—assume a good deal of technical expertise, including HTML, CGI, and hypertext skills. Although production courses cover many different issues, including topics such as World Wide Web design, according to Selber "their primary instructional goal in terms of computing is often skill building: they rarely address literacy or humanistic concerns in a substantive way" (p. 369).

The second category in Selber's survey focuses on computer literacy and courses that attempt to "broaden students' understanding of computers as they relate to theories of reading, writing, and textuality" (1994, p. 370). Courses in this category, which often focus on information design, visual communication, and technical communication for high-tech environments, move beyond basic and advanced computer skills. In this way, computer literacy courses have more of a rhetorical focus than production courses, asking technical communication students to consider not just audience and purpose issues, which production courses often do, but also how readers make meaning in online information space. This view of computer literacy moves beyond the learning of functional skills and into the realm of interpretation. In these courses, technical communicators are concerned, to a large extent, with the kinds of issues Haas (1996) outlines related to the role of technology in the cognition of literacy: the effects of computer technologies on the interpretive acts associated with writing and reading. In returning to the World Wide Web example, a computer literacy course in this area would assume the technical skills outlined earlier in the production course, but would devote its focus to heuristics, principles, and procedures for writing and reading in electronic spaces, including issues of organization, link structure, navigation, perception, usability, and aesthetics.

Selber's third category of computer-related courses distinguished by primary instructional goals is the situating of technology in historical, political, and social contexts. Courses in this area, according to Selber, "help students realize that implementing and using computer technologies are fundamentally ideological acts—acts that have broad and varied implications for those engaged in their use as well as for our society in general" (1994, p. 376). This perspective—most often found at the graduate level and in courses on technology and discourse, technology assessment, technology as a social and political phenomenon, and alternative technologies—aligns in close ways with Knoblauch's view of critical literacy, as well as with Feenberg's critical theory of technology discussed later in the chapter. Seeing the development and use of computers in historical, political, and social terms requires that students and teachers "identify reading and writing abilities with a critical consciousness of the social conditions in which people find themselves" (Knoblauch, 1990, p. 79), and the complex ways in which computers, as artifacts of an industrial culture, are connected to these social conditions and discursive exercises of power. Computer technologies, from this view, do not simply and unproblematically aid the work of technical communication. Rather, as with other literacy technologies, their constructions and uses are influenced by a wide range of nontechnological factors and ideological interests. Returning to the World Wide Web example one final time, a course in situating technology in historical, political, and social contexts in this area might, for example, consider the human problems associated with introducing a web site in an organization; the policy issues associated with creating an intranet, including issues of access, privacy, and power; the changes in communication patterns encouraged by a WWW-based communication system; and the representations of people, processes, and products instantiated in that system.

As one can imagine, the number and type of courses available in and across these three categories in any one technical communication program depends, at least in part, on a wide range of pedagogical, institutional, and geographical factors—among them program level and orientation, faculty interest, and local industry. But in general terms, when asked why they incorporate computers in technical communication curricula, the 39 faculty responding to Selber's survey provided five key reasons: "to increase students' marketability in business and industry, build skills important to technical communication activities, improve the quality of students' writing, promote collaboration, and provide faculty with an opportunity to research issues related to computers and technical communication" (1994, p. 378).

These five reasons provided by technical communication teachers for using computers in the classroom help illuminate some of the tendential forces influencing our constructions and uses of technology—among them:

- functionalist perspectives in science and engineering contexts, and, in turn, technical communication programs;
- narratives of progress equating advances in computing with better writing and communication;
- social perspectives on technical communication encouraging an increase in attention paid to collaborative activities and social and political arrangements;
- and movements in the discipline, aided by an information-based economy, to expand the roles of technical communicators.

Although this list is limited, it helps account, at least on some level, for the complex and contradictory forces that help mold computer technologies along particular axes of ideological interest in technical communication programs.

The first two reasons offered for using computers in technical communication classrooms are straightforward, connected, and help account for the large number of production courses often offered in technical communication programs at both the undergraduate and graduate levels. The alignment of the discipline with science and engineering, particularly in terms of student employment, encourages the teaching of computer skills that support the work of technical communication in those disciplines and contexts. What is at issue here for teachers and students is not the teaching of advanced word processing, desktop publishing, or HTML coding, nor the supporting of technological work, but a functionalist perspective often found in science and engineering contexts and, in turn, technical communication programs, that can encourage the learning of computer skills in isolation of their rhetorical and social implications. The concern of such a perspective, as Knoblauch explains, "is the efficient transmission of useful messages in a value-neutral medium" (1990, p. 76). The logic of such a pragmatic view, while appealing on a variety of levels, is far too one-dimensional for the responsible preparation of students for work in a complex technological society. As a number of edited collections in the area of technical communication are beginning to explain in pedagogical and professional contexts (Duin & Hansen, 1996; Selber, 1998; Sullivan & Dautermann, 1996), and as we argue elsewhere in this chapter, technologies of writing are far from "value-neutral" in both cognitive and social terms. And yet, far too often functionalist perspectives remain a strong force encouraging skills-based approaches to the teaching of technical communication in electronic environments.

Another reason identified by teachers for using computers in technical communication classrooms is to "improve" student writing. This reason is consistent with survey work done by Hawisher and Selfe, who reported that, in the late 1980s, many teachers preferred teaching writing with computers based on the following claims: with the use of hardware and software, students spent more time working on their writing; peer teaching was common; classes became more student-centered; one-on-one conferences between teachers and students increased; opportunities for collaboration increased; students shared more with other students and teachers; and communication features provided more direct access to students, thus allowing teachers to get to know them better (1991, p. 59). The concern here is not that computers cannot be associated with productive change in the teaching of writing and communication in technological contexts—in fact, we think they can. Rather, what is problematic are the causal connections sometimes constructed between advances in computer technologies and improvements in writing. Postman (1995) explores such connections in schooling contexts, claiming that "nowhere do you find more enthusiasm for the god of Technology than among educators" (p. 38). The "god of Technology" Postman refers to here is a set of culturally determined narratives of progress, encouraging technical communication teachers to equate improvements in student performance with increased network speed and improved writing and editing environments. Winner (1986) summarizes the logic of these narratives in general terms: "the almost religious conviction that a widespread adoption of computers and communications systems along with easy access to electronic information will automatically produce a better world for human living" (p. 105). Missing from such a view, however, are some of the very real factors that help determine the success, or failure, of students in classroom and professional settings, including teachers, classmates, class sizes, support systems, pedagogical orientations, and personal histories, just to name a few. But although a belief in the god of Technology is problematic, according to Staudenmaier (1995), nowhere can we find such a "master narrative so deeply entrenched in popular imagination and popular language as in the mythic idea of progress, particularly in technological progress" (pp. 262–263).

One of the most popular reasons reported by teachers for using computers in technical communication classrooms is to promote collaboration. In recent years, collaboration has become an important topic in the field for a number of reasons, but one important reason is a move to social perspectives on writing and reading in nonacademic contexts. Faigley (1985) was one of the first to encourage a social perspective on technical communication, and since then others have articulated the explanatory power of this view for both professional and educational contexts. Thralls and Blyler (1993a), for instance, map out three different approaches to a social perspective—social construction, ideologic, and paralogic hermeneutic—discussing how issues of community, knowledge and consensus, discourse conventions, and collaboration are viewed in each approach. In terms of social construction and the ideologic approach, two popular perspectives in the field, Thralls and Blyler provide the following summary:

> The social constructionist approach focuses on community, viewing communal entities as the sources of knowledge maintained by consensual agreement; as the repositories of discourse conventions by which communities are defined and shaped; and as the bodies to which nonmembers must—through collaboration—be acculturated. The ideological approach focuses on political issues downplayed in constructionists' ways of conceptualizing community, knowledge and consensus, discourse conventions, and collaboration. (pp. 13–14)

In a variety of ways, technical communication teachers have used computers to both challenge and support these social views. For example, recent work suggests that computer technologies can be used to challenge traditional notions of authorship (Bolter, 1991; Johnson-Eilola, 1994; Landow, 1992), promote constructive dissensus in collaborative teams (Selfe, 1992), and flatten organizational hierarchies that impede information flow and productive group work (Schrage, 1990). Moreover, as Selfe (1996) identifies, computer technologies have also been connected to a wider range of social and collaborative claims, including that computers can, under the right conditions, support democratic, pluralistic, and egalitarian communication; support open discussion, research, and education about civic issues; allow more people to access information and to communicate issues and topics; support rich possibilities for collaborative communication and discursively based collective action; and eliminate or modify the nonverbal paralinguistic cues which contribute to the differential exercise of power in face-to-face communication (pp. 257–258).

Providing faculty with an opportunity to research issues surrounding computers and technical communication is a final rationale provided by teachers for using computers in the classroom. This rationale includes the widest range of issues, but one issue that is congruent with recent work in the field is the use of computers to expand our view of—and assumptions about—technical communication. The domain of the discipline, as it was at one time, is no longer primarily concerned with the traditional tasks associated with writing, editing, and managing print-based publications, though these tasks remain important in some areas. Rather, as Johnson-Eilola (1996) argues, in an information versus industrial economy, the field is slowly beginning to rearticulate its value away from service and support roles toward more meaningful and central work, and technology, if understood and used in rhetorically sophisticated ways, can help the field accomplish this rearticulation. Along these same lines, Shirk (1997) describes some role expansion trends in the workplace encouraged by computer technologies, arguing that technical communicators, with a rhetorical perspective, are well positioned to assume the roles of interface design experts, product designers, information architects, customer trainers, process facilitators, media selection consultants, interpersonal communication advisors, and usability specialists.

PERSPECTIVES ON TECHNICAL COMMUNICATION AND TECHNOLOGY INFORMED BY RHETORICAL THEORY

In American colleges and universities of the last century, approaches to the teaching and practice of technical communication—and, to a great extent, our thinking about and use of technology within such programs—have been linked consistently at both philosophical and practical levels to the more general study and practice of rhetoric and composition in humanistic disciplines (Adams, 1993; Dombrowski, 1994; Kynell, 1996; Miller, 1979; and Ornatowski, 1992).

This close relationship results from a number of factors. First, at a practical level, the majority of technical communication teachers are now educated primarily as English composition teachers—within composition studies programs influenced by the historical and conceptual frameworks of rhetorical theory and history. Similarly, the directors of writing programs who often help shape the curriculum of technical communication programs, the editors of professional journals focused on technical communication, the reviewers of these journals, and the authors who write for these journals are all educated within, or in close proximity to, rhetoric and composition programs, and, thus, influenced by the current thinking of, and the cultural and intellectual capital within, these closely related fields.

At another, less immediate—but nonetheless important—level, the thinking of technical communication teachers and composition specialists converge because they draw from a common, albeit, complex set of prevailing theories about language and language practice that are part of the intellectual zeitgeist of a particular time. This thinking grows out of, and continually helps fashion, a common set of social formations that exert tendential force on the directions, and practices, of both technical communication and composition/rhetoric studies.

Acknowledging that such a linkage exists suggests the importance of examining some of the salient rhetorical theories that have come to characterize the teaching of writing in the last century—and that have also proved influential during that period in technical communication classrooms. Such an examination may also illuminate the ways in which technical communication, and the uses of technology within such programs, have been both shaped and limited.

In undertaking this examination, the work of James Berlin (1987) is useful. Berlin, in his classic work, *Rhetoric and Reality: Writing Instruction in American Colleges 1900–1985*, outlines three major conceptual approaches to rhetorical theory in the past centuryCthe objective, the subjective, and the transactionalCeach of which, he notes, has had distinct influences on the teaching, practice, and understanding of writing in American colleges. We contend that these conceptual approaches have also been highly influential in determining the direction of instruction in technical communication, and, importantly, the ways in which technical communication teachers have come to understand and use technology in their classrooms, and think about technology within workplace settings.

We offer a caution, as does Berlin (1987), that the taxonometric overview offered by these conceptual categories is useful not because the approaches it describes are exclusive, exhaustive, or monolithic, but rather because they make a discussion of theoretical approaches to the teaching of written composition—which are always multiply and contradictorily defined—manageable and generative.

Berlin's first category of approaches to teaching written communication, the objective theories, are, in his words, "based in positivistic epistemology" and assert that the "real is located in the material world," apart from individuals and their ability to perceive it, and that reality is "empirically verifiable." This set of theoretical approaches, closely aligned with the project of science, views language as a "sign system, a simple transcribing device for recording that exists apart from the verbal"; truth, in other words, precedes language and "[t]he business of the writer is to record this reality exactly as it has been experienced so that it can be reproduced in the reader" (Berlin, 1987, p. 7).

Chief among the approaches to writing instruction influenced by this theoretical strand, Berlin adds, is "current-traditional rhetoric" (1987, p. 7), a conceptual framework growing out of Scottish Common Sense Realism, which actively influenced the American mindset during the 19th century. According to this conceptual framework, with its roots in Aristotelian thought, truth is determined through observation, and "the responsibility of the observer [writer]" is to use language to describe reality in a way that eliminates "distortion" (p. 8). "Language must thus be precise," Berlin recounts, in order to "demonstrate the individual's qualifications as a reputable observer" (p. 9).

The connections between objective theories of rhetoric and our profession's approach to the understanding and teaching of technical communication are relatively direct and unambiguous, and they need little elucidation here, as they have been detailed quite thoroughly—in historical, philosophical, and practical terms—elsewhere (Dobrin, 1983; Driskill, 1989; Kynell, 1996; Miller, 1979). Both grow directly out of the related projects of Science and Positivism; both stress language that should be, above all, clear, free of error, accurate, and transparent in meaning; both subscribe to Truth as a verifiable entity, separate from any distortion introduced by human beings. Such connections, for example, inform—to varying degrees—approaches to technical communication that stress the importance of technical writers' accurate observations of mechanisms and phenomena; the completeness of technical writers' recording of complex processes; the clarity, simplicity, and directness of technical writers' language.

At another level, these connections also inform empirical studies on the effectiveness of various organizational and stylistic features of technical documents (Faigley & Witte, 1983; Teklinksi, 1992), examinations of readability (Huckin, 1983) and calculations of readability formulae (Selzer, 1983); and studies of document cycling within corporations (Dorff & Duin, 1989)

Importantly, these approaches to technical communication are also linked closely to technology and technology use: both grow out of the project of Science, and both depend on objective approaches to representing reality. If computers are a tool that science can use to identify the Truth of the natural world—by calculating accurate distances between stars, solar systems, and galaxies; by modeling the precise genetic workings of DNA and RNA; and by tracing the unambiguous molecular combinations and movements of natural and artificial elements, then these machines can also be used by writers and technical communicators to craft and convey meaning more precisely and directly through language. Within such a framework, computers provide writers with a tool for recording their most accurate observations of complex and multipart technical processes, for identifying complete instruction sets, for documenting the workings of complex mechanisms in objective and understandable terms, and for identifying precisely the risk factors associated with various hazards and materials.

According to this perspective, word processing packages offer writers the means of revising language until it approaches crystal clarity and becomes unambiguous for readers; page layout software allows for the most accurate rendition of information; graphics packages provide the opportunity to convey specific data in a variety of representational forms; and the World Wide Web provides an environment in which all of these modes of expression can be used to communicate, thoroughly and accurately, information about the human genomes, the Hubbell telescope data, and the Visible Man project. Such objective frameworks also, at least to some extent, motivate those technical communication scholars and educators who propose modeling the use of computers in educational settings directly on the needs of practitioners in business and industry (Allen, 1996; Johnson, 1996; Penrose, Bowman, & Flatley, 1987), and those who propose using computers to improve grammar and punctuation performance for writers (Smith, 1986).

Berlin's second category, subjective approaches to rhetorical theory, is most closely aligned with expressionism in the teaching of composition. Subjective theories of rhetoric, Berlin (1987) notes, "locate truth either within the individual or within a realm that is accessible only through the individual's internal apprehension" (p. 11). However, if the most effective expression of truth is located only in the individual's unconscious mind, it is not easily expressed by writers—rather, it takes a great deal of effort, usually multiple attempts to seek the most effective way of expressing concepts, and multiple rounds of identifying false or ineffective ways of communicating about specific topics. As Berlin adds, subjective approaches to writing rest on the assumption that "[t]he business of rhetoric...is to correct error as one speaker engages in a dialogue with another, each sharing a dialectical interchange in which mistaken notions are exposed" (p. 12).

In the teaching and practice of technical communication, this theoretical strand results in such common pedagogical approaches as peer-editorial groups, designed to help writers hone their individual perceptions of truth into more effective representations by engaging in discussions with other writers. Within such groups, technical communication students learn to rely on dialectical exchanges with other members of their peer-writing group—in editing and revision sessions—to help them discover the problematic expression of ideas due to such distortions as "bureaucratic voice, clinched language, and the like" (Berlin, 1987, p. 14). In related ways, subjective theories of language are also linked with user-testing as a process by which technical communicators gauge their own internally derived representations of a mechanism, an instruction set, an interface, or a procedural approach against those of users engaged in a specific task.

Additionally, in technical communication classrooms, subjective theories of rhetoric support students' use of original and effective metaphorical language, which allows them to represent their rich unconscious and tacit understanding of the world in the most effective forms—using processes of "condensation and displacement" (Berlin, 1987, p. 14). Thus, technical communication students are encouraged to help readers connect highly technical and unfamiliar processes with known and familiar processes (e.g., the unfamiliar operation of an optical scanning device with the familiar operation of human eyesight, or a simple camera; the unfamiliar context of a computer's hard disk storage system with the more familiar context of paper-based documents, files, and filing cabinets).

Subjectively-based conceptions of language also influence the uses, and understanding, of technology in technical communication programs. From such a framework, technology is seen not simply as a set of tools for the precise expression of ideas and the communication of these ideas through language, but, rather, as a fertile environment within which individual writers can explore, refine, and express their own internally derived representations of phenomena, processes, policies, or mechanisms. From this perspective, for example, computer-supported writing environments such as hypertext can be understood as writing spaces that allow individuals to construct increasingly robust representations of the world (Bolter, 1991; Johnson-Eilola, 1997a; Snyder, 1996) in terms that approximate or mirror their own cognitive structures or visions. Similarly, computer networks can provide environments within which individuals can contact and use the feedback from peer-editing groups to help them develop and test their own internally persuasive representations of the world (Cooper & Selfe, 1990). Finally, the subjective framework for understanding language and the world justifies the frequent use of metaphorical language (Johnson-Eilola, 1995; Selber, 1995; Shirk, 1991b) to explain technology itself.

The third rhetorical framework Berlin (1987) identifies involves transactional theories. In his words, "transactional rhetoric is based on an epistemology that sees truth as arising out of the interaction of the elements of the rhetorical situation: an interaction of...subject, object, audience, and language" (p. 15). Within such a framework, knowledge of the world does not precede language, rather, truths about the world are multiple because they are constructed, variously, within the full range of human language experiences. Hence, meaning is constructed through active debate among individuals and groups from different cultures, exchanges between teachers and students who hold different perspectives, documents shared by clients and technical communicators who may operate in disparate environments, and texts constructed by writers and readers with multiple or contradictory purposes—all elements functioning to simultaneously construct and be constructed by language.

Transactional theories of language have, in part, provided the impetus for an increasingly complex understanding of both technical communication and the responsibilities of technical communicators. Such a framework, for example, supports a vision of technical communicators as professionals who are not only influenced by social situations, but who function actively, and with purpose, as social agents themselves—participating in the creation of meaning within a full range of civic arenas that depend on, and are constructed by, discursive exchanges. In comparison with objective or subjective frameworks for understanding language, communication, and the creation of meaning, the transactional framework attempts to take into account many of the social and political complexities that have been identified as factors bearing on the work of technical communicators (Miller, 1979; Stotsky, 1996; Winsor, 1996).

Operating from within a transactional framework, for example, teachers of technical communication have begun to explore how their own educational responsibilities relate to the expanded program of democratic and social responsibility identified by critical pedagogy, liberatory pedagogy, rhetorical theory, composition studies, and cultural studies (Miller, 1991; Sullivan, 1990; Waddell, 1990). As a result, many technical communication programs now include the study of ethics, responsible civic decision making, social policy formation, rhetorical theory, or risk communication. Programs that include such studies are designed to explore the direct connections between the professional practice of technical communication and the social/political decision making in such cases as the Challenger disaster (Moore, 1992; Winsor, 1988) and recombinant DNA research (Waddell, 1990). Additional transactional considerations have been raised as scholars and teachers of technical communication explore the construction of various kinds of knowledge within, and among, collaborative writing groups (Battalio, 1993; Gerson, 1993; Spilka, 1993); the making of meaning—and the recognition of truth—within various scientific communities (Bazerman, 1988; Berkenkotter & Huckin, 1995; Swales, 1990); the social construction and recognition of authority and responsibility in corporate documents associated with large-scale disasters (Sauer, 1994); and the socially and culturally defined practices affecting communication in specific science and technology fields (Pinch, 1994; Taylor, 1994; Thralls & Blyler, 1993a, 1993b; Winsor, 1996). Transactional rhetorical theory, similarly, has motivated the work of scholars such as Linda Flower, John Hayes, and Heidi Swarts (1983), as they examined the cognitive requirements of audiences reading functional documents, and the work of John Gribbons (1992) and Stephen Bernhardt (1986) on cueing and page formatting techniques that increase the readability of documents for various audiences.

Transactional rhetorical theories also clearly inform the understanding and use of computer technology within technical communication programs. Such theories have, for instance, motivated a range of investigations on e-mail lists as professional environments for the social construction of technical knowledge (Selfe & Selfe, 1996); the use of computers as support systems for the collaborative writing of documents and small-group decision making (Arms, 1984; Duin, 1990; Selber, McGavin, Klein, & Johnson-Eilola, 1996); gender-based approaches to computer use and understanding (Lay, 1996; Turkle & Papert, 1990); and the advisability of technical communication curricula that focus on the social, economic, and political issues surrounding technology, as well as on the practical issues associated with training students to use such technology in completing their writing assignments (Selber, 1994).

EXPANDING AND CONTRACTING PERSPECTIVES FOR UNDERSTANDING TECHNICAL COMMUNICATION AND TECHNOLOGY

The expanding and contracting model of information design and use (Johnson-Eilola & Selber, 1996) can also offer a generative framework for thinking about the teaching and practice of technical communication, especially in connection with technology. At one level, Johnson-Eilola and Selber use the paired terms to refer to a broad range of possible characteristics for individual pieces of communication. A text characterized as contractive at its most extreme, they note, "cuts off discussion and reflective thinking—the text offers, perhaps instantly, one, and only one 'correct' chunk of information" (pp. 125–126). A text characterized as expansive at its extreme, the authors add, "offers no unqualified answers and a huge number of navigational choices (but few hard and fast rules for users to distinguish which choice to make)" (p. 126).

It is important to note, however, that Johnson-Eilola and Selber consider the terms "expansive" and "contractive" as the poles of a wide range of "social and technological possibilities" (1996, p. 125) within which texts are designed and used. Hence, the model describes not only the characteristics of individual texts, but also "how those characteristics are taken up, channeled, defined, and defied by people." As a result, the model is extended in a way that "allows us to get at not merely the functional characteristics of an isolated technology, but also at the social and political contexts in which technologies are developed, used, maintained, and reconstituted" (p. 124). Hence, texts of all kinds—both those tending toward the contractive and the expansive ends of the continuum—can be used within the context of contractive communication situations, those characterized by a corporate value on hierarchy, efficiency, and the goal of "frictionless, noise-free" (p. 125) exchanges between senders and receivers, as suggested by Shannon and Weaver's classic model of communication. In contrast, all kinds of documents—both those with expansive and contractive tendencies—can be used within the context of expansive communicative situations, those which value writing and reading as complicated activities that involve thinking, constructing and deconstructing meaning and recognizing the contradictions inherent in texts.

Given this intellectual groundwork, mapping some of the major landmarks of the "expansive and contractive" model onto the practice of technical communication is relatively straightforward. The contractive end of the model, Johnson-Eilola and Selber (1996) point out, provides an impetus for those who identify technical communication as clear, precise, error-free language that accomplishes the efficient exchange of information. Such a conception, Johnson-Eilola and Selber point out, is frequently associated with those corporate settings that are overly rigid and that value hierarchy and streamlined processes, environments that, at their extreme, reduce communication to a "mere function, a component necessary...but not of great importance (except when it fails)." From this perspective, the authors further note, "engineers and scientists...have the most power, technical communicators, as well as readers and users, are somewhat like acolytes" (p. 132).

This perspective on information design for contractive purposes—with its emphasis on clarity and correctness, on the efficiency of transmitting information, and on hierarchy and streamlining—is situated most clearly in earlier conceptions of technical communication as the clear and effective transmittal of objective information. This historically based conception, as Kynell (1996) points out, grew out of the utilitarian values that informed the engineering professions and the military-industrial complex after the American Civil War. However, the goals of contractive information design also continue to be influential within the current contexts of increasing corporate competitiveness, downsizing, and accountability.

These "deeply sedimented" utilitarian values, as Anthony Giddens (1984, p. 22) might refer to them, have several interesting effects. Such values provide the foundation for a common-sense understanding of technical communication that can serve to mask alternative ways of understanding and creating meaning (Dobrin, 1983). This naturalized set of assumptions, for instance, is represented in most technical communication textbooks which stress as key attributes of technical writing the characteristics of clarity, accuracy, accessibility, efficiency, and correctness (cf., Lannon, 1997; Lay, et al., 1995; Markel, 1996) and it encourages technical communicators to adopt the goal of producing efficient, clear, and effective memos, graphics, and documents (Cooper, 1996)—habits that they carry into the workplace upon graduation and continually reproduce in their own writings and exchanges with other employees. Utilitarian values also encourage technical communication scholars to seek the most effective strategies for accomplishing efficient and effective communication among writers and readers. Thus, technical communication scholars focus on readability formulae and derive strategies for increasing readability (Huckin, 1983; Selzer, 1983), describe how to create and use scenarios within functional documents (e.g., loan applications, insurance claims, drivers' tests) to increase readers' comprehension (Flower, Hayes, & Swarts, 1983); employ syntactically-based principles of topical focus as a way of increasing the efficiency of transmitting information to readers (Faigley & Witte, 1983); and create computer-based communication formats that attempt to "deemphasize prose and hence reading time...as a way of compensating for poor employee writing" (Tebeaux, 1996, p. 43) with the goal of increasing the speed of work flow within an organization (Dzujna, 1991).

In contrast, an expansive conception of technical communication recognizes that discursive acts have the "potential to make qualitative changes within a corporation, down to the level of [an] individual worker" (Johnson-Eilola & Selber, 1996, p. 134). Expansive environments for communication, Johnson-Eilola and Selber note, are characterized by some of the same elements that Kanter (1989, p. 88) attributes to "postentrepreneurial" corporations, including "a greater number and variety of channels" for workers to "take action and exert influence," a shift in power relations from "vertical to horizontal," and "diminished" distinctions between "managers and those managed." In such environments, technical communicators recognize the potential for agency and change inherent in language and see their work as an important form of meaning creation within the organization.

Expansive conceptions of technical communication are linked closely with related social theories of language use—Marxist, social constructivist, feminist, epistemic—and draw clearly on the work of scholars in a range of disciplines in anthropology, social sciences, rhetorical theory, composition theory, and sociolinguistics (cf., Duin & Hansen, 1996; Kynell, 1996; Sullivan & Dautermann, 1996). Expansively-based notions of technical communication, as Johnson-Eilola and Selber (1996) describe them, stress the need for change within corporate organizational contexts—the ability of technical communicators to act as social agents who can affect change—and the reciprocal effects of discourse shaping social situations and social situations influencing discursive practices.

Expansive themes focus on the discursive practices of social agents such as technical communicators and posit, moreover, that such agents can and do make qualitative differences within corporate contexts. Such assumptions have broadly influenced technical communication scholarship—shaping scholarship and practitioners' reports that, for example, describe the need for linguistic diversity and multicultural communication in workplace writing and in modern corporate contexts; explore the active role that individual's prior language experiences, theories, and knowledge structures can play in shaping the form and content of technical data and reports, and the nature of problem solving within corporate settings (Dombrowski, 1994; Gerson & Gerson, 1994); and indicate the need for "critical interpretivist research" (Blyler, 1995, p. 306) in the field of technical communication, research which contributes to increased levels of participatory involvement within corporations and that has as a primary goal the aim of involving increased numbers of workers within an organization (including technical communicators) in active decision making at all levels (pp. 305–306).

The models of expansive and contractive communication contexts can also prove generative in thinking about technology use. As Johnson-Eilola and Selber (1996) explain, technological applications—and, as a specific case in point, hypertext—supports both contractive and expansive communication environments. For instance, an online manual designed in a rigid hierarchical fashion in an attempt to streamline a worker's access to information, and thus increase their efficiency without expanding their power to determine how to make use of information, would provide an example of a form of electronic communication that is contractive in both design and use. Similarly, a piece of online documentation that is designed to provide individuals with many choices of paths for accessing and organizing information, that offers readers and writers alternative ways of making and using meaning within a communication situation, and that allows control to remain in the hands of users could be identified as expansive in both design and use. These descriptions, of course, are simplistic in that they represent examples of communication design and use that tend toward the extreme ends of the expansive/contractive continuum; most communication, of course, can be characterized by multiple and contradictory combinations of design and use.

Importantly, as Johnson-Eilola and Selber (1996) point out, the power of language associated with such technology applications in technical communication environments shape social realities, as well as reflect them. Contractive hypertext design and use, for example, may reflect, support, and reproduce a value on the efficient transmittal and use of information within a corporate setting. Expansive hypertext design and use in a corporate, technical communication setting, in contrast, may not only place "more value on the roles of writer and reader [because both are encouraged to make authoring decisions in connection with these texts], but also might help drive shifts in the corporate structure at large that also increase the perceived value of new, expanding methods of communication, not merely in documentation but in all phases of corporate life" (p. 135).

INSTRUMENTAL, SUBSTANTIVE, AND CRITICAL THEORIES OF TECHNOLOGY

Technology critic Andrew Feenberg (1991) sketches the broad outlines of a "critical theory of technology" designed to overcome some of the decontextualizing and dehumanizing aspects of technology development and use. Working from a Marxist perspective, Feenberg begins by critiquing two contemporary methods of thinking about technology: the instrumental and the substantive.

In the instrumental view, technology is considered a simple instrument for achieving predetermined ends. For example, a pencil might be used to make marks on paper; paper and pencil both are tools (in the instrumental view) for communicating. Such systems nest together in complicated but ultimately determinable ways to attain an end-goal (Horton, 1990; Price & Korman, 1993; Woolever & Loeb, 1994).

In technical communication theory and practice, instrumental views of technology frequently surface in relation to linear models of communication that prioritize the transmission of information from sender to receiver. This model, drawn from the 1940s-era telephony of Bell Laboratories engineer Claude Shannon, posits the process of communication in terms of packets of information being transmitted from a sender to a receiver down a channel.

The history of technical communication makes the transmission model particularly attractive (Slack, Miller & Doak, 1993). The engineering and scientific contexts in which many technical communication programs and departments developed valued the apparent mathematical precision of the transmission model (Connors, 1982). Readability formulas, as we have already mentioned, rely on the idea that meaning inheres in text and can be straightforwardly calculated. In addition, this model prioritized the content being transmitted (technical information) and the end-goal (efficient technological operation) over creative action (or "indeterminacy").

In such contexts, language and communication are considered primarily as tools; at best, they are transparent and effortless so that the end-product can be achieved more quickly and efficiently (Horton, 1990).

Completing the binary pair is the substantive perspective. Substantive views of technology invert the "tool" metaphor and insist that technologies are all powerful (and, in a sense, that people have become tools of technological systems). Although certainly less popular than the instrumental view, popular proponents include Luddites, neo-Luddites, and similar antitechnological movements. As one might expect, it is difficult to find substantive models of technical communication because it is difficult to imagine a technical document that was, itself, antitechnological (although some particularly badly designed documents might seem to qualify as substantive attempts to subvert technology use). In important senses, however, some of the more optimistic perspectives on technology use (for example, Negroponte, 1995) propose utopian substantive theories by insisting that technology use will bring with it automatic and positive benefits, apart from the concrete intentions of developers and users.

In critiquing these two common views, Feenberg (1991) develops a critical theory of technology that mediates between each in productive ways. According to Feenberg, the capitalist mode of production has stalled technological development and use at the moment of primary instrumentalization. The key characteristics of this stalled moment revolve around different aspects of industrialization, especially those that decontextualize workers and users from their contexts. In the primary moment of instrumentalization, Feenberg identifies aspects of decontextualization (taking an object out of context), reduction (stripping away secondary aspects of the object so it can be more easily used), autonomization (breaking systems of feedback so an operator is not affected by changes in the object), and positioning (the occupation of a strategic position so that value can be drawn from an object).

Although each of the four terms describes complex and rich methods for considering technologies, there is not enough space here to fully describe each, so we will limit ourselves to a brief description and example of the first aspect of primary instrumentalization, decontextualization. This aspect is easily seen in much contemporary technical communication, especially computer documentation. Producers of documentation construct fragmented, isolated cases and situations that are then taken by users. But as any documentation user knows, the need for documentation frequently arises out of unique, problematic situations that are not adequately addressed by decontextualized instructions. James Paradis (1991), in an analysis of deaths and injuries related to the use of an industrial stud-gun fastening system, found that instructions for these high-powered construction tools oversimplified cases with grave consequences. The coding system used in the documentation to distinguish different powered charges used to drive studs into materials relied on colors—which to most users are difficult to rank—rather than foot-pounds of force or some other contextualized (but more complex) system. Operators of the stud guns had a difficult time telling how much more powerful one charge was than another. As a result, in at least one case, an operator chose a charge much more powerful than necessary, driving a fastener completely through a wall like a bullet and killing someone in an adjacent room (see also research in Cooper, 1996; Johnson-Eilola, 1996; Mirel, 1996).

The decontextualization model has been particularly popular in the design of computer documentation. Such communications are typically constructed in an attempt to help users address an impasse (Carroll, 1990; Duffy, Palmer, & Mehlenbacher, 1992) or impediment to their current ("real") work. In the online help system for Microsoft Word 6.0, for example, information about lists discusses how to change from a bulleted to a numbered list (Figure 10.1), but not about the appropriate contextual reasons for doing so (if ordering of list items is relevant, then numbering should be considered).

Unlike the substantive view, a critical theory of technology does not reject technology for the problems of primary instrumentalization, but insists, in Marxist fashion, that technological development follows an evolutionary course in which capitalism is a necessary but flawed step. Technological development must be completed with *secondary* moments of instrumentalization that recuperate the earlier, primary moments. This moment includes four aspects: *concretization* (developing and using objects in ways that value their contexts), *aesthetic investment* (considering secondary qualities), *vocation* (learning work as a craft rather than as a simple

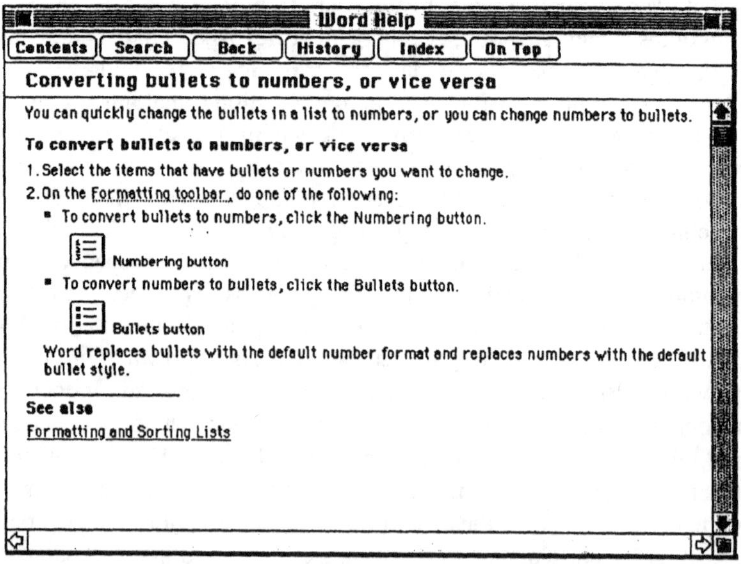

FIGURE 10.1. Decontextualized online help in Microsoft Word 6.0.

skill), and *collegiality* (working with others, including both co-workers and user/consumers, in contexts).

To return to the stud gun example discussed previously, a concretization might address the problems of decontextualization. Whereas the original (decontextualized) instructions attempted to remove context in order to make the documentation simpler, a concretized manual would explicitly address context in order to make the work safer and more effective. Instead of coding load strengths by color, the text might specify loads in variables related to materials being fastened. Even better, the text might involve workers in the process of mapping load and material vectors on a map that also illustrated the locations of other workers potentially in the line of fire. From a decontextualization standpoint, this manual appears somewhat inefficient; from a concretization standpoint, such contextualizing moves are a normal part of work.

Although moving from primary to secondary instrumentalizations can be difficult for technical communicators—because that work is still frequently cast as secondary and instrumental—computer networks may facilitate critical approaches to technology development and use. In a version of Microsoft Word online help developed under the precepts of secondary instrumentalization, users would not be offered fragmented, decontextualized, isolated assistance designed to get them back to their "real" work as soon as possible. Instead, they would enter into a fluid, collaborative, interactive learning space supporting a wide, and changing, variety of approaches. As computer-based work becomes increasingly networked, work spaces open up to broader communication possibilities (secondary instrumentalizations tend to resemble Zuboff's informed spaces). The start of such work is evident in the sometimes infrequent ability of users to move from short-term, decontexualized help to short tutorials. More ambitious versions might offer users the ability to enter into synchronous or asynchronous discussion spaces with other users of the program (Selfe, Selber, McGavin, Johnson-Eilola, & Brown, 1992) or educational spaces that deal with the context in which the users are working—users writing a technical report learning fundamental principles of technical writing or budding novelists learning about plot development and dialogue writing (Johnson-Eilola, 1997b).

Such secondary instrumentalizations resonate with Killingsworth's (1992) reconceptualization of instrumental discourse in the framework of communicative action.[1] Following Habermas's theory of communicative action, as well as Beale's work on pragmatic discourse, Killingsworth argues that instrumental discourse can function pragmatically to provide readers with a road map for action and intervention in the world. Indeed, far from being automatically repressive, instrumentalist discourse can operate as a method for articulating discourse to the world. "The dialectic could be summarized thus: the conjunction of the thesis 'subjectivity' with the antithesis 'instrumental rationality' produces the synthesis 'intersubjective or communicative rationality'" (1992, p. 177). As both Killingsworth and Feenberg (1991) point out, instrumentalism can support ambivalent, contradictory outcomes—social contexts that engender fragmenting, depersonalizing instrumentalist discourses tend to alienate and oppress users; social contexts can make explicit the connections, and possibilities, among thought, discourse, society, and action.

For technical communicators, a critical theory of technology invokes a paradox both important and dangerous. Because so many technical communication contexts currently prioritize primary instrumentalizations, introducing secondary politicizations can appear, or be labeled, as wasteful or even "political" (in the negative sense of the term). So such work must always be undertaken carefully and critically. But as Feenberg's project (1991) illustrates, secondary instrumentalizations are necessary in order to reconstruct technological development and use in ways that are both productive and humane. And technical communicators occupy a key role in this project because communication about technologies helps to construct their methods of use—from proposals to tutorials and online documentation, people rely both implicitly and explicitly on technical communication to help them understand ways to think about and use technologies.

CONCLUSION

In this chapter, we offer four different perspectives on the relationships between technical communication and computer technologies, situating these perspectives in a wide range of historical, educational, rhetorical, and theoretical contexts. The first set of perspectives considers the influence of a multitude of factors—among them, functionalist perspectives in science and engineering programs, narratives of progress, and expanded roles for writers in the workplace—on the developments and uses of computer technologies, and, in turn, how these factors help construct the pedagogical and curricular directions often seen in technical communication programs. The second set of perspectives considers the influence of three different rhetorical frameworks—objective, subjective, and transactional—on our thinking about written composition in general, and computer-based technical communication in particular. The third set of perspectives examines the ways in which expanding and contracting models of text can privilege particular ways of knowing and working, ways that can be reproduced and/or resisted in the design and use of online information spaces. And the fourth set of perspectives focuses on technology critique and how instrumental, substantive, and critical theories can encourage widely different views of the relations between technical communication, computer technologies, and human users.

These different perspectives—pedagogical, rhetorical, spatial dynamic, and critical—encourage teachers, researchers, students, and practitioners to situate the developments and uses of computer technologies environmentally, to adopt a vision that not only recognizes how the field has come to depend so heavily and so thoroughly on technology—both as a tool for technical communication instruction and study in the academy, and as a medium for technical communication practice within the workplace. These perspectives also consider the dominant cultural formations that help mold the development and use of computer technologies along particular axes of ideological interest. Such a vision is critical in moving beyond one-dimensional views that tend to locate or construct a privileged method for understanding and tracing the role of computers in technical communication, views that often fail to encourage meaningful and ethical contributions to the study and practice of communication in technological contexts.

NOTE

[1] We would point readers to Killingsworth's (1992) particularly lucid and informative article that traces the roots of instrumental discourse to Habermas, Beale, Pierce, and others, in an attempt to reconstitute its usefulness as a discourse valuable to current notions of social agency and informed citizenry.

REFERENCES

Adams, K. H. (1993). *A history of professional writing instruction in American colleges: Years of acceptence, growth, and doubt.* Dallas, TX: Southern Methodist University Press.

Allen, N. (1996). Gaining electronic literacy: Workplace simulations in the classroom. In P. Sullivan & J. Dautermann (Eds.), *Electronic literacies in the workplace: Technologies of writing* (pp. 216–237). Urbana, IL and Houghton, MI: National Council of Teachers of English and Computers and Composition Press.

Arms, V. M. (1984) The computer: An aid to collaborative writing. *The Technical Writing Teacher, 11*(3), 181–185.

Battalio, J. (1993). The formal report project as shared-document collaboration: A plan for co-authorship. *Technical Communication Quarterly, 2*(2), 147–160.

Bazerman, C. (1988). *Shaping written knowledge: The genre and activity of the experimental article in science.* Madison, WI: Wisconsin University Press.

Berkenkotter, C., & Huckin, T.N. (1995). *Genre knowledge in disciplinary communication: Cognition/culture/power.* Hillsdale, NJ: Lawrence Erlbaum.

Berlin, J. (1987). *Rhetoric and reality: Writing instruction in American colleges, 1900–1985.* Carbondale, IL: Southern Illinois University Press.

Bernhardt, S. A. (1986). Seeing the text. *College Composition and Communication, 37*, 66–78.

Blyler, N. R. (1995). Research as ideology in professional communication. *Technical Communication Quarterly, 4*(3), 285–313.

Bolter, J. D. (1991). *Writing space: The computer, hypertext, and the history of writing.* Hillsdale, NJ: Lawrence Erlbaum.

Carroll, J. (1990). *The Nurnberg funnel.* Cambridge, MA: MIT Press.

Connors, R. J. (1982). The rise of technical writing instruction in America. *Journal of Technical Writing and Communication, 12*, 329–352.

Cooper, M. M. (1996). The postmodern space of operator's manuals. *Technical Communication Quarterly, 5*(4), 385–410.

Cooper, M. M., & Selfe, C. L. (1990). Computer conferences and learning: Authority, resistance, and internally persuasive discourse. *College English, 52*(8), 847–869.

December, J. (1996). An information development methodology model for the World Wide Web. *Technical Communication, 43*(4), 369–375.

Dobrin, D. (1983). What's technical about technical writing? In P. V. Anderson, R. J. Brockmann, & C. R. Miller (Eds.), *New essays in technical and scientific communication: Research, theory, and practice* (pp. 227–250). Farmingdale, NY: Baywood.

Dombrowski, P. M. (1994). Challenger through the eyes of Feyerabend. *Journal of Technical Writing and Composition, 24*(1), 7–18.

Dombrowski, P. M. (Ed.). (1994). *Humanistic aspects of technical communication*. Amityville, NY: Baywood.

Dorff, D. L., & Duin, A. H. (1989). Applying a cognitive model to document cycling. *The Technical Writing Teacher, 16*(3), 254–259.

Driskill, L. (1989). Understanding the writing context in organizations. In M. Kogen (Ed.), *Writing in the business professions* (pp. 125–145). Urbana, IL: National Council of Teachers of English and the Association for Business Communication.

Duffy, T. M., Palmer, J. E., & Mehlenbacher, B. (1992). *Online help: Design and evaluation*. Norwood, NJ: Ablex.

Duin, A. H. (1990). Terms and tools: A theory and research-based approach to collaborative writing. *ABC Bulletin, 52*(2), 45–50.

Duin, A. H., & Hansen, C.J. (Eds.). (1996). *Nonacademic writing: Social theory and technology*. Mahwah, NJ: Lawrence Erlbaum.

Dzujna, C. C. (1991). Speeding the work flow with electronic forms. *The Office, 113*(6), 56–57.

Faigley, L. (1985). Nonacademic writing: The social perspective. In L. Odell & D. Goswami (Eds.), *Writing in nonacademic settings* (pp. 231–248). New York: Guilford Press.

Faigley, L., & Witte, S. P. (1983). Topical focus in technical writing. In P. V. Anderson, R. J. Brockmann, & C. R. Miller (Eds.), *New essays in technical and scientific communication: Research, theory, and practice* (pp. 59–68). Farmingdale, NY: Baywood.

Feenberg, A. (1991). *Critical theory of technology*. New York: Oxford University Press.

Flower, L., Hayes, J.R., & Swarts, H. (1983). Revising functional documents: The scenario principle. In P. V. Anderson, R. J. Brockmann, & C. R. Miller (Eds.), *New essays in technical and scientific communication: Research, theory, and practice* (pp. 41–58). Farmingdale, NY: Baywood.

Gerson, S. (1993). Teaching technical writing in a collaborative computer classroom. *Journal of Technical Writing and Communication, 23*(1), 23–32.

Gerson, S. J., & Gerson, S. J. (1994). Meeting corporate needs: How technical writing can prepare students for today's changing work place. *Journal of Technical Writing and Communication, 24*(2), 197–206.

Giddens, A. (1984). *The constitution of society: Outline of the theory of structuration*. Berkeley, CA: University of California Press.

Gribbons, J. (1992). Organization by design: Some implications for structuring information. *Journal of Technical Writing and Communication, 22*(1), 57–75.

Haas, C. (1996). *Writing technology: Studies on the materiality of literacy*. Mahwah, NJ: Lawrence Erlbaum.

Hawisher, G. E., & Selfe, C. L. (1991). The rhetoric of technology and the electronic writing class. *College Composition and Communication, 42*(1), 55–65.
Hayhoe, G. F. (1996). Winning the race for cyberspace. *Technical Communication, 43*(4), 327–329.
Horton, W. K. (1990). *Designing and writing online documentation: From help files to hypertext.* New York: John Wiley and Sons.
Huckin, T. N. (1983). A cognitive approach to readability. In P. V. Anderson, R. J. Brockmann, & C. R. Miller (Eds.), *New essays in technical and scientific communication: Research, theory, and practice* (pp. 90–108). Farmingdale, NY: Baywood.
Johnson, R. (1995). Romancing the hypertext: A rhetorical/historiographical view of the 'hyperphenomenon.' *Technical Communication Quarterly, 4*(1), 11–22.
Johnson, R. (1996). Tales from the crossing: Professional communication internships in the electronic workplace. In P. Sullivan & J. Dautermann (Eds.), *Electronic literacies in the workplace: Technologies of writing* (pp. 238–252). Urbana, IL and Houghton, MI: National Council of Teachers of English and Computers and Composition Press.
Johnson-Eilola, J. (1994). Reading and writing in hypertext: Vertigo and euphoria. In C. L. Selfe & S. J. Hilligoss (Eds.), *Computers and literacy: The complications of teaching and learning with technology* (pp. 195–219). New York: MLA.
Johnson-Eilola, J. (1995). Accumulation, circulation, association: Economies of information in online spaces. *IEEE Transactions on Professional Communication, 38*(4), 228–238.
Johnson-Eilola, J. (1996). Relocating the value of work: Technical communication in a post-industrial age. *Technical Communication Quarterly, 5*(3), 245–270.
Johnson-Eilola, J. (1997a). *Nostalgic angels: Rearticulating hypertext writing.* Norwood, NJ: Ablex.
Johnson-Eilola, J. (1997b). Wild technologies: Computer use and social possibility. In S. A. Selber (Ed.), *Computers and technical communication: Pedagogical and programmatic perspectives* (pp. 97–128). Greenwich, CT: Ablex.
Johnson-Eilola, J. D., & Selber, S. A. (1996) After automation. Hypertext and corporate structures. In P. Sullivan & J. Dautermann (Eds.), *Electronic literacies in the workplace: Technologies of writing* (pp. 115–141). Urbana, IL and Houghton, MI: National Council of Teachers of English and Computers and Composition Press.
Kanter, R. M. (1989, November-December). The new managerial work. *Harvard Business Review*, 85–92.
Killingsworth, M. J. (1992). Realism, human action, and instrumental discourse. *Journal of Advanced Composition, 12*(1), 171–200.
Knoblauch, C. H. (1990). Literacy and the politics of education. In A. Lunsford, H. Moglen, & J. Slevin (Eds.), *The right to literacy* (pp. 74–80). New York: MLA.
Kynell, T. C. (1996). *Writing in a milieu of utility: The move to technical communication in American engineering programs 1850–1950.* Norwood, NJ: Ablex.
Landow, G. P. (1992). *Hypertext: The convergence of contemporary critical theory and technology.* Baltimore: Johns Hopkins University Press.
Lannon, J. M. (1997). *Technical writing* (7th ed.). New York: Longman.
Lay, M. (1996). The computer culture, gender, and nonacademic writing: An interdisciplinary critique. In A. H. Duin & C. J. Hansen (Eds.), *Nonacademic writing: Social theory and technology* (pp. 57–80): Mahwah, NJ: Lawrence Erlbaum.

Lay, M., Wahlstrom, B. J., Doheny-Farina, S., Duin, A. H., Little, S. B., Rude, C. D., Selfe, C. L., & Selzer, J. (1995). *Technical communication*. Chicago: Irwin.

Markel, M. (1996). *Technical communication: Situations and strategies* (5th ed.). New York: St. Martin's Press.

Miller, C. R. (1979). A humanistic rationale for technical writing. *College English, 40*, 610–617.

Miller, T. P. (1991). Treating professional writing as social praxis. *Journal of Advanced Composition, 11*(1), 57–72.

Mirel, B. (1996). Writing and database technology: Extending the definition of writing in the workplace. In P. Sullivan & J. Dautermann (Eds.), *Electronic literacies in the workplace: Technologies of writing* (pp. 91–114). Urbana, IL and Houghton, MI: National Council of Teachers of English and Computers and Composition Press.

Moore, P. (1992). When politeness is fatal: Technical communication and the Challenger disaster. *Journal of Business and Technical Communication, 6*(3), 269–292.

Negroponte, N. (1995) *Being digital*. New York: Alfred A. Knopf.

Ornatowski, C. (1992). Between efficiency and politics: Rhetoric and ethics in technical writing. *Technical Communication Quarterly, 1*(1), 91–103.

Paradis, J. (1991). Text and action: The operator's manual in context and in court. In C. Bazerman & J. Paradis (Eds.), *Textual dynamics of the professions: Historical and contemporary studies of writing in professional communities*. Madison, WI: University of Wisconsin Press.

Pearsall, T. E., & Warren, T. L. (1996). The Council for Programs in Technical and Scientific Communication: A retrospective. *Journal of Technical Writing and Communication, 26*(2), 139–146.

Penrose, J. M., Bowman, J. P., & Flatley, M. E. (1987). The impact of microcomputers on the Association for Business Communication (ABC) with recommendations for teaching, writing, and research. *Journal of Business Communication, 24*(4), 79–91.

Pinch, T. (1994). Cold fusion and the sociology of scientific knowledge. *Technical Communication Quarterly, 3*(1), 85–100.

Postman, N. (1995). *The end of education: Redefining the value of school*. New York: Alfred A. Knopf.

Price, J., & Korman, H. (1993). *How to communicate technical information: A handbook of software and hardware documentation*. Redwood City, CA: Benjamin/Cummings.

Sauer, B. (1994). The dynamics of disaster: A three dimensional view of documentation in a tightly regulated industry. *Technical Communication Quarterly, 3*(4), 393–419.

Schrage, M. (1990). *Shared minds: The new technologies of collaboration*. New York: Random House.

Selber, S. A. (1994). Beyond skill building: Challenges facing technical communication teachers in the computer age. *Technical Communication Quarterly, 3*(4), 365–390.

Selber, S. A. (1995). Metaphorical perspectives on hypertext. *IEEE Transactions on Professional Communication, 38*(2), 59–67.

Selber, S. A. (Ed.). (1997). *Computers and technical communication: Pedagogical and programmatic perspectives*. Greenwich, CT: Ablex.

Selber, S. A., McGavin, D., Klein, W., & Johnson-Eilola, J. D. (1996). Issues in hypertext-supported collaborative writing. In A. H. Duin & C. J. Hansen (Eds.), *Nonacademic writing: Social theory and technology* (pp. 257–280). Mahwah, NJ: Lawrence Erlbaum.

Selfe, C. L. (1992). Preparing English teachers for the virtual age: The case for technology critics. In G. E. Hawisher & P. LeBlanc (Eds.), *Re-imagining computers and composition: Teaching and research in the virtual age* (pp. 24–42). Portsmouth, NH: Boynton/Cook.

Selfe, C. L. (1995). Lest we think the revolution is a revolution: Images of technology and the nature of change. *Proceedings of the Council for Programs in Technical and Scientific Communication Annual Meeting*, 1–7.

Selfe, C. L. (1996). Theorizing e-mail for the practice, instruction, and study of literacy. In P. Sullivan & J. Dautermann (Eds.), *Electronic literacies in the workplace: Technologies of writing* (pp. 255–293). Urbana, IL and Houghton, MI: National Council of Teachers of English and Computers and Composition Press.

Selfe, C. L., & Selfe, R. J. (1996). Writing as democratic social action in a technological world: Politicizing and inhabiting virtual landscapes. In A. H. Duin & C. J. Hansen (Eds.), *Nonacademic writing: Social theory and technology* (pp. 325–358). Mahwah, NJ: Lawrence Erlbaum.

Selfe, R., Selber, S., McGavin, D., Johnson-Eilola, J., & Brown, C. (1992). Online help: Exploring static information or constructing personal and collaborative solutions using hypertext. *ACM SIGDOC '92 Proceedings*, 97–102.

Selzer, J. (1983). What constitutes a 'readable' technical style? In P. V. Anderson, R. J. Brockmann, & C. R. Miller (Eds.), *New essays in technical and scientific communication: Research, theory, and practice* (pp. 71–89). Farmingdale, NY: Baywood.

Shirk, H. N. (1991a). 'Hyper' rhetoric: Reflections on teaching hypertext. *The Technical Writing Teacher, 18*(3), 189–200.

Shirk, H. N. (1991b). Hypertext and composition studies. In G. E. Hawisher & C. L. Selfe (Eds.), *Evolving perspectives on computers and composition studies: Questions for the 1990s* (pp. 177–202). Urbana, IL and Houghton, MI: National Council of Teachers of English and Computers and Composition Press.

Shirk, H. N. (1997). New roles for technical communicators in the computer age. In S. A. Selber (Ed.), *Computers and technical communication: Pedagogical and programmatic perspectives* (pp. 353–374). Greenwich, CT: Ablex.

Silker, C. M., & Gurak, L. J. (1996). Technical communication in cyberspace: Report of a qualitative study. *Technical Communication, 43*(4), 357–368.

Slack, J. D., Miller, D. J., & Doak, J. (1993). The technical communicator as author: Meaning, power, authority. *Journal of Business and Technical Communication, 7*(1), 12–36.

Smith, C. (1986). Attitudinal study of graphic computer-based instruction for punctuation. *Journal of Technical Writing and Communication, 16*, 267–272.

Snyder, I. (1996). *Hypertext: The electronic labyrinth*. Carlton South, Victoria, Australia: Melbourne University Press.

Souther, J. (1989). Teaching technical writing: A retrospective appraisal. In B. E. Fearing & W. K. Sparrow (Eds.), *Technical writing* (pp. 2–13). New York: MLA.

Spilka, R. (1993). Collaboration across multiple organizational cultures. *Technical Communication Quarterly, 2*(2), 125–145.

Staudenmaier, J. M. (1995). Rationality versus contingency in the history of technology. In M. R. Smith & L. Marx (Eds.), *Does technology drive history?: The dilemma of technological determinism* (pp. 259–273). Cambridge, MA: MIT Press.

Stotsky, S. (1996). Participatory writing: Literacy for civic purposes. In A. H. Duin & C. J. Hansen (Eds.), *Nonacademic writing: Social theory and technology* (pp. 227–256). Mahwah, NJ: Lawrence Erlbaum.

Sullivan, P., & Dautermann, J. (Eds.). (1996). *Electronic literacies in the workplace: Technologies of writing.* Urbana, IL and Houghton, MI: National Council of Teachers of English and Computers and Composition Press.

Sullivan, D. (1990). Political-ethical implications of defining technical communication as a practice. *Journal of Advanced Composition, 10*(2), 375–386.

Swales, J. (1990). *Genre analysis: English in academic and research settings.* Cambridge, England: Cambridge University Press.

Taylor, C. A. (1994). Science as cultural practice: A rhetorical perspective. *Technical Communication Quarterly, 3*(1), 67–84

Tebeaux, E. (1996). Nonacademic writing into the 21st century: Achieving and sustaining relevance in research and curricula. In A. H. Duin & C. J. Hansen (Eds.), *Nonacademic writing: Social theory and technology* (pp. 35–55). Mahwah, NJ: Lawrence Erlbaum.

Teklinski, B. (1992). Style analysis of award winning technical manuals. *Journal of Technical Writing and Communication, 22*(4), 415–423.

Thralls, C., & Blyler, N. R. (1993a). The social perspective and professional communication: Diversity and directions in research. In Thralls, C., & Blyler, N. R. (Eds.), *Professional communication: The social perspective* (pp. 3–34). Newbury Park, CA: Sage.

Thralls, C., & Blyler, N. R. (1993b). The social perspective and pedagogy in technical communication. *Technical Communication Quarterly, 2*(3), 249–270.

Turkle, S., & Papert, S. (1990). Epistemological pluralism: Styles and voices within the computer culture. *Signs, 16*(1), 128–157.

Waddell, C. (1990). The role of pathos in the decision-making process: A study in the rhetoric of science policy. *Quarterly Journal of Speech, 76,* 381–400.

Winner, L. (1986). *The whale and the reactor: A search for limits in an age of high technology.* Chicago: University of Chicago Press.

Winsor, D. A. (1988). Communication failures contributing to the Challenger accident. *IEEE Transactions on Professional Communication, 31*(1), 101–107.

Winsor, D. A. (1996). *Writing like an engineer: A rhetorical education.* Mahwah, NJ: Lawrence Erlbaum.

Woolever, K. R., & Loeb, H. M. (1994). *Writing for the computer industry.* Englewood Cliffs, NJ: Prentice Hall.

Zakon, R. H. (1993). *Hobbes' Internet timeline v2.5* [On-line.]. Available: http://info.isoc.org/guest/zakon/Internet/History/HIT.html

11

Refining a Social Consciousness: Late 20th Century Influences, Effects, and Ongoing Struggles in Technical Communication

Jo Allen
East Carolina University

Over the years, the interest in writing the history of technical communication has been alternately fever-pitched and frustrated. The most obvious problem with writing such a history is determining what kind of history to chronicle. Should the work explain the rise of technical communication as a career; as an academic field of inquiry; or as a centuries-old endeavor that went without a name until the 20th century? Should the work examine the subjects, the concept, or the writers of technical communication? And which writers should it examine—those who practiced technical communication or those who have studied it and helped it emerge as a legitimate field of inquiry?

As in any historical accounting, the second consideration must be one of perspective—*how* to describe the events and circumstances in such a way as to create the proper sense of development. Although most of us in Western cultures have been taught to think of history as a tidy continuum, writers such as Thomas Kuhn (1970), Richard Rorty (1979), and others have helped reshape our thinking, pointing out that minor adjustments in our understandings may lead to major changes in overall perspectives (what Kuhn calls "revolutions") and that those changes tend to follow a more helter-skelter pattern, characterized by stagnant periods and episodic upheavals, rather than the tidy progression our textbooks, literature reviews, and pedagogical approaches typically present.

Our problems in writing our history could be easily dismissed were it not for the role that history will undoubtedly play in shaping our future. In her article contemplating the future of technical communication, Carolyn Miller (1990) writes that "we need to conceive of technical communication broadly and generously, not as a career path or job description based on current or past experience, but as a network of social practices and needs" (p. 108). In response, this work addresses one portion of our history by offering an evolutionary perspective on the social nature of technical communication, showing how ideas about major issues have evolved over the years; some significant historical, ideological, and theoretical forces behind those changes; and the consequences of those changes within certain areas of our profession. Less concerned with the absolute starting points of these issues than with the representational evolution of the discussions, the purpose of the work is to explore the development of social consciousness that serves as a significant indicator of the maturation of technical communication as a discipline.

A couple of caveats helps establish a context for this discussion and its conclusions. First, in no way do I mean to suggest that the individuals who populate our profession have only recently developed a social consciousness; what I am suggesting is that—as a discipline—we began our explorations of popular topics (clarity, conciseness, and accuracy, for instance) with little or no incorporation of social consciousness into our work. Now, at the end of the 20th century, we are more aware, perhaps, of the social components and consequences of our work. Second, the historical, ideological, and theoretical issues on which I base this argument are not to be interpreted as isolated or monolithic movements; rather, they are representative of the kinds of changes that can and have affected our profession. Most significant for the profession is that social awareness has enabled us to find a place that transcends the limiting view of our profession as an extension of, and perhaps even co-conspirators in, what many would consider a precarious 21st century leaning toward technocracy (see, for instance, Dobrin, 1985; Miller, 1990; Sullivan, 1990). Instead, we are now better positioned to clarify what is "valuable"—in its most humanistic sense—about our profession and to guard against encroachments on that humanism, simply defined as an awareness of the best that humans have to offer: flexibility, resiliency, critical thinking, a sense of ethics or morality, and potential for growth. The interrelatedness, therefore, of historical, humanistic, and theoretical concepts support the social perspective on knowledge and its concommitant responsibilities.

In his article, "Post-Modernism as the Resurgence of Humanism in Technical Communication Studies," Paul M. Dombrowski (1995) asserts that four areas of study—rhetoric of science, social constructionism, feminist critiques of science, and ethics—demonstrate the humanistic resurgence in technical communication.

I agree, and in fact, I take a similar approach in this essay to extend Dombrowski's argument. I believe we can take technical communication out of a theoretical, postmodernist perspective and into a historical movement (the Civil Rights Movement) to initiate an alternative positioning of our profession within humanism. Next, I hope to show that humanism itself—especially as characterized in a 1996 debate about instrumental/nonrhetorical discourse and totalizing rhetoricism—is very much a goal of technical communication, although the actual debate seems to lose sight of humanism rather quickly. Finally, I show that social constructionism, rather than merely demonstrating the postmodernist/humanist position, actually undergirds the positions established in historical (civil rights) and ideological (humanist) frontiers. Then, following Dombrowski's lead, I show the influence of these historical, ideological, and theoretical influences in discussions of four subjects—collaboration, corporate culture, gender studies, and intercultural communication—as they evolve in our publications. My point is to show how the discussion of these issues has evolved, reinforcing the social contingency of our research and scholarly discourse.

Tracing the development of thought about social issues in our field is crucial: First, because that development reflects the ongoing vitality of exploratory research and scholarship in technical communication that works to serve the profession; second, because the insistence on that development prevents misinterpretations of current research and practice as "the way things have to be" regarding any aspect of our work; and third, because a chronicle of such developments helps us reevaluate historical beliefs and practices in light of newer developments, encouraging us to reexamine old ideas that may yet prove fruitful, while lessening the chances that we will squander research time and resources on unproductive approaches and ideas that have been rightfully abandoned.

Equally important, perhaps, is the role that social awareness has played in reshaping the image of the technical communicator. As stereotypical as the scientist in the ubiquitous lab coat, the image of the technical communicator in the windowless, sterile cubicle, totally out of touch with the world outside has plagued many potential technical communicators who opt for other fields because of that aloof, antisocial image. Indeed, the highly technical and technological subject matter of much technical communication lends credence to that sterile, passive image. Our concerns with technology, production budgets, and other issues clearly at the root of much of our work have perpetuated that apparent schism between rhetorical humanism and nonrhetorical objectivity in technical communication (see Hagge, 1996; Kreth, Miller, & Redish, 1996; Moore, 1996a and 1996b, discussed more thoroughly below). But sociological influences, and the resulting studies that investigate their causes and consequences, have called attention to the grander scheme of technical communication as a profession that owes allegiance to the communities of readers it serves, as well as to the professionals who—in one way or another—carry out its mission.

HISTORICAL INFLUENCES: THE CIVIL RIGHTS MOVEMENT

The historical influences and developments within scientific and technical communication have been well traced by others in this collection of essays, and especially by William Rivers in his useful bibliographical account. Unlike those and other works, however, the focus on particular manifestations of understandings within the scientific and technical communication community (the plain style versus the ornate style, specific authors, the attention to rhetorical patterns and influences, etc.), I turn to a prominent historical movement within America as a mobilizing factor in some ideological aspects of the technical communication profession.

Throughout its history, and like many of our global partners, America has been characterized, challenged, and anguished by its social agendas. Balancing, precariously, the principles of the earliest navigators and founders with those of the indigenous peoples, the search for religious freedom with the mounting concern over cultism, and the capitalistic foundations of the nation with the inclusive principles of equality, the United States fights an ongoing battle with its history and, in a generous interpretation, perhaps a growing concern for making things right.

Although certainly not the first effort to remove significant societal barriers, the Civil Rights Movement of the 1960s stands as one of the finest testimonies to a sustained concern for making things right—for preserving a heritage that should be preserved (most often founded on principles of equality) and for overturning the elements of its heritage that should be eradicated (most often founded on practices of inequality). And although other movements challenged laws and, for instance, empowered Blacks and women to vote, it was the Civil Rights Movement that most significantly questioned attitudes, long accepted as the primary motivation for action. That movement has—however eloquently or imperceptibly—pervaded much of what (and how) we in academe study and teach, and what (and how) others in the profession practice.

If we acknowledge that the heart of our profession is the mission to relay knowledge, we can also see that our mission itself points to the same equalizing effects demonstrated in the Civil Rights Movement. When A holds the same knowledge as B, the scene is right for equality. Of course, equality comes about from far grander actions and attitudes than the mere transference of knowledge, but that transference is surely a key component in the equation. (Reciprocal knowledge, on the other hand, takes the concept of transference into larger social realms and is discussed later.) The inclusion of women, people of color, and those without money and power in the considerations of the governance of the nation are mirrored in the concerns for these same groups within our profession and within our readership. Making information available to those groups, in short, removes a primary element of powerlessness and disenfranchisement: lack of information or ignorance.

The struggle, of course, is ongoing, and many would argue that only modest (if not token) accommodations have been made. Clearly, our profession is still dramatically dominated by Caucasians, if a quick glimpse at the population of our professional conferences is any indication at all. And although we have made strides in creating opportunities for women within the profession, the penalty—lower proportional salaries than when men dominated the field—is disheartenngly severe (see Stoner, 1988).

The awareness of the issues, however, leads us to ponder the role of information—and the roles that technical communicators play as deliverers of that information—within our communities. Consequently, our studies of readers, interpersonal relationships, collaboration, management, gender, race, class, ethics, economics, and power—both internal and external to our profession—are increasingly attentive to the very concepts raised during the Civil Rights Movement. And if, as we so often contend, information is power, then the parallels become clear. Who has access to power and information? Who determines what kind or how much power/information to dispense? Who benefits from providing or hoarding power and information? Who determines how much money power and information are worth—and how do they determine their value? How do power and information become more accessible? Should or can we control how people can use power and information?

More specifically, in what ways does our society limit information (via education, scholarship, opportunity, or other institutionalized "systems") from women, from the poor, and from people of color? How can we encourage more minorities to enter the profession? And how will their entrance affect our profession? The answers to these questions will clearly reflect a social agenda with a potential for disturbing assessments of the balance between readers and their rights, holders and dispensers of information, and the powers of industrialized societies.

IDEOLOGICAL INFLUENCES: THE PERPETUATION OF HUMANISM

Serving in many ways as the ideology supporting the Civil Rights Movement, humanism represents the second source of social influence, calling attention to the rather tenuous relationship between the ideological foundations of capitalism and technologization and humanism. Again, America's foundations as a capitalist society necessitate an ever-present concern for the bottom line—a concern that has played a significant role in movements of both racial and gender equality, as well as a concern which in innumerable instances has led to untenable breaches of ethics among businesspeople, politicians, and even writers and, thus, an austere turn from the principles of humanism.

Fueling the capitalistic engine, of course, is the technologization of America; the *ability* to make faster, better, snazzier anythings becomes the *justification* for production and mass marketing. (The 1997 congressional debates over cloning—especially regarding its ethical and unethical uses—exemplify this notion within our society.) That is not to say, of course, that capitalism is necessarily evil (although many philosophers and some of my colleagues may disagree); it does mean, however, that capitalism demands an equally strong dose of humanism to correct its tendency toward tilt: the abandoning of social consciousness and the good of most people in favor of despotism, recklessness, and the good of the (very) few.

Of course, humanism is hardly a new movement; its roots in the liberal arts hearken to its place in ancient societies as the curriculum of freed citizens. As Paul Dombrowski (1994) notes, "The pursuit of these liberal arts was thought to be freeing in itself, elevating the practitioner above baser impulses while cultivating a noble, civic mentality" (p. 3). The reference to elevated pursuits reminds us that technical communicators are, as humanists, historically attuned to the very issues raised in our social inquiries: matters of language, interpretation, meaning, and manipulation (in its neutral connotation) for the good of others—a charge that directly positions us to address issues of inclusion, representation, and interpersonal abilities and responsibilities.

In technical communication, therefore, it should be no surprise that humanism has taken its grandest swipes at the very subject of the profession: technology. Recognizing the dangers of losing sight of the many, humanism most eloquently supports the mission of the profession: information exchange for the benefit of all, a goal best characterized in the spirit of the liberal arts. Paul Dombrowski's work *Humanistic Aspects of Technical Communication* (1994) situates the humanist's role as one concerned "with the humanity of the sender and receiver of the communication model and with the wide social impact of the communication" (p. 1). In her article, "A Humanistic Rationale for Technical Writing," Carolyn Miller (1979) argues that the certainties of positivist science and technology have infringed on the flexibility of language, leading professional communicators to believe that there can be "one and only one way" to communicate clearly (see Britton, 1965). The humanist perspective is the alternative to this kind of positivism, recognizing that language and interpretation are defiantly complex and irreducible to formulaic notions of expression and communication.

Perhaps no more interesting an argument has appeared in our professional journals in the past few years than the Humanism debate in the last two issues in 1996 of the *Journal of Business and Technical Communication*. Opposing Carolyn Miller and others who have argued that technical communication (in order to be humanistic) is necessarily rhetorical in nature and practice, Patrick Moore (1996a) posits that instrumental discourse (for example, computer documentation) is essentially nonrhetorical (i.e., nonpersuasive) because the writer does not have to *persuade* the reader to hook up the computer or press "Escape" to return to the main menu. Challenging the opposing view, requiring some aspect of rhetoricism (concern for ethically persuading the reader, based most often on the reader's best interest) in humanism, Moore goes on to contend that the mere absence of rhetorical suasion does not correlate to the absence of humanistic concerns. On the contrary, he states that perhaps the greatest evidence of the technical communicator's humanistic concern is his or her reliance on persuasionless (nonrhetorical/objective) prose, especially in instances where rigid explanations may prevent the reader's harm.

Supporting Moore's position (and extending it in a few instances), John Hagge (1996) writes that the essential difference between Moore and Miller is their contrasting view of "the relationship between words and the world" (p. 462), with Miller accepting no connection between the two, while Moore accepts a representational reliance of words to a correspondening world concept. Espousing a relativist point of view, Hagge suggests that the debate is not nearly as monolithic as Miller and other advocates of the staunchly nonpositivist perspective propose—that, indeed, middle ground exists that allows for rhetorical sensitivities amidst rule-bound realities. In fact, Hagge argues, "It is instrumental discourse, with its commitments to a realist linguistic ontology in which words connect to real-world consequences, and not discourse conceived rhetorically...that has true ethical implications for people's daily lives" (p. 472).

Citing Moore's distinction between rhetoric and instrumental discourse as false, Miller (Kreth, Miller, & Redish, 1996) argues the more useful distinction resides between those who see rhetoric as a type of discourse versus those who see it as an element of all discourse. Adopting the second interpretation, Miller, as well as Kreth and Redish, argues that the existence of rhetoric, in all forms of discourse, enriches language and helps technical communicators perform their missions of accommodating audiences and their needs. Moore's final rejoinder (1996b) differentiates between the content, audience, purposes, genres, invention, and empowerment of instrumental versus rhetorical discourse, concluding that these distinctions necessitate distinctions in our classrooms and in our approaches to these forms of discourse.

The dismissal of "humanism" fairly early in this argument is far more evident than the mere absence of the word in the discussions' latter pages. The debate modifies the famous Wittgenstein anecdote in which a student laughs at the silliness of people who believed that the sun traveled around the Earth. In response, Wittgenstein reportedly asks the student to imagine how different things would look if the sun, indeed, did travel around the Earth—the point being that it would look exactly the same. Similarly, I cannot help but wonder whether the inclusion or absence of humanistic concern would alter the appearance (or substance) of the rhetorical/nonrhetorical argument. In other words, when both parties argue the humanistic concern at the heart of their rhetorical/nonrhetorical debate that also pits inaccurate charges of necessarily positivist instrumental discourse against relativist socially constructed discourse, one must at least consider the possibility that *both* versions of the endeavor either promote or deflate humanism. In the tenor of the argument, at least, one would almost have to argue that the latter is the case.

A reasonable, humanistic insertion into the argument would hold that technical communicators have to acknowledge humanistic concerns—for their own welfare and ethics, as well as for their readers' safety, cultural understandings, and needs or uses for information—regardless of whether those concerns arise in instrumental or persuasive discourse and regardless of whether that discourse survives on a consciously social, rhetorical, or even coercive plane. The responsibility, in short, falls to the technical communicator to have at least a hand in regulating or massaging the message and its consequences as dictated by whatever source in power. The concerns continue to rest on issues of who is in charge and what agenda they serve—issues that clearly resonate with the concerns over technocracy and power discussed earlier.

Of course, humanism does not have to play Dr. Jeckyl to technology's Mr. Hyde (or vice versa, depending on your point of view). Nor does it have to be the counterconcern to industrial interests. Indeed, while the very existence of humanism as an ideological influence on technical communication has increased the attention to the social issues embedded in our discipline, that attention is welcomed by those who understand the kind of responsibilities we have identified, and are continuing to identify, as crucial to information exchange. And even the confirmed misanthrope may appreciate the humanist's attention to these issues, as least as a step toward identifying the fodder for lawsuits. The longstanding duality of "technical" versus "communication" mimics the science-humanism schism and dissolves in the face of critical abilities for uniting technology with humanism, without lessening the potency and value of either.

Countering that leaning toward technocracy in many respects, and as perhaps our clearest ongoing gesture of social awareness, technical communicators have always borne a keen sense of responsibility to readers, but the view of readers as "consumers" or "receivers" of information has necessitated a distance that could arguably be interpreted as both condescending and even primitive. The notion of "helping the reader" implies a hierarchical (and even patriarchal) stance that has been at least modestly challenged by recent works on audience (see Allen, 1989; Ede & Lunsford, 1984). Thus, as social awareness within our profession has increased, new ideas have emerged about many topics in our field, and they have introduced, added to, reshaped, or undermined some of the once-held beliefs about those issues. In short, the evolution of social consciousness has drastically altered the technical communication profession, moving us beyond the "I" of being good writers toward the "we" of creating meaning and significance that sustains a progressive professional and social community.

Thus, the humanist perspective has made us far more accountable to readers because it situates them as pivotal elements in the technical communication "story." Then, to borrow from social constructionism, it contends that the readers' interpretations are every bit as valid as the writer's intentions. Even if they are not accurate from the writer's point of view, those interpretations are not necessarily changeable, at least not without sustained dialogue between writer and reader or within communities of readers seeking to interpret our texts (see Kent, 1989). That kind of legitimacy, as well as integral binding, between the writer and the reader clarifies the growing sense of sociological commitment and responsibility within the profession. If we cannot dismiss our readers as "thick-headed" (because *we* are *they*, after all), then we must reevaluate our understanding of our roles as communicators. And our responsibilities to our communities of readers increase by virtue of our philosophical stance toward that audience as deserving of knowledge.

THEORETICAL INFLUENCES: THE RISE OF SOCIAL CONSTRUCTIONISM

Following on the heels of the historical and ideological movements of the 1960s and beyond, the events of our experiences naturally find their way into our profession's theoretical perspectives as well. One of the most significant developments in our thinking about technical communication has resulted from the works of Thomas Kuhn (1970), Richard Rorty (1979), Lester Faigley (1985), Lee Odell (1985), Kenneth Bruffee (1986), Karen LeFevre (1987), Rafoth & Rubin, (1988), and others who pointed out—in various ways and within various communities—that writers not only write *to* readers, but also write *with* them. In short, readers make meaning based on their interpretations, experiences, and needs, making reading a task of situational convenience as well as a task of information gathering. In turn, writers may present the technical message but recognize that the reader's interpretation of that message is not a matter of simple clarity (see Miller, 1979), but rather a matter of the reader's experiences, expectations, and needs dovetailing with the writer's. If sociology is the study of people coming together, writing and reading is clearly a sociological activity. (See Dombrowski, 1994, pp. 81-96, for an excellent review of literature on social constructionism.)

Citing his own failed attempts to publish works that refer to some kinds of technical communication as instrumental, rather than rhetorical, Patrick Moore (1996b) points out that works that fall outside currently faddish views of technical communication do not get published—a testimonial validating the theory of social constructionism as socially or communally negotiated perspectives. The parallels between our own technical communication community and the scientific community, as Thomas Kuhn (1970) describes it, are clear: When the community reaches a (negotiated) concensus on an issue, those who do not ascribe to that concensus fall to the outskirts of the profession.

As a community, we seem satisfied with the view of technical communication as a discipline that embraces socially constructed knowledge, although a great deal of criticism can be rather easily hurled at that satisfaction (see the conclusion of this work). Of course, the view of technical communication as socially constructed is useful in our studies and characteristically humanistic but means little if the view is wrongheaded. But few challenges to social constructionism remain, suggesting we are more confident of the "rightness" of that view as time has passed. For example, in challenging totalizing rhetoricism, Moore (1996a) focuses on instrumental discourse, which he pointedly situates within a socially constructed context, and adopts social constructionism as an underlying principle for his arguments.

The significance of social constructionism resides primarily in our attitudes toward our readers. Knowing that readers find significance from works that (1) coalesce with, or at least acknowledge, their experiences and (2) hold up a particular kind of social conversation, means that writers have to take a different view of readers: not so much as clients or students anymore, but as partners. The designation of "partner" is important, for it clearly suggests a much more intimate, knowledgeable, and responsibility-laden relationship than the "consumer" or "receiver" designation of previous communication theories and models (see, for instance, Shannon & Weaver, 1949). Further, the disdain with which writers may have treated some readers who had difficulty fathoming their messages decreases with this view of readers as partners.

As the writer's partner, the reader contributes to shape what is written, how it is written, how the information is used, and how it is modified by various groups and situations. The primitive determination of the reader's mere familiarity with technical subjects—a practice that hearkens to positivist, static analyses of audiences and their needs—becomes the topsoil that is quickly blown away by graver issues of relevance, cognitive abilities, and even, to a large extent, place and social determinism.

Again, the limitations of this view are challenged in the conclusion, but that sense of equality between partners and the humanistic roles invoked between the partners echoes the sociological patterns and movements already described; a closer look at how these influences have played out in our profession clarifies the extension of ourselves into our communities.

COLLABORATION

One of the most immediately recognizeable forms of socialization in technical communication is the collaborative nature of the technical communication workplace. Kenneth Bruffee's article, "Collaborative Learning and the 'Conversation of Mankind'" (1984), is one of the earliest theoretical works portraying the interaction between writers and their communities. A study conducted by O'Donnell, et al. (1985) provides some of the earliest empirical evidence that collaboration is, in itself, a likely improvement over isolated writing. In addition, the 29th International Technical Communication Conference included at least one presentation on integrated teams and members' roles in proposal writing at Martin Marietta Denver Aerospace (Hoerter & Brenner, 1982).

Of course, technical communicators have long been practitioners of collaboration, in its most elementary sense, as they receive information from subject matter experts and then shape that information into an understandable form for a particular audience. But advancing our ideas about collaboration, beyond the assembly-line image of adding and shaping information from subject matter expert to writer to editor, has required observation and characterization in dialogues that not only clarify the processes of collaboration, but that also mimic the social constructionism that shapes the discussion. The best evidence of the difficulties in describing (and defining) collaboration may be the finding that 87% of the writers who write with others do not necessarily see their work as collaborative (Ede & Lunsford, 1990). Implicit in this assessment, perhaps, is an understanding that collaboration requires something besides proximity and even interchange to qualify as true collaboration. Theory and research have taken even longer to address this function of technical communication, with early acknowledgments that it happens, but no actual studies emerging until the middle to late 1980s.

Studies of collaboration have flourished over the past decade, breaking the silence, with the clear message that we are no longer asking whether it happens, but *how* it happens—and happens successfully. Addressing matters of interpersonal relations and conflict (see, for instance, Burnett, 1991 and 1993), management of collaborative teams (see, for instance, Bosley, 1991), and predictors of collaborative successes (see Williams & Sternberg, 1988), we are increasingly engaged in understanding the social nature of collaboration, with its products a matter of participatory inclusiveness, rather than mere division of labor (Killingsworth & Jones, 1989).

A brief look at how this discusssion has developed sociologically characterizes the very process of social constructionism. Ede and Lunsford (1990) write from personal experience that their own collaborative process is truly an integrated one, with both sitting at the computer and crafting their prose. Lay (1989) acknowledges the potential influences from other disciplines, calling on gender studies as an informant for her discussion on interpersonal relationships that strengthen (or harm) collaborative teams. In Lay and Karis's (1990) collection of essays on collaboration, contributors extend the discussion by recognizing pedagogical and industry-based difficulties and resolutions with collaborative writing. And Burnett (1993) clarifies the kinds of interpersonal conflicts that most often affect collaborative teams. This process of personal observation, disciplinary inclusiveness, and fleshing out clearly imitates social conversation, and our understanding of collaboration is all the richer for that conversation.

Numerous other works, of course, contribute to expand this discussion (see Debs, 1991). But all work to refine, reshape, counter, and negotiate an understanding of collaboration and its role in the work of technical communicators.

CORPORATE CULTURE

The dominant workplace image of the 1950s and 1960s is unquestionably the image portrayed in movies, television, and even comics and children's books, showing Daddy (rarely ever Mommy—see notes on gender, below) going off to work for "the company." If Daddy changed jobs, it was because of some evil external force or Daddy's own role as a misfit worker. Daddy hardly ever changed jobs because of moves across the country, increased industrialization and technolization, acquisitions and mergers, or downsizing. Estimates that American workers will change jobs 9–10 times and careers 3–4 times in their lifetimes clearly shatter those early perceptions. And because our jobs, roles, and responsibilities do indeed change, we must also change—hence, the awareness of corporate (or organizational) cultures and their impact on technical communicators.

Matters such as ethics, responsibilities, and expectations are inextricably bound in our perceptions of corporate cultures and the socialization required to fit into any culture to which we may aspire. Scholars have consistently shown that corporate culture requires a highly sophisticated process for understanding the "rules" of fitting in and getting along and then deciding whether to accommodate those rules (see, for example, Hofstede, 1991; Lipson, 1986 and 1987; Parsons, 1987; Rawlins, 1988; and Southard, 1990). Personal experiences bear out the research, as we exchange stories about our own work environments at our professional conferences and over e-mail, while noting the effects of corporate culture in ethnographic and other research publications.

As unpopular as corporate culture may be in some quarters because of its perpetuation of the *status quo*, which may easily lead to dilemmas with ethics and philosophical sellouts (see Sullivan, 1990), the truth may well be, as James Zappen (1989) writes, that our understandings of our work and the culture that inspires and supports it means that we "need...not only to communicate within the context of several discourse communities but also, and especially, to develop the ability to step outside the boundaries of particular discourse communities and to participate in conversations with others on problems of mutual interest and concern" (p. 9).

Our discussions regarding corporate culture have moved from earliest works that characterize the complexity of the concept (see, for instance, Geertz, 1973) to works that describe how corporate culture plays out as a major influence on the work that professional communicators do (see, for instance, Southard, 1990; Lipson, 1986 and 1987; Bosley, 1993). The conclusion that the workplace is a highly complex sociological community with its own set of governing rules becomes even clearer when we understand that it, like the larger society, has its own *laws* (often corresponding with those of the larger society, but occasionally organizationally-specific, meriting either legal prosecution or dismissal), its formal or informal *code of ethics* that informs the morality of the organization, and even its own *standards of etiquette and sensitivity* that may or may not affect the work that is done, but that certainly affect the sociological environment in which that work takes place.

Our studies of corporate culture, in short, have helped create an awareness of the system of rules that govern our workplace and, by extension, our work. Social awareness and responsibility easily fall to communicators who either advance, modify, or deconstruct that culture through their writings. Yet, the potential for challenging the culture, widely seen as suicidal throughout the history of the American workplace, has found remarkably fertile ground in the past 40+ years, perhaps as a direct result of the Civil Rights Movement, which has awakened a number of questions about the status of the *status quo* and the rights of those in power to maintain and perpetuate it.

Technical communicators who acquiesce to management pressure to write more (Winsor, 1996), or to say it differently (Ornatowski, 1995), begin to understand not only the ethical dilemmas they face, but also the more complex nature of writing with others—this time, with managers who have special interests to protect, whether their own or those of their bosses or consumers. Therefore, the community of writers that operates within the community of the organization and, inevitably, within the community of the whole finds additional grounds for investigating and understanding the sociological impact of its work.

Our theoretical and research-based inquiry into corporate culture, again mirrors a socially constructed view of the workplace—a necessary evil, in some aspects, and an indefatiguable reality in others. The sense of what is best about a particular organization's culture relies heavily on a sense of workers' rights (in this case, the technical communicator's rights) to craft both a message and a mission for the organization, meaning that the communicator is far more than a mere conduit for the culture, but also maintains, erodes, or refines that culture.

GENDER

Discussions about gender may characterize most clearly the social evolution of a topic in technical communication because of the way they adopt the larger society's discussions of gender. Early publications on the role of gender in technical communication argue for gender-neutral language: ridding our vocabularies of the generic "he" and its concomitant understandings that men are the practitioners and doers in our society (see, for instance, Christian, 1986; Dell, 1990; Vaughn, 1989; Veiga, 1989).

These early admonitions against offensive language lay the groundwork for more sophisticated understandings of what and how attitudes about gender affect the profession and its work. Mary Lay (1989) introduces gender into discussions of interpersonal collaborative relationships; Elizabeth Tebeaux (1990) offers exploratory investigations into the myth of women's sensitivity in writing; and Jo Allen (1991) addresses the effects of gender in the profession itself, in our research, and in our pedagogy.

In her later publication, however, Louise Rehling (1996) offers a number of findings regarding collaborative teams that contradict earlier findings (or suppositions) of gender-based difficulties in collaborative settings. Perhaps most interesting is Rehling's observation that (1) groups tend to stereotype men as technical experts and (2) subdivisions of the group tend to self-segregate by gender. Set within the context of the classroom, one could always argue that this generation's students are more enlightened about traditional forms of sex stereotyping (assigning clerical duties to women; investing in hierarchically structured management of the team; engaging in gender-based conflict; and valuing product over process—behaviors that Rehling's study finds statistically insignificant in these groups), which could necessitate a tribute to the social consciousness of these students, as well as for communicators who have identified these behaviors.

What these students may not be enlighted about, however, are these other forms of sex-based behaviors. Is there anything pernicious in these behaviors? Does it matter that men are awarded special dimensions of intellect based on their gender? Does it matter if men and women segregate if it seems to be on their own initiative? The answers are arguable, but what is significant for our discussion here is that the issue of difference has clearly taken a new line in our conversation, moving us once again beyond superficial understandings and into deeper waters of inquiry.

Again, the evolution of the discussion has clarified and extended our understandings of the history, as well as the sociological effects, of a gendered profession and a gendered readership. The role that women have played in that history and its sociological development are well documented (see Tebeaux, 1993; Tebeaux & Lay, 1992; and "Historical Contributions," 1997, which places women even more profoundly within the history of our profession.) The result, I should hope, is that we are no longer surprised by the roles women have played in our profession, just as we have moved beyond mere invectives against the age-old practice of referring to a writer, reader, or any other active agent as "he," and toward identifying the more subtle forms of sexism in society and in the workplace. The sociological effect of this evolving discussion has created a more inclusive view of our profession and our readers, an inclusion that mirrors the kind of communal understanding our larger society has also increasingly acknowledged about the roles of women, as well as the effects of sexist language and assumptions.

GLOBALIZATION

As a final example of the increasing understanding of the social implications of professional communication, globalization (and its partners multiculturalism, multiethnicism, and desegregation) also follows a sociological pattern of development. Although only a dunderhead would "forget" that the rest of the world exists, many professional communicators have experienced the luxury of writing exclusively for readers who share their cultures and nationalities. Raising our consciousness about ways cultures differ and how those differences affect communication has been a primary concern in a number of professional organizations and publications that see changes ahead for these communicators (see, for instance, Scott, 1990; Terpstra & David, 1991).

Technology in the form of the Internet, as well as more obvious global mechanisms (such as transportation), brings together information and product, resulting in worldwide access and responsibility. The social implications of writing for a global community in general are compounded, therefore, by the additional filter of technology. The issue of technology and, more specifically, access, is far more complex than it seems, of course, resulting in far less panaceic effects than we might hope. Matters of origination of and culpability for information are complicated by matters of economics, availability, interference, and use. In other words, *who* puts information on the Internet (or any other "worldwide" format) and their motivations are complicated by cultural and national senses of perspective, responsibility, and ethics, as well as by simpler questions of who actually owns computers and the software to make information available, and who monitors or even censors that information. Cavalier assumptions that everyone can participate are swept aside by the realities that many countries do not have access or that it is heavily censored. These kinds of obstacles foster concern about the misperception of a global community with universal access to information.

Technical communicators are increasingly aware of this dilemma. Our studies have moved from interesting accounts of language and the practice of professional communication in or between countries (see, for instance, Greene, 1990; Johnston, 1988; Kinosita, 1988; and Varner, 1988) to the more sophisticated explorations of the ethnocentric assumptions we hold about the rest of the world, the effect of particular rhetorical strategies on messages and readers, and the expansion of our ethnocultural assumptions about work and communication (see, for example, Boiarsky, 1995; Bosley, 1993; Carson, Carson, & Irwin, 1995). As a result of the evolution of these investigations and speculations, we have significantly elevated our studies of the sociological ramifications of technology and globalization beyond the investigation of language differences and beyond the cavalier assumption of global villages and into the realm, again, of issues of power, control, ownership, and, thus, social significance.

CONCLUSION

In conclusion, we need to look at a number of issues that technical communication's social consciousness raises about the profession and its practice: In short, what kinds of issues characterize a strong society? And how do those characteristics apply to the society of professionals who teach, study, and practice technical communication? I would argue that some of the most important elements are still to be addressed fully by our profession and, if for no other reason than because of their sociological and communal import, merit concerted attention from both researchers and scholars. In light of that argument, I offer a few observations, positions, and questions regarding the sociological makeup of our profession:

Balance. In the interest of thwarting the tyrannies of same-mindedness, regardless of Kuhn's estimation that such same-mindedness (or concensus) is inevitable, we must continue to challenge our ideas about how communication succeeds, fails, and/or changes in light of disparate audiences, contexts, and purposes. Thus, in spite of the current popularity of social constructionism, can we conceive of what technical communication would look like if it were not socially constructed? The return to positivism is unattractive and intellectually indefensible, but how do we go forward from this point?

Inclusiveness. As a profession, technical communication will benefit from being more inclusive of minorities and the powerless as readers and as colleagues. Various perspectives and experiences can simultaneously expand and clarify information, and our interests in participants in information development and exchange demand a more inclusive gathering. Our research needs to ask how inclusive we are of readers from other cultures—within, and outside of, the United States? How inclusive are we of writers from other cultures? How inclusive are we of men and women within collaborative groups? As readers? How do we respond to subject matter experts, editors, illustrators, and other contributors to our writing projects from other cultures? To women of other cultures? In short, the work of our professional organizations to target these audiences may well signal a major commitment to minority interests—provided we continue to push our organizations to uphold that commitment.

Equity. Technical communicators are uniquely positioned, as information sources, to help level the playing field among various segments of the population regarding information access and use. Accepting this role will likely mean ethical dilemmas, economic challenges, and a great deal of soul-searching; refusing the role will undoubtedly mean the same things.

This work, therefore, briefly traces some of the historical, ideological, and theoretical social influences on our field, as they play out in four inherently social issues—collaboration, corporate culture, gender, and globalization—in essence answering three primary sociological questions: Who? does what? in what setting? Answers will most likely show that the isolationist, technocratic view of our work and our profession is keenly obsolete and even harmful.

The result of the study of these evolutions suggests additional directions our research might take and serves as a model of the kinds of evolutionary development that may well characterize the thinking about numerous topics in our field. Most significant, however, is that these conversations continue to evolve, suggesting that (1) we deeply care about these issues, (2) we recognize the relationship between the development of a social awareness and the future of our profession, (3) we acknowledge the open-ended potential of technical communication as a discipline, and (4) we can analyze the correlation between this social consciousness and our profession to confirm who and what we as professional communicators are, as well as to craft the view we will take of ourselves into the 21st century.

REFERENCES

Allen, J. (1989). Breaking with a tradition: New directions in audience analysis. In B. E. Fearing & W. K. Sparrow (Eds.), *Technical writing: Theory and practice* (pp. 53–63). New York: MLA.

Allen, J. (1991). Gender issues in technical communication studies: An overview of the implications for the profession, research, and pedagogy. *Journal of Business and Technical Communication, 5*, 371–392.

Boiarsky, C. (1995). The relationship between cultural and rhetorical conventions: Engaging in international communication. *Technical Communication Quarterly, 4*, 245–259.

Bosley, D. S. (1991). Designing effective technical communication teams. *Technical Communication, 38*, 504–512.

Bosley, D. S. (1993). Cross-cultural collaboration: Whose culture is it anyway? *Technical Communication Quarterly, 2*, 51–62.

Britton, W. E (1965). What is technical writing? *College Composition and Communication, 16*, 113–116.

Bruffee, K. A. (1984). Collaborative learning and 'the conversation of mankind.' *College English, 46*, 635–652.

Bruffee, K. A. (1986). Social construction, language, and the authority of knowledge: A bibliographical essay. *College English, 48*, 773–790.

Burnett, R. E. (1991). Substantive conflict in a cooperative context: A way to improve the collaborative planning of workplace documents. *Technical Communication, 38*, 532–539.

Burnett, R. E. (1993). Conflict in collaborative decision-making. In N. R. Blyler, & C. Thralls (Eds.), *Professional communication: The social perspective* (pp. 144–162). Newbury Park, CA: Sage.

Carson, K. D., Carson, P. P., & Irwin, C. (1995). Enhancing communication and interactional effectiveness with Mexican-American trainees. *Business Communication Quarterly, 58*(3), 19–25.

Christian, B. (1986). Doing without the generic he/man in technical communication. *Journal of Technical Writing and Communication, 16*, 87–98.

Debs, M. B. (1991). Recent research on collaborative writing in industry. *Technical Communication, 39*, 476–484.

Dell, S. A. (1990). Promoting equality of the sexes through technical writing. *Technical Communication, 37*, 248–251.

Dobrin, D. N. (1985). Is technical writing particularly objective? *College English, 4*, 237–251.

Dombrowski, P. M. (Ed.). (1994). *Humanistic aspects of technical communication*. Amityville, NY: Baywood.

Dombrowski, P. M. (1995). Post-modernism as the resurgence of humanism in technical communication studies. *Technical Communication Quarterly, 4*, 165–185.

Ede, L., & Lunsford, A. (1990). *Single texts/plural authors: Perspectives on collaboration*. Carbondale, IL: Southern Illinois University Press.

Ede, L. A., & Lunsford, A. (1984). Audience addressed/audience invoked: The role of audience in composition theory and pedagogy. *College Composition and Communication, 35*, 155–171.

Faigley, L. (1985). Nonacademic writing: The social perspective. In L. Odell & D. Goswami (Eds.), *Writing in nonacademic settings* (pp. 231–248). New York: Guilford.

Geertz, C. (1973). *The interpretation of cultures: Selected essays.* New York: Basic Books.

Greene, B. G. (1990). International business communication: An annotated bibliography. *The Bulletin of the Association for Business Communication, 53,* 76–79.

Hagge, J. (1996). Ethics, words, and the world in Moore's and Miller's accounts of scientific and technical discourse. *Journal of Business and Technical Communication, 10,* 461–475.

Hoerter, G. E., & Brenner, T. B. (1982). Teamwork...the key to managing proposal production. *Proceedings of the 29th International Technical Communication Conference,* 45–48.

Hofstede, G. H. (1991). *Cultures and organizations: Software of the mind.* New York: McGraw-Hill.

Johnston, J. (1988). Japanese firms in the US: Adapting the persuasive message. *The Bulletin of the Association for Business Communication, 51*(3), 33–34.

Kent, T. (1989). Paralogic hermaneutics and the possibilities of rhetoric. *Rhetoric Review, 8*(1), 24–42.

Killingsworth, M. J., & Jones, B. G. (1989). Division of labor or integrated teams: A crux in the management of technical communication. *Technical Communication, 36*(3), 210–221.

Kinosita, K. (1988). Language habits of the Japanese. *The Bulletin of the Association for Business Communication, 51*(3), 35–40.

Kreth, M., Miller, C. R., & Redish, J. (1996). Comments on "Instrumental discourse is as humanistic as rhetoric." *Journal of Business and Technical Communication, 10,* 476–490.

Kuhn, T. (1970). *The structure of scientific revolutions.* Chicago: University of Chicago Press.

Kynell, T., Tebeaux, E., & Allen, J. (Eds.). (1997). Historical contributions by women to technical communication [Special issue]. *Technical Communications Quarterly, 6*(3).

Lay, M. M. (1989). Interpersonal conflict in collaborative writing: What we can learn from gender studies. *Journal of Business and Technical Communication, 3*(2), 5–28.

Lay, M. M., & Karis, W. M. (Eds.). (1990). *Collaborative writing in industry: Investigations in theory and practice.* Amityville, NY: Baywood.

LeFevre, K. B. (1987). *Invention as a social act.* Carbondale, IL: Southern Illinois University Press.

Lipson, C. (1986). Technical communication: The cultural context. *Technical Writing Teacher, 13,* 318–322.

Lipson, C. (1987). Teaching students to "read" culture in the workplace: Reply to Gerald Parsons. *Technical Writing Teacher, 14,* 267–270.

Miller, C. (1979). A humanistic rationale for technical writing. *College English, 40,* 610–617.

Miller, C. (1990). Some thoughts on the future of technical communication. *Technical Communication, 37,* 108–111.

Moore, P. (1996a). Instrumental discourse is as humanistic as rhetoric. *Journal of Business and Technical Communication 10,* 100–118.

Moore, P. (1996b). A response to Miller and Kreth. *Journal of Business and Technical Communication, 10*, 491–501.

Odell, L. (1985). Beyond the text: Relations between writing and social context. In L. Odell & D. Goswami (Eds.), *Writing in nonacademic settings* (pp. 249–280). New York: Guilford.

O'Donnell, A., Dansereau, D. F., Rocklin, T., Lambiotte, J. G., Hythecker, V. I., & Larson, C. O. (1985). Cooperative writing: Direct effects and transfer. *Written Communication, 2*, 307–315.

Ornatowski, C. M. (1995). The writing consultant and the corporate/industry culture: How to learn the lingo, mind-set, and issues. *Journal of Business and Technical Communication, 9*, 446–460.

Parsons, G. M. (1987). The elusiveness of workplace culture: Response to "Technical communication: The cultural context." *Technical Writing Teacher, 14*, 265–266.

Rafoth, B. A., & Rubin, D. L. (Eds.). (1988). *The social construction of written communication.* Norwood, NJ: Ablex.

Rawlins, C. (1988). Changes in corporate culture and organizational strategy: The effect on technical writers. *Technical Writing Teacher, 15*, 31–36.

Rehling, L. (1996). Writing together: Gender's effect on collaboration. *Journal of Technical Writing and Communication, 26*, 163–176.

Rorty, R. (1979). *Philosophy and the mirror of nature.* Princeton, NJ: Princeton University Press.

Scott, J. C. (1990). An annotated reference list of publications relating to international business communication: Expanding horizons beyond traditional business communication sources. *The Bulletin of the Association for Business Communication, 53*, 72–76.

Shannon, C.E., & Weaver, W. (1949). *The mathematical theory of communication.* Urbana, IL: University of Illinois Press.

Southard, S. G. (1990). Interacting successfully in corporate culture. *Journal of Business and Technical Communication, 4*(2), 79–90.

Stoner, R. B. (1988). Economic consequences of feminizing technical communication. *Proceedings of the 35th International Conference on Technical Communication*, 108–110.

Sullivan, D. (1990). Political-ethical implications of defining technical communication as a practice. *Journal of Advanced Composition, 10*, 375–386.

Tebeaux, E. (1990). Toward an understanding of gender differences in written business communication: A suggested perspective for future research. *Journal of Business and Technical Communication, 4*(1), 25–43.

Tebeaux, E. (1993). Technical writing for women of the English Renaissance. *Written Communication, 10*, 164–199.

Tebeaux, E., & Lay, M. M. (1992). Images of women in technical books from the English Renaissance. *IEEE Transactions on Professional Communication, 35*, 196–207.

Terpstra, V., & David, K. (1991). *The cultural environment of international business* (3rd ed.). Cincinnati, OH: Southwestern.

Varner, I. I. (1988). A comparison of American and French business correspondence. *Journal of Business Communication, 25*(4), 55–65.

Vaughn, J. (1989). Sexist language—still flourishing. *The Technical Writing Teacher, 16*, 33–40.

Veiga, N. E. (1989). Commentary: Sexism, sex stereotyping, and the technical writer. *Journal of Technical Writing and Communication, 19*, 277–283.

Williams, W. M., & Sternberg, R. J. (1988). Group intelligence: Why some groups are better than others. *Intelligence, 12*, 351–377.

Winsor, D. (1996). Writing well as a form of social knowledge. In A. H. Duin & C. J. Hansen (Eds.), *Nonacademic writing: Social theory and technology* (pp. 157–172). Mahwah, NJ: Lawrence Erlbaum.

Zappen, J. P. (1989). The discourse community in scientific and technical communication: Institutional and social views. *Journal of Technical Writing and Communication, 19*, 1–11.

part IV
Bibliography in the History of Technical Communication

12

Studies in the History of Business and Technical Writing: A Bibliographical Essay*

William E. Rivers
University of South Carolina

Scholarship in the history of business and technical writing has made tremendous strides, especially over the past 15 years. Not only have the number of articles that deal with business, technical, and scientific writing from the past increased dramatically, but they have also become much more sophisticated. Early essays tended to be short and anecdotal; they called attention to and catalogued qualities in older writers, usually (and appropriately) with an eye on the classroom. In the 1960s and 1970s we were, for the most part, literature-trained teachers of business and technical communication who took to writing about historical examples because of how we were trained. We were also, in a piecemeal, tentative way, beginning an important search for our roots. Even in the very early studies, one finds a sense that the study of business and technical documents from the distant and recent past could help us better understand how language both influences and is influenced by the

---

* This essay is an expansion of an article that originally appeared in the *Journal of Business and Technical Communication,* 8(1), pp. 6–57. The original essay was greatly improved through the suggestions of Elizabeth Tebeaux and M. Jimmie Killingsworth. For this version I am indebted to Michael Moran and Teresa Kynell for their editorial support and patience and to Brad Bostian, Doug Wedge, and Robert Williford for their valuable help as I searched and sorted.

technical and scientific disciplines and institutional cultures. As the studies have become more sophisticated, more scholarly, and more interdisciplinary, that sense of the value of historical studies has grown.

Although we continue to see—and still need—the more informal, anecdotal essays that call attention to particular authors and works, increasingly the studies that deal with the history of business and technical writing:

- place works and authors in their intellectual, cultural, and linguistic settings to determine what forces shaped their writing;

- trace patterns of influence and change over time; or

- analyze individual works and groups of works using techniques and insights from different disciplines and special areas of interest, including composition, rhetoric, speech, anthropology, philosophy, literary criticism, linguistics, and history.

The purpose of this bibliography is to show both the richness of accomplishment and the richness of opportunity for those interested in the history of business and technical communication.

In the analysis and bibliography that follow, I have tried to include all items published in business and technical writing journals over the past 20 years. Athough I have included some studies from journals and books from other disciplines (e.g. speech, philosophy, history, literature), I have not tried to be exhaustive. The recent studies listed here, however, provide enough bibliographic information to facilitate research in any of the many fields of study that so fruitfully contribute to our understanding of business and technical writing.

BIBLIOGRAPHIES AND STATEMENTS ABOUT METHOD

One sure sign that an area of study is coming of age is the appearance of bibliographies on the subject. During the 1980s, three annotated bibliographies and three bibliographical essays on the history of business and technical writing were published, all of which made contributions. However, several of these studies, especially the bibliographical essays by Moran and Zappen, went well beyond simply listing what others had written; they began to define new directions for research. Together with this study, they record the brief history of an area of study that is still growing, maturing, and defining itself.

John Brockmann provided the first bibliography in 1983. His valuable annotated bibliography of 36 items focused only on technical writing. Although his introductory remarks are brief, Brockmann there makes an astute observation: The "central promise" of research in the history of technical communication as he saw it in 1983 was in its continued movement from mainly essays that narrowly focus on famous writers to more studies that take into account "a broad spectrum of celebrated and uncelebrated writers" (p. 155).

Michael G. Moran's 1985 bibliographical essay, "The History of Technical and Scientific Writing," in *Research in Technical Communication*, revealed that the broader, more sophisticated scope Brockmann wished to see in studies of the history of technical communication was beginning to develop. However, in the introduction to his essay, Moran identified another way in which the history of technical and business communications needs to grow when he said, "we have only scattered pieces of scholarship that, when fitted together, do not yet make up a complete picture" (p. 25). One of the major contributions of Moran's essay is that he helped us get a sense of the size and composition of this picture by grouping the 65 studies he discusses into historical periods or other categories. Moran's bibliography was also valuable in that it broadened our perspective by including not only studies that focused on technical or business communication from the point of view of those who teach technical and business writing, but it also included studies written by scholars in literary history and criticism. Although most of these studies are discussed in his section on 17th-century prose, their appearance serves to remind us that our research must reach out and include scholarship from other disciplines.

Debra Hull's 1987 book, *Business and Technical Communication: A Bibliography 1975–1985*, includes a list of 24 items under the heading "Influences from the Past" (pp. 1–6). Although she picks up a few items not listed in the other bibliographies, hers is not as helpful as the others because of its limited chronological range and its major gaps.

Two recent bibliographical studies focus on early technical writing in America and Canada. The more substantial of these contributions is Margaret W. Batschelet's *Early American Scientific and Technical Literature: An Annotated Bibliography of Books, Pamphlets, and Broadsides* (1990). In this valuable bibliographical tool, Batschelet lists and annotates over 800 works published in the America between 1639 and 1820. Divided into three major sections (Medical Titles, Technical Titles, and Natural and Physical Science Titles), her bibliography not only reveals the quality and quantity of scientific and technical writing in America, but also provides a resource for making its study a less daunting task. (Deborah Journet's 1991 review of this bibliography, in the *Journal of Business and Technical Communication*, provides a more detailed overview of its resources and contributions to scholarship.) Although her bibliographical contribution is smaller in comparison to Batschelet's, Jennifer J. Connor's "History and the Study of Technical Communication in Canada and the United States" (1991) is significant. After a brief review of essays devoted to the history of technical writing, she says a need still exists for "focused historical studies...particularly in North America" (p. 3). Connor then goes on to list several recent microfilm projects and reference guides that now make this rich array of primary materials easily accessible to researchers. After illustrating the value of these bibliographcial tools by discussing several documents she found using them, she ends her essay by examining the pedagogical and professional reasons for studying the history of technical and scientific writing.

Finally, *Medieval and Rennaissance Letter Treatises and Form Letters: A Census of Manuscripts in Part of Western Europe, Japan, and the United States of America* (1994), by Emil J. Polak, will prove valuable to those whose research focuses on the letters of these periods. This volume includes collections in the British Isles, the Low Countries, Scandinavia, Switzerland, the Iberian Peninsula, and Greece, as well as Japan and the United States. (A review of this volume, by Judith Rice Henderson, appears in *Rhetorica* [Winter 1996].) An earlier volume covers collections in Central and Eastern Europe, and a planned third volume will include material held in France, Italy, Austria, and Germany.

Two recent articles discuss methods for conducting research in the history of business and technical communication. Kitty Locker was joined by Scott L. Miller, Malcomb Richardson, Elizabeth Tebeaux, and JoAnne Yates for a "conversation" article in 1996 that focuses on studies in the history of business communication. In turn, they address questions about identifying topics and research questions, finding archival and published sources, interpreting texts and contexts, and dealing with special problems in historical research. At the end of the article, they each define several areas and issues in the history of business communication in which additional research is needed.

Although the focus of her article is exclusively on medical texts, Jennifer J. Connor's 1993 essay, "Medical Text and Historical Context: Research Issues and Methods in History and Technical Communication," elaborates a methodology that can be successfully applied to the study of documents from a variety of technical and scientific disciplines. Connor builds her essay on her extensive work in Canadian and American medical history, but ranges through European and English works (especially those by Harvey) to illustrate her points. Her 116-item bibliography is in itself a valuable resource.

THE RHETORIC OF SCIENCE

The two essays on method previously mentioned both stress the interdisciplinary nature of research in business and technical communication. The burgeoning publications on the rhetoric of science illustrate that point well, but also place before us insights and approaches that will continue to inform our study of institutional and scientific writing. A number of separate items within various other sections of this bibliography fit into the category of texts and theories explored by rhetoricians of science; however, I have chosen to add a separate section to call special attention to this body of work.

James P. Zappen's provocative bibliograhpical essay, "Historical Studies in the Rhetoric of Science and Technology" (1987), performed the valuable task of identifying a new direction in studies of the rhetoric of science. He demonstrated that whereas earlier studies of the rhetoric of science emphasize its difference from classical rhetoric, more recent studies emphasize the parallels, and in so doing, call attention to the "political/ethical and probabilistic character of contemporary science and technology and of the contemporary discipline of technical communication" (p. 258). To support that thesis, Zappen presents and discusses a 115-item bibliography that included not only articles from technical communication, but also studies from philosophy, rhetoric, and speech communication. His heavy reliance on scholarship from other disciplines makes his bibliographical essay an excellent example of how studies in the history of technical communication are, and should be, building on a broader and more sophisticated scholarly base. His article also provides an excellent bibliography from which those interested in the rhetoric of science might build.

I have tried to include in this bibliography most of the works on the rhetoric of science that directly focus on the history of technical and scientific communication. However, anyone wishing to conduct research in this area should consult Zappen's bibliography, and also should examine at least three of the book-length studies of the rhetoric of science that have recently appeared: Charles Bazerman, *Shaping Written Knowledge: The Genre and Activity of the Experimental Article in Science* (1988); Lawrence J. Prelli, *A Rhetoric of Science: Inventing Scientific Discourse* (1989); and Alan G. Gross, *The Rhetoric of Science* (1990). Both Bazerman and Gross contain easily accessible and valuable bibliographies at the ends of their studies. (There are two reviews of Bazerman's work, one authored by David S. Kaufer, and the other by Carolyn R. Miller; Gross's work also has two reviews, one by Jeffrey Jones and another by Carolyn D. Rude.) Prelli's bibliography is embedded in endnotes after each chapter. His study is the most theoretically oriented of these studies and, therefore, contains only one chapter that focuses on specific rhetorical situations and works. (Prelli is reviewed by James P. Zappen and by M. Jimmie Killingsworth.) Bazerman and Gross consistently ground their theory in discussions of specific writers, works, and trends; their works, therefore, contain a number of chapters that are independently coherent. To facilitate access to these discussions, I deal with them separately in the appropriate sections of this essay. For the convenience of those who would like to consult longer evaluations of each book, each review mentioned is also listed in the reference section.

To these single-author books we should also add the three collections of essays discussed by Mary Lay in her review essay, "Rhetorical Analysis of Scientific Texts: Three Major Contributions" (1995): Peter Dear's *The Literary Structure of Scientific Argument: Historical Studies* (1991); Murdo William McRae's *The Literature of Science: Perspectives on Popular Scientific Writing* (1993); and Jack Selzer's *Understanding Scientific Prose* (1993). Written by historians of science, the essays in Dear's volume cover scientific materials from the 17th through the 19th centuries and are discussed separately in the bibliography. As Lay points out, most of these articles are very detailed analyses of materials that will be new to scholars in composition or even technical communication but which illustrate the central role of texts in creating scientific knowledge.

The collections edited by McRae and Selzer force us to address two questions concerning what materials constitute the history of business, scientific, and technical writing: How old does a text have to be to warrant study as a historical document? Should texts about science written specifically for the popular market be included among those written specifically for professional audiences? Essays in McRae's collection focus exclusively on popular scientific writing, including several writers who are still active (e.g., Oliver Sacks, Richard Selzer); thus, they raise both questions. I have included that collection, and several other studies of "popular" science writing, later in the bibliography because these "popular" texts are involved in a larger "conversation" about science, technology, society, our culture, and our interior and exterior environments. The analysis of those texts helps us see distinctions that elucidate the rhetorical dynamics of both pure scientific and popular science writing. Recent history is still history with lessons to teach—thus, even the most recent history deserves our study. The problem for the bibliographer becomes how much and what of it to include. Bascially, I have listed items written by scholars working in the rhetoric of science or the history of scientific and technical communication in the hope that, though not inclusive, the entries here will suggest lines of reading and research that individual researchers can build upon. The essays in McRae's collection are a valuable resource for scholars interested in exploring the rhetorical intersections and interactions between popular and scientific cultures.

The title of Selzer's collection, *Understanding Scientific Prose* (1991), was chosen, according to his introduction, not so much because it is indicative of the book's contents, as it is indicative of the book's goal, one parallel to that of Brooks and Warren in *Understanding Poetry*. The goal of Selzer and his contributors was "to demonstrate and domesticate new methods of practical criticism in order to 'open up' another kind of literature, scientific discourse" (p. 3). They seek to achieve that goal by applying different approaches of rhetorical analysis to a single text, "The Spandrels of San Marco and the Panglossian Paradigm: A Critique of the Adaptationist Programme," written by Stephen Jay Gould and R. C. Lowontin and presented at the 1978 Royal Society of London Symposium. Lay has reservations about the collection: Among other things, she points out that "The Spandrels" is a unique text designed to attack a major evolutionary theory before a broad audience and thus has characteristics that make it much like popular science literature. For the whole collection to work as its authors intended, we must assume that "traditional and popular scientific texts must have more similarities than differences" (1995, p. 297). Is the "richness" of this essay—the "richness" that enables the "richness" of the analyses in the collection—typical of what we traditionally call scientific literature? Lay's reservations and endorsements (she recommends it as a text for an introductory graduate class in methodology) are interesting and well founded. Gould certainly has earned a place in the history of scientific writing and even if this particular essay had not heretofore deserved mention among his historically significant works, it has certainly now gained that status simply because of the attention focused on it by the influential rhetoricians writing in *Understanding Scientific Prose*.

The rhetoric of science has also generated controversy. In his review of Marcello Pera's *The Discourses of Science* (1994), Alan Gross takes Pera to task for his limited claim for rhetoric—a claim which allows him to separate scientific rhetoric from its "traditional concerns with ethical and pathetic proofs and with style or arrangement," despite Pera's view that the "place of rhetoric in science is definite, constitutive, and substantial" (p. 252). This is a statement, Gross asserts, "one would expect a philosopher to make." That explains, according to Gross, why the collection of essays Pera coedited with William Shea, *Persuading Science: The Art of Scientific Rhetoric* (1991), does not include a single rhetorician of science. Pera has, in Gross's words, kept "his discourse pure for philosophy by avoiding any intellectual engagement with what is now a substantial literature of science" (p. 252). In his "reply" to Gross's review, Pera asserts that the label "'rhetoric of science' covers two kinds of people: those who are interested in *rhetoric* and those who are interested in *science*" (p. 255). Pera puts Gross in the first category, himself in the second. Those in the former group, he writes,

> share a "deconstructivist" approach aiming to dilute science into a sort of speech or narration or tale which is effective as long as it is enjoyable to its audience. That is why they take rhetoric as being mainly concerned with questions of style and arrangement, or ethos and pathos. That latter have a "foundationalist" or "reconstructivist" approach which aims to find the discursive constitution of science qua knowledge of nature, the grounds of its specific discourse, the reasons for its typical effectiveness, and the ways of it proper rationality. That is why they take rhetoric as being concerned with questions of *logos*. (p. 255)

I have quoted freely from both Gross and Pera to demonstrate the intensity of the opinions held by those who stand on either side of this major divide about the nature of the rhetoric of science. It would seem that, at least for now, even the powers of rhetoric itself are unable to bridge this gap.

A second debate was played out in the Summer 1993 issue of the *Southern Communication Journal*. That issue centers on a lengthy essay by Dilip Parameshwar Gaonkar in which he calls into question the ability of rhetoric to function as a discipline because, he argues, rhetoric does not have a theory "thick" enough for criticism. In particular, Gaonkar focuses on the attempt by rhetoricans to globalize rhetoric—to show that it is a hermeneutical metadiscourse that can "'swallow up' any other discourse" (Keith, 1993, p. 256). Gaonkar chooses to make his argument against the assertion that rhetoric is global in character by looking at the rhetoric of science, in particular the work of John Campbell, Alan Gross, and Larry Prelli. Following Goankar's lengthy critique of the rhetoric of science are responses by

Campbell, Gross, and Prelli, as well as Steve Fuller and Michael Leff. The debate is highly theoretical and complex, but whether they agree or disagree with Goankar, or qualify his view, all the respondents review their work in the rhetoric of science in interesting ways.

PEDAGOGICAL APPLICATIONS OF HISTORICAL MATERIAL

In his bibliography, Moran (1985) comments that all the work on the history of business and technical writing has had "little effect on teaching" (p. 25). In a sense, this is true: As Moran points out, textbooks do not draw upon historical scholarship; furthermore, only a handful of articles have been written whose central purpose is to address the connections between historical studies and pedagogy. Many of the earliest essays that dealt with the history of technical or business writing do, however, have a strong pedagogical emphasis. Walter J. Miller's "What Can the Technical Writer of the Past Teach the Technical Writer of Today" (1961/1975) presents 12 writers, from classical times through the 20th century, whose works can help students gain a fuller sense of technical writing skills. Robert Masse and Patrick M. Kelley, in "Teaching the Tradition of Technical and Scientific Writing" (1977), report on how they use major works in the history of science to enhance their students' awareness of the value of clear communication. Stephen L. Gresham's two frequently cited essays on the history of technical writing also have a strong pedagogical emphasis. In "When Technical Communicators Face the Past" (1978), Gresham reports on a survey of technical writers which reinforces the idea that placing technical writing in a historical context is important and should be given more emphasis in courses. In a later essay, "From Aristotle to Einstein: Scientific Literature and the Teaching of Technical Writing" (1981), Gresham explains how he uses the works of these early writers in his classroom. Jane Allen, in "The Literature of Science and the Technical Writing Curriculum" (1984), builds on Gresham's work by reporting on her classroom applications of early texts.

Two high-profile issues discussed in technical writing publications—the rhetoric of science and the ongoing debate over technical writing as a humanities course—have stimulated studies that emphasize classroom concerns. Daniel R. Jones's article, "A Rhetorical Approach for Teaching the Literature of Scientific and Technical Writing" (1985), argues that classic works from the history of science can be used not only to teach style, but also to teach science as rhetoric, thus, challenging student assumptions about the objectivity of science. He has his students read original texts and rhetorical analyses of those works, for example, John Angus Campbell's essays on Darwin (discussed under "Late 17th Century and the 18th Century," later in the chapter). In "History, Rhetoric, and Humanism: Toward a More Comprehensive Definition of Technical Communication" (1991), Russell Rutter argues that technical communication should be returned "to the mainstream of rhetoric and the liberal arts" (p. 148). That tradition, Rutter believes, should lead us to reexamine our definitions of technical writing and shift our modes of teaching to more clearly reflect liberal arts ideals.

In addition to the studies that focus directly on how historical works can be used in the classroom, many other essays presented in this bibliography contain very clear pedagogical implications or include whole sections devoted to how historical materials might be used by the technical or business writing teacher. However, even as I stress that the connections are there, I must emphasize that we need to think more carefully, and more often, about how the knowledge we gain from the past can be used effectively in today's classrooms, especially as the studies become more specialized and theoretical.

THE HISTORY OF TEACHING BUSINESS AND TECHNICAL WRITING

The best article-length overview of the history of teaching technical writing is Robert J. Connors' "The Rise of Technical Writing Instruction in America" (1982). Covering the period from 1895 through 1980, Connors not only traces trends in instructional techniques and texts, but also examines the forces that influenced technical writing instruction, especially over the 40-year period when technical writing instruction experienced its "coming of age."

Two recent book-length studies of writing instruction have made major contributions to our knowledge of the evolution of technical and business writing instruction. In her formidable 1993 book entitled, *A History of Professional Writing Instruction in American Colleges: Years of Acceptance, Growth, and Doubt*, Katherine H. Adams examines the history of all upper-level writing instruction (including creative writing, journalism, and technical writing for business, engineering, and agriculture) in America from the beginnings of these courses in the early 19th century through to the present decade. Based on extensive archival work, Adams shows in detail how these courses emerged from the traditional curriculum and then evolved in different directions to become housed, in many institutions, in separate departments. Adams also documents the interesting and complex relationship between writing instruction and the professions. Students of business and technical writing will be especially interested in Chapter 7, "'Professional Writing' in Agriculture, Engineering, and Business." Adams' overall thesis is that the fragmentation of writing instruction into different departments has been unfortunate—in that writing instructors in different contexts have not been able to share knowledge and techniques effectively. (In his largely complimentary review in 1994, David Russell offers a different conclusion—that the divisions, and consequent diversity, have been a good thing. Russell deals with business and technical writing at several points in his 1991 book, *Writing in the Academic Disciplines, 1870–1990,* particularly in Chapter 4, "Writing and the Ideal of Utility: Composition for the Culture of Professionalism.")

Teresa C. Kynell's 1996 book, *Writing in a Milieu of Utility: The Move to Technical Communication in American Engineering Programs 1850–1950*, focuses on the emergence of technical writing out of the engineering curriculum in America. In the first part of her study, Kynell largely documents the process that established engineering as a profession as it moved from its pre-1850 status as a craft learned through an apprenticeship to its evolution by the end of the century to a discipline entered mainly through a four-year technical college curriculum. English and writing instruction in these early years were included in the curriculum as "culture" courses designed to "humanize" members of the young profession and give them "status" much like that of physicians or lawyers. Their experience in English classes had little to do with the actual writing they would do on the job. During the first decades of the 20th century, however, professional engineers and engineering faculty began to criticize the writing skills of engineering graduates, calling for better, more applied writing instruction. This pressure lead schools to experiment with writing courses, thus beginning a slow but sure development toward technical writing as we know it. Changes in course purpose and design also generated debate about where the course should be taught—in English departments or in engineering schools. The 1940s brought war, rapidly evolving technology, and, in the second half of the decade, a dramatically increasing student population; all of these factors placed new demands on technical writing and influenced how it was thought of and taught. The sophisiticated technology developed during and after the war also generated the need for people with special expertise in technical writing and thus began the development of technical writing as a new profession. In her interesting narrative, Kynell documents this institutional history carefully and extensively, providing a clear view of the "turbulent" early history of technical writing as it grew out of need and technological change, through its search during the first half of the 20th century for intellecual and curricular space somewhere between engineering and English. Kynell resisted suggestions that she extend her study through the 1990s, arguing that "the period 1850–1950 demonstrates *the move* to technical communication in American engineering programs," whereas, "the period 1950 to the present will likely demonstrate the move of technical communication into our English and Humanities programs as a distinct major" (p. xi). Clearly, a major divide occurred around 1950 that justifies her decision to stop at that point in time. However, Kynell's tantalizing preview of the major professional and pedagogical changes that occured in the second half of the 20th century (changes that many of us know, at least in part, from our own experience) argues strongly for a second book from Kynell to complete this important study—a study that will help us better understand where we are and where we might be going as a profession.

Rhonda Carnell Grego's article, "Science, Late Nineteenth-Century Rhetoric, and the Beginnings of Technical Writing Instruction in America" (1987), supplements Connors' review of the early period of technical writing instruction by looking carefully at the evolution of technical writing at Penn State University. Whereas Connors' history focuses on engineering, Grego looks at both engineering and English faculty and their reasons for teaching technical writing as they did. She demonstrates that the English faculty were influenced by late 19th-century rhetoricians' interest in the inductive methodology of "pure" science—an emphasis that made conclusions self-evident and therefore minimized the importance of argument. Engineering faculty became dissatisfied with this approach because of their students' need to persuade their superiors that their applications of abstract, scientific laws did indeed fit the specific situations presented by specific clients. Thus, dissatisfaction of engineering faculty came not from English instructors' attempts to "humanize" students through the study of literature, but from their using examples from "pure" rather than "applied" science.

Other essays on technical writing instruction focus on teachers and texts. "The First Textbook on Technical Writing" (1977), by Richard W. Schmelzer, is a tribute to Dr. Ray Palmer Baker of Rensselaer Polytechnic Institute for his early technical writing text *The Preparation of Reports* (1924). Mary Rosner's "Style and Audience in Technical Writing" (1983) offers an analysis of 17 technical writing texts from the 1910s through the 1930s, as well as a comparison of those texts with three "modern" texts (one from 1979; two from 1980). In her 1995 article, "English as an Engineering Tool: Samuel Chandler Earle and the Tufts Experiment," Teresa Kynell examines Earle's 1911 presentation to the Society for the Promotion of Engineering Education to demonstrate his pivotal role in the development of writing courses for engineers, particularly his encouragement of dialogue between English (writing) teachers and engineering faculty. In "Frank Aydelotte: AT&T's First Writing Consultant, 1917–1918" (1995), Mike Moran reveals the contribution of another important early player in the development of technical writing by examining Aydelotte's adaptation of his MIT technical writing course to fit the needs of new employees at AT&T during World War I. And Teresa Kynell, in "Considering our Pedagogical Past through Textbooks: A Conversation with John M. Lannon" (1994), interviews Lannon and evaluates five editions of his textbook, *Technical Writing,* as a way to see how technical communication responded to the social realities of the 1980s.

Studies of the history of business writing instruction in America mainly deal with its beginnings in the late 19th century. Arthur B. Smith, Jr., began the serious investigation of business writing instruction with "Historical Development of Concern for Business English Instruction" (1961). His analysis focuses on activity shortly before and after the turn of the century. Carter A. Daniel's "Sherwin Cody: Business Communication Pioneer" (1982) argues that Sherwin Cody, who is remembered for his high-profile, highly commercialized "School of English," must be regarded as a founder of modern business communication. The most recent and most scholarly study is "The Spurious Paternity of Business Communication Principles" (1989) by John Hagge. Hagge rejects the claims of Smith, Daniel, and others and argues that the beginnings of business writing instruction originated much earlier, especially with the work of J. Willis Westlake, beginning in 1876. But even Westlake, Hagge points out, was working in a 2,000-year old epistolographic tradition.

Finally, Vanessa Dean Arnold's 1989 article, "A Twenty-five Year Perspective on the Pedagogy of Business Communication," provides a quick overview of the evolution of interests among business writing teachers, as seen through a selection of their comments in the *ABC Bulletin* and the *Journal of Business Communication* between 1963 and 1988.

ANCIENT AND CLASSICAL LITERATURE

The dramatic increase in publications on the history of business and technical writing that center on classical literature is one of the clearest indications that scholars are now seeing, and building upon, the connections between traditional rhetoric and business and technical writing. In 1985, Moran could mention only Elbert W. Harrington's 1948 book-length study, *Rhetoric and the Scientific Method of Inquiry*, and two articles that touched on classical writers as part of a survey (i.e., those by Miller [1961] and Shulman [1960]). Since 1985, a number of articles—most of them quite scholarly—have appeared. These studies are perhaps only the beginning of what is obviously a rich ground for those interested in the history of business and technical writing.

Most of the essays in this group deal with connections between classical rhetoric and contemporary business and technical writing. Although he was not the first to point out the influence classical rhetoric had on modern business and technical writing practices, Edward P. J. Corbett's essay, "What Classical Rhetoric Has to Offer the Teacher and the Student of Business and Technical Writing" (1989), brings to this line of inquiry a higher profile because of his well deserved reputation as an expert on classical rhetoric, as well as its implications for composition studies. Corbett's purpose is to show that "business and professional communication has its own rhetorical system...that has been shaped not only by the natural demands of...verbal exchange among human beings, but also by the enduring principles of classical rhetoric" (p. 66). Corbett's short essay discusses a number of connections, including structure, style, audience awareness, and even physical appearance.

Other essays that follow the pattern found in Corbett's essay are more narrowly focused, but also more detailed in their treatment of both classical rhetoric and contemporary applications. Rosemary L. Gates, in "Understanding Writing as an Art: Classical Rhetoric and the Corporate Context" (1990), examines two classical rhetorical concepts (*kairos*, or "appropriateness," and *aitia*, or "cause"), showing how they function in modern business, scientific, and technical writing. She then offers ten recommendations for teaching technical writing based on these two rhetorical concepts. John Hagge's 1989 essay, "Ties that Bind: Ancient Epistolography and Modern Business Communication," offers a thorough and scholarly discussion of the parallels between ancient letter writing traditions and more recent practices as seen through late 19th- and 20th-century textbooks. Herbert W. Hildebrandt also focuses on letters in "Some Influence of Greek and Roman Rhetoric on Early Letter

Writing" (1988). The influence Hildebrandt demonstrates is from the classical theory of oral rhetoric (seen in Aristotle, Cicero, Quintilian, and others) to English, Italian, and German practitioners of letter writing (*dictamen*) in the Medieval and Renaissance periods.

At least two essays use stylistic advice from classical Greek rhetoric to argue for a more complex—and therefore more versatile and persuasive—prose style in business communications. Craig and Carol Kallendorf, in "The Figures of Speech, *Ethos*, and Aristotle: Notes Toward a Rhetoric of Business Communication" (1985), focus specifically on the figures of speech as presented by Aristotle in his *Rhetoric* and argue not only that these stylistic devices can (and should) be used in contemporary business writing, but also that business communication constitutes a fourth, and modern, rhetoric to supplement the classical distinctions of judicial, deliberative, and epideictic. In "Business Prose and the Nature of the Plain Style" (1987), Michael Mendelson rejects as too narrow the standard of prose style advocated in most business writing texts and offers a broader definition based on the Greek rhetoricians. He finds in these classical sources a plain style that allows for the wider range of "dictional, syntactic, and figurative choices" (p. 3) necessary to accommodate the complex and diverse nature of modern business.

In Alan G. Gross's 1988 article, "Discourse on Method: The Rhetorical Analysis of Scientific Texts" (revised to become Chapter 1 in *The Rhetoric of Science* [1990]), he advocates a neo-Aristotelian rhetoric that is capable of "a systematic examination of the most socially privileged communications in our society" (p. 184)—scientific texts. In this essay, which ranges over the history of science to find examples, but works primarily out of classical rhetoric, Gross argues that the rhetoric of science should be recognized as a new humanistic discipline.

Individual writers have attracted the attention of modern scholars looking to ancient literature for antecedents and correctives to modern practices. As might be expected, most studies focus on Aristotle. In addition to the Kallendorfs' work mentioned above, Aristotle inspired H. W. Hildebrandt's "Aristotelian Views of the Twentieth Century" (1984)—the Association for Business Communication Presidential Address written in the form of a letter from a 20th-century Phaedrus who describes and critiques modern business communication from an Aristotelian perspective. Charles R. Fenno and Terrance Skelton, in their separate 1986 ITCC papers, show how Aristotelian rhetorical principles (especially invention and arrangement) have a place in technical writing—even in online writing and editing.

In "Cicero's Arrangement in Scientific Writing" (1988), Lori L. Alexander argues that Cicero's discussion of logical and ethical appeals can be studied profitably by modern scientific writers now that we (following Kuhn and others) accept the place of subjectivity and persuasion in science. S. Michael Halloran and Merrill D. Whitburn also look to Cicero's rhetoric as a guide for modern scientific and technical discourse. In "Ciceronian Rhetoric and the Rise of Science: The Plain Style Reconsidered" (1982), they argue that the best model for the plain language movement should not be the 17th century impetus toward the plain style associated with

Bishop Sprat and the Royal Society, but the ideas on plain language found in Cicero and other classical rhetoricians: They argue that Cicero's more sophisticated approach seeks to achieve clarity with language that is not stripped of personality or subtlety. Michael Mendelson's "The Rhetorical Case: Its Roman Precedent and the Current Debate" (1989) draws upon the Roman debate over the value of declamation and Quintilian's insistence that declamation not drift into an empty verbal game to inform the current debate over the value of problem-solving cases in business and technical writing. I use that same point from Quintilian—that rhetoric must be tested in the crucible of the forum—to support my argument that business writing fits well within the humanist tradition in "The Place of Business Writing in English Departments: A Justification" (1980).

Julius Caesar's skills as an engineer and technical writer are pointed out by Tim Whalen in "Gaius Julius Caesar, Technical Writer" (1986). Whalen also offers "A History of Specifications: Technical Writing in Perspective" (1985), a "history" that not only reviews the use of specifications in ancient and classical times, but also divides those specifications into three types: Control of Nature (e.g., civil engineering), War and Military Science, and Commerce.

At least two scholars have produced interesting studies by applying modern techniques of rhetorical analysis to ancient texts. Carol S. Lipson's "Ancient Egyptian Medical Texts: A Rhetorical Analysis of Two of the Oldest Papyri" (1990) describes the complex rhetorical dynamics operating within ancient texts that challenged traditions even as they valued them. Elisabeth M. Alford uses theories from rhetoric and sociology, as well as schema theory and reader-response theory, to inform her analysis in "Thucydides and the Plague in Athens: The Roots of Scientific Writing" (1988). She demonstrates that Thucydides' sophisticated and varied rhetorical strategies are largely responsible for the mixed reaction to his work over the centuries.

MEDIEVAL EUROPEAN

Studies in the history of business writing during the Medieval period are dominated by one scholar, Malcolm Richardson. In five essays, he has surveyed the history of business writing in the 15th century ("The First Century of English Business Writing, 1417–1525" [1985]), studied the nature and influence of business letters in the 15th century ("The *Dictamen* and Its Influence on Fifteenth-Century English Prose" [1984] and "The Earliest Business Letters in English: An Overview" [1980]), examined the relationship between business correspondence and literacy during the period ("Business Writing and the Spread of Literacy in Late Medieval England" [1985]), and provided a valuable outline of research methods for those who wish to investigate business writing in medieval times ("Methodology for Researching Early Business Writing in English" [1985]). In his work on the *dictamen*, a genre that had continuing influence through the Renaissance and after, Richardson is joined by Luella M. Wolff. While Richardson traces the influence of the *dictamen* on medieval prose,

Wolff's "A Brief History of the Art of Dictamen: Medieval Origins of Business Letter Writing" (1979) examines that form as an antecedent for modern business letter principles. More recently, Les Perelman, in his "The Medieval Art of Letter Writing" (1991), argues that the *ars dictaminis* (or, collectively, the *dictamen*) was a new form of rhetoric manual that "transformed the complex rhetorical traditions of the classical period with their emphasis on persuasion into a phatic rhetoric of personal and official relations" (p. 116). These texts, Perelman suggests, were largely responsible for establishing and stabilizing the influence of the medieval church and, thus, medieval society. The *dictamen* illustrate (in the words of Charles Bazerman and James Paradis in their introduction to the collection of essays in which Perelman's essay appears) how "rhetorical forms set standards for the structuring of human relations" (p. 6).

Studies of technical writing in the Medieval period tend to point out models of good technical writing. Barbara M. Olds presents a brief analysis of a technical passage from the Anglo-Saxon *Leechbook* (ca. 950 A.D.) in "An Anglo-Saxon Technical Writer" (1984). In "Agricola's Preface to *De Re Metallica*" (1986), Jane Allen provides a similar treatment of Agricola's "Preface" to *De Re Metallica* (1541), the first book to record mining methods. Bradford B. Broughton's "The Art of Falconry: A Surprising Manual of Rhetoric" (1989) offers a description of the rhetorical strategies used by Frederic II, the Holy Roman Emperor, in his detailed study of falconry. Finally, Max Loges offers an analysis of the first technical manual on sport fishing in *"The Treatise of Fishing with an Angle*: A Study of a Fifteenth-Century Technical Manual" (1994). Loges points out features of the text that make it effective technical writing and that help account for the apparent success of *The Book of St. Albans*, the collection in which the *Treatise of Fishing* appeared.

The medieval author who has attracted the most attention as a technical writer is Geoffrey Chaucer. Although there were a few scattered earlier mentions of Chaucer as a technical writer (for example, W. A. Freeman's "Geoffrey Chaucer, Technical Writer" [1961]), a number of scholars "discovered" Chaucer's technical writing skills in the 1980s. Three of these studies—George Ovitt's "A Late Medieval Technical Directive: Chaucer's *Treatise on the Astrolabe*" (1981), Edmond A. Basquin's "The First Technical Writer in English: Geoffrey Chaucer" (1981), and Carol S. Lipson's "Descriptions and Instructions in Medieval Times: Lessons to be Learnt from Geoffrey Chaucer's Scientific Instruction Manual" (1982)—all appeared within a few months of each other. All these studies are careful to point out Chaucer's skill in solving technical writing problems with few, if any, models to guide him. Lipson's study—the most scholarly of the essays in this group—includes a comparison of Chaucer's text with his Arabian source, thereby clearly demonstrating his innovations. More recently, in 1993, Peter J. Hager and Ronald J. Nelson offer a long, detailed study, "Chaucer's *A Treatise on the Astrolabe*: A 600-year-old Model for Humanizing Technical Documents," in which they argue that *The Astrolabe* is a model for modern technical writers because in it Chaucer merges his logico-rational self with his humanistic self to produce a text that is not only technically accurate but also accessible.

Two more recent studies of Chaucer as technical writer were designed primarily for readers whose concerns do not include the teaching of technical communication. Sigmund Eisner's essay, "Chaucer as a Technical Writer" (1985), examines Chaucer's *Treatise* from the perspective of a literary scholar, and offers a comparative study of Chaucer's text and other technical documents of the time to reveal the quality of his work. In 1987, George Ovitt offered a second essay on Chaucer, entitled "History, Technical Style, and Chaucer's *Treatise on the Astrolabe*." Here, he argues that the astrolabe and Chaucer's *Treatise* occupy a special place within the history of technology and ideas because the instrument, paired with Chaucer's procedures for its construction and use, helped to create a context in which scientific exercises (mathematical analyses based on accurate astronomical observations) could be performed.

Finally, John Hagge, in his recent essay, "The First Technical Writer in English: A Challenge to the Hegemony of Chaucer" (1990), asserts that Chaucer was not the first technical writer in English, the explicit and implicit claims of previous scholars notwithstanding. Hagge backs up his claim by pointing to a technical writing tradition in Old and Middle English that was well underway before Chaucer. Hagge's work here, and the additional articles on technical writing in Old and Middle English he plans, suggest that the Medieval period will continue to grow as an active area for study in the history of business and technical writing.

THE RENAISSANCE: ENGLAND AND THE CONTINENT

Since Moran's bibliography appeared in 1985, scholarly attention focused on business and technical writing in the Renaissance has increased phenomenally. These studies are fairly evenly balanced between those that trace the evolution of genres or prose style and those that focus on the contributions of individual authors. Three scholars contributed essays on the development of Renaissance business correspondence to the Association for Business Communication's 1985 collection, *Studies in the History of Business Writing*. In "Humanistic Influences on the Art of the Familiar Epistle in the Renaissance," Donald R. Dickson demonstrates the influence of both the Renaissance humanists (e.g., Erasmus, Vives, Ascham) and later compilers of books on letter writing practices in England (e.g., Day, Fullwood). He also documents a shift from a more complex Ciceronian style to the more curt Senecan style and structure. Herbert W. Hildebrandt's "A 16th Century Work on Communication: Precursor of Modern Business Communications" centers on Angell Day's *The English Secretary or Method of Writing Epistles and Letters*. Hildebrandt briefly sets Day's work into context and then offers a detailed summary and analysis of his contribution to business communication. Kitty O. Locker's "The Earlier Correspondence of the British East India Company (1600–19)" is an intriguing look at the actual letter writing practices of East India Company employees who had to rely almost entirely on writing to conduct the business of that far-flung enterprise. Elizabeth Tebeaux's recent essay, "Renaissance Epistolography and the Origins of Business Correspondence, 1568–1640" (1992), further expands our understanding of how letters were approached in the 16th and 17th centuries. She locates the earliest influence on letter writing in the *dictamen* produced by Erasmus and others, showing how this rhetorical, process-oriented approach flourished in the 16th century, then faded in favor of a more rigid, formulaic, product-oriented approach by the early 17th century.

Tebeaux's essay on business correspondence in the Renaissance is but one of several important contributions she has made to our knowledge of the writing produced in that period. In fact, in a series of articles and a recent book on the subject of technical writing between the years 1475 and 1640, Tebeaux has established herself as the most prolific and influential scholar working in this period, or perhaps any other. In "The Evolution of Technical Description in Renaissance English Technical Writing, 1475–1640: From Orality to Textuality" (1991), she traces the shifts in style and format as writers modified their approaches to technical description to better take advantage of opportunities inherent in print media. This movement was characterized by more reliance on illustration. Thus, verbal description in texts moved from an oral tradition to a text-based tradition in response to new and more effective visual techniques. In three other essays, Tebeaux demonstrates that the better Renaissance technical writers reveal a clear sense of audience awareness, especially in their attention to visual elements. In "Books of Secrets—Authors and Their Perception of Audience in Procedure Writing" (1990), she surveys a number of "how to" works on agriculture, navigation, medicine, and other similar topics and finds that these writers were sensitive to readers' present needs, their previous knowledge, and their reading context. Tebeaux also found, and illustrates, through facsimiles of several pages, that writers were consciously working to enhance readability through page design. Tebeaux examines page design further in "Visual Language: The Development of Format and Page Design in English Renaissance Technical Writing" (1991). Her survey of representative works between 1475 and 1640 reveals an evolving sophistication in the use of design features, including partitions, headings, visual aids, enumeration and listing devices, and font changes. Tebeaux takes this interest in visual design a step further in "Ramus, Visual Rhetoric, and the Emergence of Page Design in Medical Writing of the English Renaissance: Tracking the Evolution of Readable Documents" (1991). Here, she argues convincingly that one of the major *philosophical* influences behind the sudden interest in page design in the late 16th century was the work of Peter Ramus, who used brackets to group information and to show relationships of groups to larger concepts. Again, Tebeaux provides numerous facsimile reprints of pages to show how Renaissance writers used white space, headings, and brackets to reveal hierarchies, to show relationships, and to emphasize key points. The only other study that touches on the visual element in Renaissance technical communication is "John White: Renaissance England's First Important Ethnographic Illustrator" (1990), by Michael Moran. Moran's interesting account of White's drawings of Native Americans, however, deals only with illustration, not the visual elements of texts. Clearly,

Tebeaux has taken the lead in the study of the relationships between text and design—a lead that, we hope, will stimulate other studies of the visual element in other pre-20th century texts. (See also Olsen's essay on the evolution of visuals in technical writing, under "The 19th Century: England and the Continent.")

Tebeaux, along with Mary Lay, broke ground in another important area of study for historians of technical and scientific writing: the place of women in a culture as seen through technical documents. In "Images of Women in Technical Books from the English Renaissance" (1992), Tebeaux and Lay examine a number of technical books written for women (mostly in the middle class) and conclude that these women were very active in their culture and, based on the style and presentation in these texts, were considered as literate as men. They also detect in these texts evidence that literacy among women increased dramatically during the Renaissance. And, even though the writers of Renaissance technical books addressed to women did consider women as literate as men, they also developed distinct ways of presenting information to women that modern writers might well consider. Tebeaux pursues her investigation of Renaissance writing for women in a subsequent article, "Technical Writing for Women of the English Renaissance: Technology, Literacy, and the Emergence of a Genre" (1993). These documents not only indicate the nature of the work done by women, but also suggest, among other things, a pattern of improving reading skills among women between the early 16th century and the 17th century, as the documents shifted from an oral style to a more analytical style that also reflected the growth of knowledge during this period.

In the 1992 essay, "Expanding and Redirecting Historical Research in Technical Writing: In Search of Our Past," Tebeaux and M. Jimmie Killingsworth have produced a valuable touchstone piece for scholars working in the Renaissance, or in any period. In this important essay, they first address the problems and opportunities inherent in studying the history of technical writing and then offer guidelines for scholars to consider as framing questions for such studies. They stress the importance of studying trends within periods and of making distinctions between technical and scientific writing. Tebeaux and Killingsworth go on to illustrate the approach they suggest by applying their guidelines to the technical writing done in the Renaissance.

Tebeaux's book-length study, *The Emergence of a Tradition: Technical Writing in the English Renaissance, 1475–1640*, builds upon, but goes well beyond, these articles to provide an insightful overview of the rise of technical writing in the 16th and early 17th centuries. The book is copiously illustrated and contains chapters that expand Tebeaux's previous work on format and page design, the responses of writers to their audiences, and the move from orality to textuality in English culture. Of particular interest, however, is a chapter in which Tebeaux argues that the plain style long attributed by literary scholars to the influence of Bacon and his followers had long existed in the noncannoical technical texts written and printed in, and before, the 16th century. As Tebeaux points out, highly respected scholars seriously misread and misunderstood one of the most prolific periods in European cultural history because they did not take into account what she calls "pragmatic discourse." Thus, Tebeaux dramatically illustrates the potential value of all scholarship in the history of business and technical writing and validates the argument recently made by some literary scholars that *all* texts produced within a culture must inform our understanding of that culture, in particular, and human communication and knowledge, in general. Tebeaux ends her study with a chapter that not only sums up her conclusions, but also offers a valuable list of research suggestions for those interested in working on the history of technical communication, particularly in the Renaissance period.

Another scholar who has made several valuable contributions to the study of Renaissance technical and scientific writing is James Stephens. His two essays on style in Renaissance science are fascinating studies of the very difficult rhetorical problems faced by scientists such as Paracelsus, Copernicus, Galileo, Kepler, and Bacon. In "Rhetorical Problems in Renaissance Science" (1975), Stephens demonstrates these writers' acute awareness of the importance of stylistic choices when presenting their new theories and discoveries to skeptical, even antagonistic readers. The solutions they found were varied and sophisticated, as these writers were often trying to discourage less intelligent readers who might not understand and approve of their ideas, even as they were attempting to engage wise, well-read readers. Stephens takes this study of style an intriguing step further in "Style as Therapy in Renaissance Science" (1983), singling out one element common to many of these scientists—their attempt to include in their writing enough of the old, familiar "truths" to "alleviate the pain of confronting new truths" (p. 187–188). Alan Gross works a similar theme in Chapter 7 of *The Rhetoric of Science* (1990), "Copernicus and Revolutionary Model Building." Gross's title is a bit misleading in that his focus in this chapter is more on the rhetorical strategy and influence of Rheticus' *Narratio Prima*, the first printed work supporting Copernicanism. Published in 1540, three years before Copernicus' *De Revolutionibus*, the *Narratio Prima* does more, according to Gross, than just summarize the heliocentric argument: it justifies Rheticus' choice of the "Copernican over the Ptolemaic hypothesis....by going beyond argument and evidence" into a conversion narrative, a rhetorical model Gross calls "rational conversion" (p. 98), and traces through Brahe, Maestlin, and Kepler. The "rational conversion" model was, Gross argues, a necessary, but unscientific, rhetorical device used to create "modern exact science," a fact not realized until after the work of Newton.

Because he was so influential a thinker and so prolific a writer, Francis Bacon has attracted the interest of scholars for many years. James P. Zappen began this trend in the 1970s with two essays that focused on Bacon's special contributions to modern scientific rhetoric. In "Francis Bacon and the Rhetoric of Science" (1975), Zappen shows that, in addition to criticizing the flowery, Ciceronian prose common in the Renaissance, Bacon developed a sophisticated theory of rhetoric that turned on subject and audience analysis. In "Francis Bacon and the Topics" (1977), Zappen argues that Bacon's criticism of the topics led to the modern understanding of topics not as invention tools, but as schemes for organizing information.

Although he is, with reason, usually thought of as one of the first advocates of a plain, unadorned style in scientific writing, several recent studies of Bacon's writing focus on the complexity of his own style. In "Francis Bacon and the Technology of Style" (1988), Christopher Baker examines on Bacon's use of aphorisms as a device for expressing scientific facts. In her essay, "Francis Bacon and *Plain* Scientific Prose: A Reexamination" (1985), Carol Lipson argues that Bacon's own style is often at odds with the plain prose ideal for which he himself argued. Furthermore, his rhetorical method obscured his ideas enough so that the ideas the Royal Society picked up and transmitted to posterity are often at variance with his theory.

Readers wishing to study in detail Bacon's ideas about rhetoric might examine essays by Cogan, Harrison, and Wallace, as well as book-length studies by Briggs, Stephens, and Wallace. These works (listed in the references but not all discussed here), approach Bacon's contributions from a literary or philosophical perspective and, thus, build on complex links between philosophy, rhetoric, science, and the communication of scientific truths—links we must be willing to explore as we try to understand the history of technical writing. Marc Cogan, for example, presents in "Rhetoric and Action in Francis Bacon" (1981) an interesting analysis that focuses on Bacon's "striking innovation" in rhetorical theory. That innovation, that "rhetoric is...determined by certain operations of the mind..., rather than by the requirements of certain sorts of issues, the shape of certain sorts of arguments, or the effects of certain sorts of language" (p. 214), gives rhetoric an "inward function and movement" and thus, gives "an explicitly rhetorical dimension to action." Dietrich Rathjen's article, "The Problem of Synonymy: Bacon's Third Idol Explained" (1987), first points out Bacon's emphasis on the importance of words, on the necessity of accurate, specific definitions, and the proper use of words, then uses Bacon's pronouncements as a starting point to argue for rigorous accuracy in contemporary practice. "The integrity of language," he quite rightly points out, "determines the integrity of science, society, and politics" (p. 383).

Whereas Bacon's work has been approached by scholars interested in the history of business and technical writing, as well as by philosophers, historians, rhetoricians, and literary historians, Galileo's work has generated only one article in a business and technical writing journal—a brief, but interesting piece by Barbara Smith entitled "Audience Analysis in the 17th Century" (1984), which explains how Galileo designed his manual on the operation of his military compass to meet the needs of his readers. However, Galileo has been the subject of several very interesting and extensive contributions to the history of science that deal with Galileo as a rhetorician. For example, in *Galileo and the Art of Reasoning: Rhetorical Foundations of Logic and Scientific Method* (1980), Maurice A. Finocchiaro provides a detailed analysis of the *Dialogue Concerning the Two Chief World Systems*, in which he argues that Galileo's logic breaks down occasionally because he gave rhetorical needs precedence. Interestingly, reviews of the book vary radically in their assessment of its value. Ch. Perelman, writing in *Philosophy and Rhetoric*, finds it an important book. However, Floyd D. Anderson, in a review in the same issue of the same journal, argues that the book is deeply flawed because of its "inadequate conception of rhetoric" (p. 137). In "The Interplay of Science and Rhetoric in Seventeenth Century Italy" (1989), an essay published in *Rhetorica*, Jean Dietz Moss discusses the impact of the "new science" on rhetoric, as seen through the writings of Galileo, Guiducci, and Grassi. Their writings in support of the Copernican theory—a theory that could not be proven through traditional direct demonstrations based on common experience—reveal a reliance on rhetoric that grows throughout the controversy. Moss expands on this thesis in her 1993 book on the Copernican debate, *Novelties in the Heavens: Rhetoric and Science in the Copernican Controversy*. Here, Moss examines the role of rhetoric as the Copernican issue evolved from the time of *De Revolutionibus* in 1543 through the publication of Newton's *Principia* in 1687. Her main focus is on Galileo's contributions, as it should be, and she argues that he was the "father of the revolution in the use of rhetoric" to support scientific arguments. Although they sometimes question aspects of her thesis, reviewers of Moss's book (Marie J. Secor, Alan Gross, and Maurice A. Finocchiaro) praise its scholarly thoroughness and evenhandedness, and see it as a major contribution to our knowledge of this pivotal moment in the history of science and rhetoric. In "Narrative, Anecdotes, and Experiments: Turning Experience into Science in the Seventeenth Century" (1991), Peter Dear, a historian of science, examines the controversy between Galileo and Jesuit scholastic scholars. He argues that Galileo's "experiments" were not experiments in our current sense, but held an "irredeemably linguistic dimension" (p. 8) driven by the then-current rhetorical practice.

In "Thematic Repetition as Rhetorical Technique" (1991), Jo Allen examines William Harvey's *On the Motion of the Heart and Blood in Animals* to determine how he tried to overcome long held contemporary misconceptions about the body and its functions. She argues that Harvey used circular references, metaphors, and organizational techniques not only to support his thesis that blood circulates, but also to disarm his critics by suggesting "the circular pattern as part of God's natural order for the universe" (p. 29). In their 1992 article, "Commentary on Rhetorical Analysis of William Harvey's *De Motu Cordis* (1628)," J. T. H. Connor and Jennifer J. Connor take Allen to task for not being aware of studies of Harvey's writing (including the circle metaphor) in major journals in the history of medicine; for ignoring the probable influence of Aristotle on his thinking about the circle; for giving Copernican ideas precedence over Ptolemaic ideas when it should be the other way around; for mislabeling his target audience (since the work was in Latin, Harvey would not have had a "public" audience in mind); and for using only one translation (and that a problematic one) upon which to base her rhetorical analysis. In her "Commentary: A Response to J. T. H. Connor and Jennifer J. Connor's Analysis" (1992), Jo Allen, while admitting that the Connors' response does raise a few important points (e.g., the difficulties inherent in what she calls "crossover" studies—studies that cross lines between, for example, technical communication and the history of medicine), she effectively defends her research and writing against most of the major objections raised by the Connors. The exchange is enlightening, not only for what it shows us about Harvey's work, but also for what it shows us about the high scholarly standards that must now be applied as we study and write about the history of technical and scientific writing.

Other studies single out the work of four other important Renaissance writers and thinkers: Thomas Hobbes, Sir Thomas Browne, Ben Jonson, and Leonardo da Vinci. Joan Bennett, in "Science and the Plain Style" (1971), points out that the philosopher and mathematician Thomas Hobbes distrusted rhetoric and called for an exact, even mathematical use of language, thus fitting in with the 17th century movement toward the plain style. In "The Rhetorical Principles of Sir Thomas Browne" (1979), Michael H. Markel finds in the physician's works three rhetorical principles for technical writers: the importance of careful audience analysis, the need to balance overview statements and details, and the writer's ethical responsibility to convey information accurately. Ronald J. Nelson points out in "Ben Jonson's *Timber*: A Compilation of Verities" (1990) that the famous playwright also offers, in this collection of quotations, paraphrases, and Jonson's own random observations, valuable advice on the nature of language, the elements of style, and effective speech. In another of Nelson's essays, "Leonardo da Vinci as Technical Writer" (1988), he briefly notes the detail and accuracy of Leonardo's descriptions of the human body.

Finally, Stijn A. Verrept reminds us of the international nature of Renaissance business in "Review Essay: *Colloquia et Dictionariolum Septem Linguarum...Antverpiae* 1616: A Starting Point for Systematic Research on Cross-Cultural (Business) Communication" (1985). The book Verrept describes—one of many published during the time—not only contains a seven-language word list, a survey of grammar, and a pronunciation guide, but also includes sample phrases to use in conversation in other languages, sample letters in several languages, and advice on how to conduct oneself in different cultures. In 1616, this anonymous author chose Dutch, German, English, French, Latin, Spanish, and Italian as the languages essential for the international businessperson. Verrpt argues—quite rightly—that we today would do well to look at our need for cross-cultural studies of business communication, especially since our list of languages would not have the linguistic and cultural coherence of the Renaissance list.

LATE 17TH CENTURY AND THE 18TH CENTURY

The 17th century has long been recognized as a pivotal period in the history of English prose style. Writers in all areas were beginning to leave behind the more ornate patterns common in the Renaissance and adopt a less formal, more direct, "plainer" mode of expression. The pressures that drove these changes were many, ranging from intellectual influences (e.g., the essays of Montaigne, Bacon, and Dryden) to political and economic changes that enlarged the reading public and stimulated the advent of the periodical press. The single 17th-century institution that gets the most credit for this style shift, however, is the Royal Society. Its stylistic ideals, known chiefly through Thomas Sprat's *History of the Royal Society*, exerted a strong and complex influence on an English culture trying to come to grips with the major scientific, economic, political, and social changes of the time. Because of its broad cultural impact, the Royal Society, and the plain style it advocated, have attracted scholarly attention for many decades. (For alternative accounts of the origin of the plain style, see the article by Halloran and Whitburn [under the section "Ancient and Classical Literature"] and Elizabeth Tebeaux's book [in the preceding section, "The Renaissance: England and the Continent"].)

In three essays, "Science and English Prose Style in the Third Quarter of the Seventeenth Century" (1930), "Science and Criticism" (1951), and "Science and Language in England of the Mid-Seventeenth Century" (1971), Richard F. Jones argues that the plain style advocated by many Royal Society figures (e.g., John Wilkins, Thomas Sprat, William Petty) can be distinguished from the plain style of the anti-Ciceronians. For Jones, the Royal Society group was the main force behind the plain style because of its rejection of rhetorical florish. Morris Croll, who used the term *anti-Ciceronians* in *Attic and Baroque Prose Style* (1966), joined R. S. Crane in "Reviews of R. F. Jones 'Science and English Prose Style in the Third Quarter of the Seventeenth Century'" (1971) to attack Jones's thesis. They argued that the Royal Society was influenced by the earlier anti-Ciceronian movement and that Jones's position did not take into account the clear shift toward a plainer style across genres and subject areas in the 17th century. In *The Rise of Modern Prose Style* (1971), Robert Adolph rejects the arguments of both Jones and Croll as incomplete and explains the advent of the plain style to a growing utilitarian movement that began with Bacon.

Clearly, the Royal Society's emphasis on stylistic plainness was but one of many influences operating within the language in the 17th century. However, as a number of scholars have shown, its influence on writers of the time was significant and can be documented. In "A Reformed Writer in 1676" (1982), R. John Brockmann, building on Jones' 1930 study, focuses on the changes in style evident in Joseph Glanvill's 1667 revision of the *Vanity of Dogmatizing* as he sought approval from, and membership in, the Royal Society. In "The Royal Society, Henry Oldenburg, and Some Origins of the Modern Technical Paper" (1981), James G. Paradis examines the work of Oldenburg, the first editor of the Royal Society's *Philosophical Transactions*. Paradis shows how his translations and his emendations of foreign and domestic correspondence began to set the tone and style of modern scientific papers. In "Argument and Narrative in Scientific Writing" (1991), Frederic L. Holmes analyzes the writing conventions of the Royal Society by comparing them to those of the Royal Academy of Sciences in Paris around 1700. His study reveals that the Royal Society writers relied much more heavily on dense narrative, whereas the Academy writers used a more even balance between narrative and formal argument, a technique that Holmes argues is closer to modern practice in scientific papers. In "Pump and Circumstance: Robert

Boyle's Literary Technology" (1984), Steven Shapin analyzes Boyle's approach to experimentation and finds a major linguistic component in the rich circumstantial detail Boyle included in his descriptions. This technique enhanced believability by allowing the reader to become what Shapin calls a "virtual witness" to the experiment. Finally, Alan G. Gross bases his study "The Emergence of a Social Norm," Chapter 11 in *The Rhetoric of Science* (1990), on Royal Society figures, particularly on Newton. He uses the priority dispute between Newton and Leibniz as the central example in an essay that examines the impact of priority on scientific and social change, and the place of rhetoric in that process.

Newton's influence on the style and arrangement of the scientific article through his writings on the nature of light and optics has been studied by at least three scholars. In "The Literature of Enlightenment: Technical Periodicals and Proceedings in the 17th and 18th Centuries" (1987), Joseph E. Harmon surveys the content and stylistic conventions of technical periodicals from their origins in the work of the Royal Society to the end of the 18th century. He ends his essay with a short description of Newton's 1672 paper on optics. Both Alan Gross and Charles Bazerman devote chapters in their books on the rhetoric of science to detailed evaluations of Newton's evolution as a science writer, as seen through his publications and correspondence on light and optics. Together these chapters provide two different, but equally interesting, interpretations of the major shift in rhetorical approach Newton made between his largely unsuccessful paper of 1672 and his 1704 *Optics*—a work that transformed the 18th century's understanding of light. In Chapter 8 of *The Rhetoric of Science* (1990), "Newton's Rhetorical Conversion," Gross argues that Newton's first paper failed to persuade because it emphasized the radically new nature of his theory of light and color. In *Optics*, Newton adopted a very different approach: Instead of calling attention to the radical newness of his ideas, he "invented an essential continuity between his own work and the optical and scientific past" to make what was revolutionary seem (at least initially) to fit comfortably with what had been thought before (p. 112). In Chapter 4 of *Shaping Written Knowledge* (1988), "Between Books and Articles: Newton Faces Controversy," Charles Bazerman provides a longer, more detailed discussion of these same documents. Bazerman also carefully examines the responses of other scientists to his initial article and Newton's answers to them to demonstrate Newton's movement from article to book—from informal presentation of a discovery and resulting theory, to what Bazerman calls "a logical and empirical juggernaut" that carefully backs up every step in the reasoning with "experimental experience precisely related to the formal proposition" (p. 121). Through correspondence with these critics, Newton learned how to take rhetorical control.

Bazerman's chapter on Newton is the second of three in his book devoted to the emergence of the experimental article in the late 17th and 18th century. The first of these three chapters ("Reporting the Experiment: The Changing Account of Scientific Doings in the *Philosophical Transactions of the Royal Society*, 1665–1800") is an analysis of the first 135 years of the *Philosophical*

Transactions of the Royal Society of London. This study, based on Bazerman's analysis of a representative sample of articles (over 1,000) that report on experiments, documents an evolving form as scientists tried "to harness the stories of the smaller world of the laboratory to general claims about the regularities of the larger world of nature" (p. 79). That form changed as scientists sought to "satisfy the objections and desires of the growing scientific community." As Bazerman points out, the form for these articles continues to change even today as scientists grow in their "ability to formulate...objections and desires" (p. 79). The final chapter in this set ("Literate Acts and the Emergent Social Structure of Science") builds on the first two to determine "how the influence of scientific communication in journals had impact on the social structure of the scientific community" (p. 16). Thus, the movement in these three chapters is circular: he reveals how "social situations structure communication events and how forms of communication restructure society" (pp. 128–129). Tracing that circle, according to Bazerman, reveals how "the scientific community developed around the engendering and management of conflict" (p. 149).

The influence of the Royal Society, its members, and its publications on technical writing has been studied by a number of different scholars using a variety of approaches. Merrill D. Whitburn was one of the first to analyze late 17th-century scientific publications from the perspective of contemporary technical writing. As is true of most early studies in the history of technical writing, Whitburn's approach was descriptive and his purpose practical and pedagogical. In three different essays, "The Plain Style in Technical Writing" (1978), "The Past and Future of Scientific and Technical Writing" (1977), on Sprat and the scientific method, and "Personality in Scientific and Technical Writing" (1976), on Glanvill's "A Plurality of Worlds," he calls attention to effective seventeenth-century writing strategies and conventions and points out their utility for modern technical writers and teachers of technical communication. S. Michael Halloran and Annette Norris Bradford in "Figures of Speech in the Rhetoric of Science and Technology" (1984) use the Royal Society's appeal for a plain style as a point of departure to show that scientific writing—that produced and approved by the Royal Society, and that being written today—relies heavily on figures of speech, especially metaphor.

Another way to assess the impact of the rise of science and the writing it produced is to study their influence on other areas. In his 1963 book, *The Language of Science and the Language of Literature, 1700–1740*, Donald Davie surveys this influence looking in particular at the number of scientific and technical terms Samuel Johnson included in his dictionary and how the satirists used scientific language in their attacks on scientific materialism. In "Samuel Johnson: Technical Writer," William Kniskern (1986) shows that, in addition to his considerable literary talents, Samuel Johnson also produced a small but interesting body of technical writing, including essays on determining longitude at sea and on determining stresses in bridge design. These works, and a number of prefaces he wrote for technical publications of others, demonstrate "the range of Johnson's knowledge and ability" (p. 3).

Among 18th-century scientists, chemists (especially Joseph Priestley) seem to have attracted the most attention. Three studies of Priestley deal with how his writing reflects his objectivity and pattern of thought. In "Priestley's Personal Style" (1979), Della A. Whittaker shows how Priestley's detailed first-person descriptions of his experiments (including failures) give his work objectivity. Chester A. Lawson, in "Joseph Priestley and the Process of Cultural Evolution" (1954), studies Priestley's writings to isolate his patterns of thinking and research. Michael G. Moran, in "Joseph Priestley, William Duncan, and Analytical Arrangement in 18th-century Scientific Discourse" (1984), reviews Priestley's discussion of analytical arrangement in his book, *A Course of Lectures on Oratory and Criticism* (1777), pointing out the persuasive power inherent in Priestley's habit of presenting all the steps in the discovery process. More recently, in 1991, Charles Bazerman examines another aspect of Priestley's role as a scientist and writer. In "How Natural Philosophers Can Cooperate: The Literary Technology of Coordinated Investigation in Joseph Priestley's *History and Present State of Electricity* (1767)" (1991) Bazerman shows that Priestley's purpose in this work is not to stake out and defend a theory, and thus follow the pattern of "systematic competitiveness of modern science" (p. 14)—a pattern that quickly became the norm in scientific articles produced in the 17th and 18th centuries. Here, Priestley is consolidating "the scattered productions of natural philosophers into a stable and progressive knowledge structure" (p. 4) to foster cooperation, encourage new researchers, and direct researchers into productive areas and methods. He is thus trying to coordinate future study by bringing order to past study.

Maurice A. Finocchiaro, in "Logic and Rhetoric in Lavoisier's Sealed Note: Toward a Rhetoric of Science" (1977), offers an analysis of the note Lavoisier deposited with the Secretary of the Academy of Sciences in Paris, describing his discoveries about combustion and, thus, the nature of oxygen. This move by Lavoisier, to protect the priority of his work, Finocchiaro argues, provides an interesting study in logic and rhetoric.

In "Bacon, Linnaeus, and Lavoisier: Early Language Reform in the Sciences" (1983), James Paradis demonstrates the concern of early scientists to find the right language to define and explore the natural phenomena they studied. After reviewing Bacon's call for language reform, Paradis discusses attempts by 17th-century scientists to find names and classification systems, before focusing on Linnaeus' botanical nomenclature system and Lavoisier's textbook on chemistry—a book that "invented" a modern discipline by setting down a "language of chemical discourse" (p. 218). In "Scientific Nomenclature and Revolutionary Rhetoric" (1989), Wilda Anderson reveals the contemporary impact of Lavoisier's work by examining the French centralized school system's efforts to utilize his work soon after the French Revolution to take advantage of the "rhetoric-neutralizing potential of scientific nomenclature" (p. 4). Most recently, Lissa Roberts, in "Setting the Table: The Disciplinary Development of Eighteenth-Century Chemistry as Read Through the Changing Structure of Its Tables" (1991), examines a number of 18th-century chemical tables, including Lavoisier's, to show how these scientists were using language in these structures to define chemistry as a scientific discipline with rigorous, well-defined standards of proof, rather than as a more loosely defined art.

In "Gilbert White and the Personal Style" (1986), Michael G. Moran discusses the stylistic elements of White's *The Natural History of Selborne*, which may have contributed to the amazing success of this frequently reprinted collection of letters describing the natural phenomena of an isolated English parish. Geoffrey L. Scott's study, "The Scientific Poetry of Erasmus Darwin" (1982), focuses on two late 18th-century poems written in heroic couplets, *The Botanical Garden* and *The Temple of Nature*, to demonstrate technical accuracy and sound technical writing principles.

In a very different kind of study, "Eighteenth-Century Medical Education and the Didactic Model of Experiment" (1991), Lisa Rosner examines the relationship between the medical lecture-demonstration and medical dissertations produced by 18th-century students at Edinburgh to show how the rhetorical demands of the oral presentation, and its reliance on narrative, influenced written documents.

Except for Kitty Locker's survey of dunning letters written by East India Company employees between 1592 and 1873 (discussed in the next section), essays that deal with business writing in English in the 18th century have focused entirely on Lord Chesterfield's advice to his son about writing business letters. In two essays, "Lord Chesterfield on the Craft of Business Writing: The Relationship of Reading and Writing" (1979) and "'Elegant Simplicity': Lord Chesterfield's Ideal for Business Communication" (1985), I present Chesterfield's astute comments on the importance of clear writing in business correspondence. Although he offers specific advice on grammatical and stylistic matters, Chesterfield was acutely aware that a pleasing and effective style could only be developed through wide reading. Bradford B. Broughton, in "'No Man Is Allowed to Spell Ill': Modern Communication Advice from an Eighteenth Century Expert" (1985), reviews Chesterfield's advice on writing and speaking well. More recently, in "Eighteenth-Century Antecedents for Chesterfield's Concept of Style: A Reconsideration of Some Business Writing Traditions" (1991), I return to Chesterfield's life and work to identify the sources for his ideas about style. I identify five communities of conversation, reading, and writing in which Chesterfield actively participated, and which would have contributed to his sense of the power and subtleties of language and, especially, his awareness of the needs and interests of his readers.

Finally, Herbert W. Hildebrandt and Iris Varner's study, "The Communication Theory of Johann Carl May: Its Influence on Business Communication in Germany" (1985), provides a detailed description and analysis of business correspondence theory and practice as it evolved in 18th-century Germany through Christian Furchtegott Gellert's *Briefe nebst einer praktischen Abhandlung von dem Guten Geschmacke in Briefen* (1751), and, especially, Johann Carl May's *Versuch in Handlungs-Briefen* (1765). Hildebrandt and Varner's solid work with these important contributions to German business correspondence, and my brief description of Chesterfield's extensive correspondence in several languages, call attention to the need for more studies which explore cross-cultural influences on business writing.

THE 19TH CENTURY: BRITAIN AND THE CONTINENT

Despite the scientific, technical, and economic energy that characterized England during the 19th century, scholars interested in the history of business and technical writing have produced only a handful of studies. Most of these focus on the writings of individual scientists.

As might be expected, Charles Darwin has attracted the most attention, but even here one scholar, John Angus Campbell, has provided most of the studies. In "Darwin and *The Origin of the Species*: The Rhetorical Ancestry of an Idea" (1970) and "The Polemical Mr. Darwin" (1975), Campbell shows how Darwin used commonly accepted scientific, and even religious, beliefs to help make his thesis more accessible to his audience. Darwin was very much aware that his arguments supporting natural selection were problematic and, therefore, knew that his presentation of it had to be persuasive. Campbell continues his analysis of Darwin's rhetorical strategies in "Scientific Revolution and the Grammar of Culture: The Case of Darwin's *Origin*" (1986). Here, Campbell focuses on Darwin's use of natural theology and Baconian science to make his revolutionary idea more intelligible and acceptable. Barbara Warnick, through her work, "A Rhetorical Analysis of Episteme Shift: Darwin's *Origin of the Species*" (1983), also contributes to our understanding of Darwin's use of appeals (both rhetorical and nonrhetorical) to bring about a major "episteme shift," or revolution in thinking. In "The Invisible Rhetorician: Charles Darwin's 'Third Party' Strategy" (1989), Campbell argues that Darwin was the "chief architect" of the defense of *Origin of the Species*, although tradition has cast him as playing only a very minor role. Based on Darwin's correspondence, Campbell's study shows that Darwin was actively involved in organizing and supporting publications and research that would support his theory. Campbell concludes that this "third party strategy" reveals much about Darwin's "evolution from scientist to rhetorician and the ethical tension inherent in his dual commitment to persuasion and truth" (p. 5). More recently, in 1993, Campbell's "The Comic Frame and the Rhetoric of Science, Epistemology, and Ethics in Darwin's Origins" uses Darwin's distortions of his predecessors' "cognitive grounding and explanatory motives" (p. 29) to articulate in the *Origin* a "comic, or at least a tragi-comic frame for the rhetoric of science" (p. 50) as a way to confront both the ethical and epistemic problems inherent in that tactic. Campbell builds his argument, in part, on an article by Philip Kitcher, in which he uses Darwin to illustrate five ways rhetoric is essential to the work of science. (Kitcher's article, "Persuasion," appears in *Persuading Science* (1991), edited by Marcello Pera and William R. Shea.)

"Darwin and Style" (1984), by John Stephen Patterson, is a short overview of Darwin's method of composition, focusing in particular on the care and time he spent to make his prose style clear, sincere, and unpretentious. In Chapter 10 of his book, *The Rhetoric of Science* (1990), Alan Gross provides a close study of the portion of Darwin's *Red Notebook* in which he begins to work out his ideas about evolution and the origin of species. Gross's rhetorical analysis reveals a "drama of self-persuasion," in which Darwin is "driven forward by the rush of new concepts and facts," but then "held in check by the need to maintain the self as a coherent network of beliefs." Gross shows that as the theory evolves and Darwin convinces himself of its truth, he shifted from a style "in close touch with primary mental processes to one that anticipates public forums" (p. 159). Most recently, in 1996, Jeanne Fahnestock in her "Series Reasoning in Scientific Argument: *Incrementum* and *Gradatio* and the Case of Darwin," argues that series reasoning, especially *incrementum* (an ordered series) and *gradatio* (a series in which each item in the series after the first repeats the end of the previous item), are as conceptionally important in scientific reasoning and rhetoric as metaphors. Although she draws on other scientific writings to illustrate her points (including Newton, as well as several interesting examples from 20th-century publications in paleontology, biology, astronomy, psychology, and ecology), she presents Darwin's use of series in *The Origin of the Species* as the major support for her provocative thesis.

Other 19th-century figures whose works have been studied include Lyell, Snow, Davy, Flammarion, and Malthus. Arthur E. Walzer's 1987 study of Thomas Malthus, "Logic and Rhetoric in Malthus's *Essay on the Principle of Population, 1798*," is a full-length article analyzing the rhetorical strategies that helped Malthus's book become one of the most influential works of the 19th century. (The strong association of Malthus with 19th-century science—especially Darwin's work—is the major reason I have included Malthus in my discussion of 19th century writers, though this work was published at the very end of the 18th century.) A short, but interesting article by Yuri V. Novozhilov and Jacques G. Richardson, "Fifty Years After the Death of Flammarion, The Science Popularizer" (1976), calls attention to the contribution of the French astronomer Camille Flammarion through his *L'Astronomie Populaire*. Lyell, Snow, and Davy are featured on *Technical Communication*'s "Models for Technical Communicators" page. Mary M. Lay, in "A Classical Example of a Procedure" (1980), calls attention to John Snow's 1824 paper, "The Prevention of Cholera." In "Sir Charles Lyell: Geologist and Technical Communicator" (1980), Shari A. Kelley and Patrick M. Kelley examine a passage from Lyell's famous book, *Elements of Geology*. Donald W. Bush, Jr., focuses his evaluation of "The Diction of Sir Humphrey Davy" (1978) on Davy's first description of his safety lamp for use in coal mines. All these essays suggest possibilities for further research.

Three pieces in the recent collection of essays entitled *The Literary Structure of Scientific Argument* focus on genres and disciplines in the 19th century. In "J.C. Reil and the 'Journalization' of Physiology" (1991), Thomas H. Broman argues that the different genres of scientific writing actively shape, constrain, and support different kinds of knowledge. He uses Reil's journal, *Archiv*, and the changes it went through to illustrate his point. In "Writing Zoologically: The *Zeitschrift fur wissenschaftliche Zoologie* and the Zoological Community in Late Nineteenth-Century Germany" (1991), Lynn K. Nyhart explores how a zoological journal unintentionally moved from being a cooperative workshop for the discipline to an instrument that institutionalized published disputes as a way of establishing scientists' memberships in the disciplinary community. Finally, Bruce J. Hunt, in "Rigorous Discipline: Oliver Heaviside Versus the Mathematicians" (1991), studies the implications of the rejection by the Royal Society of a paper on mathematics by Heaviside. This rejection, Hunt asserts, came mainly because Heaviside did not provide a rigorous deductive argument, the "genre" required by the Royal Society. Thus, form and style were major ways mathematicians in the Royal Society had of controlling and shaping their discipline.

STUDIES OF ENGLISH THEMES, PRINCIPLES, OR PRACTICES ACROSS SEVERAL CENTURIES

The essays in this section trace a wide variety of specific practices or genres in business or technical writing over several centuries. In two of the first essays devoted to the history of technical writing, Joel J. Shulman deals with authorial questions. In "The Anonymous Technical Writer in History" (1960), he examines the problems in determining who wrote anonymous documents, as well as documents spuriously attributed to famous writers. Shulman's "Technical Writers Who Became Famous as Scientists" (1960) focuses on writers whose fame is based not on their own research, but on their synthesis of the work of others. In both essays Shulman begins with examples from the ancient world before presenting examples from later periods. Wayne A. Losano in "The Technical Writer as Naturalist: Some Lessons from the Classics" (1979), reviews the accomplishments of three well-known naturalists (William Bartram, Henry Walter Bates, and Charles Darwin) to demonstrate their ability to present clear, detailed, accurate descriptions of flora and fauna without sacrificing readability.

Two essays deal with the literary aspects of technical writing. Joseph Harmon's "Perturbations in the Scientific Literature" (1986) is a brief, but interesting survey of technical writers over several centuries, using such literary devices as anagrams, acrostics, puns, metaphors, litotes, and neologism. "Poetry at Work: Historical Examples of Technical Communication in Verse" (1988), by Jennifer J. Connor, reviews the use of poetry as a form for communicating technical information, particularly in instructions.

At least three studies focus on the nature and evolution of scientific periodicals over the past three centuries. Joseph E. Harmon continues his study of 17th- and 18th-century technical periodicals (see "Late 17th Century and the 18th Century") into the 19th and 20th centuries with "Development of the Modern Technical Article" (1989), a study that concentrates on the advent of specialized journals and the conventions they generated for research papers. In "The Structure of Scientific and Engineering Papers: A Historical Perspective" (1989), Harmon looks carefully at the "topical structure," its typical parts, and their evolution during the 19th and 20th centuries.

In "The Arrangement of the Scientific Paper," Chapter 6 of *The Rhetoric of Science* (1990), Alan Gross argues that the motives behind the regular patterns followed in scientific papers are epistemological: "they reenact the scientists' faith in the existence of a suite of methods by which the casual structure of the world can be displayed, directly or indirectly, to the senses" (p. 85). Gross traces these patterns from their beginnings in Baconian induction through Boyle's 1662 paper on the behavior of gasses, a 1961 scientific paper in molecular biology, and Einstein's papers on relativity. A recent study by Bryce Allen, Jian Qin, and F. W. Lancaster (1994) uses an analysis of the references in scientific articles published during the 325-year history of the Royal Society to show how "persuasive communities" function. ("Persuasive communities" is their term for a group of scientists whose work is cited and who constitute the audience that must be persuaded of the validity of new scientific claims.) In "Persuasive Communities: A Longitudinal Analysis of References in the Philosophical Transactions of the Royal Society, 1665–1990," these authors acknowledge that their analysis yielded very predictable results: referencing (and therefore the size of persuasive communities) increased steadily during the late 19th century and well into the 20th century, but increased exponentially in the years following World War II; during this latter period, references tended to be to more recent publications suggesting that scientific work remains influential (a part of the persuasive community) for shorter periods of time.

Two recent essays look at antecedents for our contemporary visual culture. John R. McNair, in "Ancient Memory Arts and Modern Graphics" (1991), surveys the use of memory arts and their reliance on the power of visual images from classical times through to the 16th century, then suggests theoretical and practical applications for modern graphics. In "Eideteker: The Professional Communicator in the New Visual Culture" (1991), Gary R. Olsen traces the changing importance given to visual elements in technical publications from the advent of the printing press to the present. The Renaissance saw a time of "visual efflorescence" (see also Tebeaux, under the section "The History of Teaching Business and Technical Writing") that faded during the Scientific Revolution due to the strong emphasis on abstract math and words. Recent technology and developments in scientific theory have again demonstrated the flexibility and power of visual presentations.

Several recent articles examine the history of rhetoric and science over several centuries. William A. Wallace, in "Aristotelian Science and Rhetoric in Transition: The Middle Ages and the Renaissance" (1989), traces the relationship between rhetoric, dialectics, and science from the 13th century through to the Renaissance. In particular, Wallace examines the impact on medieval science of nominalism and probabilism, especially during the Renaissance. He ends with an evaluation of Galileo's "pioneering work in mathematical physics, which paradoxically required the entry of rhetorical argument into the 'new science' to assure its acceptance and ultimate success" (p. 3).

James P. Zappen has written two studies that deal with the development of the rhetoric of science over the past three centuries. In "Historical Perspectives on the Philosophy and the Rhetoric of Science: Sources for a Pluralistic Rhetoric" (1985), he discusses the contributions of Sprat, Newton, Darwin, Einstein, and others to argue for the importance of a public, "pluralistic" rhetoric in dealing with both the threat and promise of science since Einstein. In "Scientific Rhetoric in the Nineteenth and Early Twentieth Centuries: Herbert Spencer, Thomas Huxley, and John Dewey" (1991), Zappen shows how scientific rhetorics (especially those by Spencer, Huxley, and Dewey) "reaffirm the relationship of rhetoric to inquiry and to the social community, to science and civilization" (p. 146). More specifically, Zappen finds that in their rhetorics, Spencer and Huxley give authority to society and see science as the servant to society. Dewey, however, gives rhetorical authority to science, thus placing society within the larger frame of science. Despite this basic difference in outlook, all three rhetorics, Zappen argues, show how communities use language not only to define themselves as separate entities, but also to relate to those communities that fall outside their self-definitions.

In "Experiment as Text: The Limits of Literary Analysis" (1993), Alan Gross uses the Michelson-Morley-Miller experiments, a sequence that marked the transition from classical to relativistic physics, to demonstrate the limits of literary analysis. Gross argues that in its focus on the text as its primary object of analysis, literary analysis often "emphasize(s) message over content" (p. 298) and thus deepens the split between words and things and also misses the rhetorical elements inherent in the steps of scientific experiments before texts are generated. Gross uses Habermas's theory of communicative action to argue for a theory of action (which includes a theory of rhetoric) to explain how scientists "identify sites of interpretation" (p. 299) and thus create science.

Kathryn A. Neeley's "Woman as Mediatrix: Women as Writers on Science and Technology in The Eighteenth and Nineteenth Centuries" (1992) is one of the first of what should become a growing list of studies that examine the place of women in technical writing. In this article, Neeley examines the careers of four mediatrix (Emilie du Chatelet [1706–1749], Mary Somerville [1780–1872], Jane Marcet [1769–1858], and Louise Otto [1819–1895]) and demonstrates how they, through a number of different forms that include textbooks, introductions, translations, and other works that synthesize and evaluate, handle the "crucial functions of establishing unity, order, and mutual understanding" (p. 209). Neeley's conclusion is as interesting as her subject: It is not "so much about changing the status of women as it is about changing the status of writing, especially mediative writing" which, according to Neeley, plays "an integral role in creative intellectual activity in science" (p. 216).

The one contribution to the history of business writing in this category is Kitty O. Locker's study of collection letters, "'Sir, This Will Never Do': Model Dunning Letters, 1592–1873" (1985). Locker reviews the offerings of "letter writers" (books containing model letters for use in business situations) from the 16th through the 19th century and identifies four approaches to dunning letters: The Legal Context: Debt and Imprisonment for Debt; Modest and Apologetic Letters; Comminatory and Vituperative Letters; and Brisk and Businesslike Letters.

EARLY AMERICAN WRITERS (COLONIAL THROUGH THE 19TH CENTURY)

As several of the studies mentioned in this and the next section reveal, the history of technical writing in America began early in Colonial times with the Puritans. In "Cotton Mather, America's First Great Technical Writer" (1963a, 1963b), Joel J. Shulman calls attention to Mather's standing as a scientist—he was the first American to be elected to the Royal Society. However, Mather's more significant contribution was not as a scientist but as a technical writer. He carefully edited the work of other scientists, reported on new discoveries, and introduced new statistical testing techniques. More recently, Margaret W. Batschelet has returned to the Puritans in her study of "Plain Style and Scientific Style: The Influence of the Puritan Plain Style Sermon on Early American Science Writers" (1988). After a discussion of the Puritan plain style sermon with special attention to its organizational structure, Batschelet shows the influence of this prose model on Samuel Danforth (1664) in *An Astronomical Description of the Late Comet or Blazing Star as it appeared in New-England in the 9th, 10th, 11th, and in the beginning of the 12th moneth* [sic], *1664* and John Winthrop IV (1759) in *Two Lectures on Comets Read in the Chapel of Harvard College, in Cambridge, New-England, in April 1759.*

The early American writer and scientist who has attracted the most attention from students of the history of technical and business writing is Benjamin Franklin. In 1962, Charles C. Hargis, Jr., began this study of Franklin's work with "America's First Great Technical Writer," an essay that focused on Franklin's advocacy of good technical writing and the role of his own writing in generating his international reputation as a scientist. John A. Brogan's short study, "Lessons from Benjamin Franklin" (1965), describes Franklin's technique of improving his prose style by imitating earlier writers. Steven L. Gresham, in "Benjamin Franklin's Contributions to the Development of Technical Communication" (1977), though centered on Franklin's technical writing style as a model for the modern technical communicator, shows how Franklin's practice was informed by his philosophy of technical communication and his view of the special role of the scientist in society. The two most recent studies of Franklin analyze his contributions by looking beyond specific technical writing documents. Elizabeth Tebeaux, in "Franklin's *Autobiography*—Important Lessons in Tone, Syntax, and Persona" (1981), shows that the elements that make for good technical communication, in his more technical work, are also present in the *Autobiography* and concludes that "good 'technical' style is not an isolated type of writing, but a powerful means of controlling tone and meaning" (p. 341). Mark Bernheim, in "Benjamin Franklin: Communicator Unlimited" (1981), looking at Franklin more from a business than a technical writing perspective, deals with the popular criticism of Franklin as "the epitome of smugness and American capitalist self-aggrandizement" (p. 35). After reviewing this criticism and Franklin's goals, Bernheim concludes that Franklin's goal as a communicator—to improve "the flow of ideas among all educated" people—was worthy, genuine, and, based on the longevity of his influence, achieved.

Studies of 19th-century American technical writing vary widely in topic. In "*Moby-Dick*: A Whale of a Handbook for Technical Writing Teachers" (1981), Deborah Kilgore discusses the famous whaling chapters of *Moby Dick* as a resource for contemporary technical writing teachers. Melville, Kilgore argues, provides in these passages excellent examples of the major techniques used in technical writing, as well as the stylistic energy that makes good technical writing successful.

In "A Classical Case of Poor Communication: P. G. T. Beauregard's Battle Orders and Report of the First Battle of Bull Run" (1995), Max Loges argues that the first battle of Bull Run was won despite Beauregard's poor technical writing in his battle plan, and that the many politically inept statements in his report to Jefferson Davis on the battle lead to his demotion and reassignment to the West.

Susan Wells adds to our knowledge of medical writing in the 19th century (see the contributions by Jennifer Connor in the next section) with "Women Write Science: The Case of Hannah Longshore" (1996), which examines the writing and professional experiences of several 19th-century women physicians. Wells mentions Mary Jacobi and Harriot Hunt, two other physicans who contributed to the medical literature of the 19th century, but deals mainly with Hannah Longshore's career and, in particular, Longshore's assessment of it in an 1895 speech before the Women's Medical College of Pennsylvania.

STUDIES OF AMERICAN THEMES, PRINCIPLES, OR PRACTICES OVER TIME

A group of at least 10 essays deals with changes in the practice of business and technical writing in America and, thus, the corresponding changes in attitude signaled by the new practices. Although these studies do not fit together to form a complete picture, they do deal with different aspects of professional writing produced in every time period, beginning with early efforts in the 17th century. Michael E. Connaughton's article, "Technical Writing in America" (1981), provides a valuable point of departure for further studies in the early history of technical writing in America. He looks at early technical and scientific periodicals and notes changes as contributors shifted from being amateurs in the 17th century to professional scientists in the early 19th century. Joseph W. Wenzel, in "Rhetoric and Anti-Rhetoric in Early American Scientific Societies" (1974), tracks changes in attitude toward the place of rhetoric in scientific writing. Under the influence of the Royal Society and Ramistic ideas, early American scientists regarded rhetoric as ornamental and, therefore, not appropriate for scientific prose; however, by the late 19th century, scientists had come to see rhetoric as the study of effective prose style and, thus, to see the value of rhetoric as an area of study for those involved in scientific discourse.

Two other studies have focused on particular genres within technical writing. In "Does Clio Have a Place in Technical Writing? Considering Patents in a History of Technical Communication" (1988), R. John Brockmann first argues that technical writers need a historical perspective so that they can more accurately evaluate technical writing standards. He then turns to a brief explanation of why the 200-year-old federal collection of patents is a convenient and revealing resource for studying the writing of ordinary mechanics of the 19th century. Though focused more on analysis of content and the explicit and implicit values expressed in house pattern books, "Nineteenth Century American House Pattern Books: A Rhetorical Analysis" (1981), by Deborah C. Andrews and William D. Andrews, provides a fascinating overview of this specialized genre and again suggests opportunites for further research.

The collection *Studies in the History of Business Writing* contains four essays that deal with changes in business writing practices during the 19th century and into the 20th century. In "The Etiquette of American Business Correspondence" (1985), Lynn W. Denton provides an interesting overview and analysis of the advice on business correspondence found in etiquette books popular between 1850 and 1900. George H. Douglas's "Business Writing in America in the Nineteenth Century" (1985), reviews 19th-century business correspondence, with particular attention to the relationship between technical innovation (specifically, typewritten letters replacing handwritten ones) and the trend toward impersonal tone in business letters. In "The Historical and Cultural Significance of Direct-mail Fund-raising Letters" (1985), John Pauly explores how the advent of direct-mail advertising in the late 19th century, and its growth in the 20th century, reflect changes in American community and culture. Finally, JoAnn Yates examines the surprisingly interesting implications of changes in record-keeping systems from the mid-19th century to the mid-20th century in "From Press Book and Pigeonhole to Vertical Filing: Revolution in Storage and Access Systems for Correspondence" (1985). She concludes that the revolution brought about by the typewriter, carbon paper, and vertical files had impacts on corporate culture that are parallel in many ways to those presently being felt by many companies as a result of the electronic revolution. Yates' book-length study, *Control Through Communication: The Rise of System in American Management* (1989), expands on that theme and illustrates it through an analysis of changes in communication patterns in three large American corporations.

In "Semantic Bypassing in Technical Communication: The Historical Case of *Antiseptics*" (1988), Jennifer J. Connor and J. T. H. Connor deal with the special kind of misunderstandings and confusions that occur when readers "bypass" the intended meaning of a technical term and instead use another meaning, thinking they understand when they really do not. They illustrate their point through a review of the confusion among Canadian physicians over the British physician Joseph Lister's use of the term "antiseptics." Lister's use of an old term in a new way created confusion that delayed acceptance of his system by most Canadian physicians for 30 years. Jennifer J. Connor continues her study of Canadian medical writing in a 1994 article, "Self-Help Medical Litarature in 19th-Century Canada and the Rhetorical Convention of Plain Language." Connor's examination of self-help medical literature in Canada reveals that claims for the plain style were essentially rhetorical posturings in a tradition that stretches back well before the 16th century. Thus, the advocacy of a plain style did not not mean a plain style was used. In fact, Connor's comparisons reveal few significant differences between medical textbooks of the time and the "plain style" self-help books.

In "Technical Writing's Roots in Computer Science: The Evolution from Technician to Technical Writer" (1988), Henrietta Nickels Shirk argues that the history of technical writing in the second half of the 20th century closely parallels trends in computer science. In particular, she points out that the growing use of visual elements—especially diagrams—and the metaphorical use of "user" and "module," now so common in technical writing come from computer science. She feels that with the advent of online documentation, however, technical writers will be able to become equals with computer scientists in their influence on trends in technical writing.

Charles Bazerman devotes two chapters in *Shaping Written Knowledge* (1988) to the analysis of trends in scientific publications during the 20th century. In Chapter 6, "Theoretical Integration in Experimental Reports in Twentieth-Century Physics: Spectroscopic Articles in *Physical Review*, 1893–1980," Bazerman argues that the large-scale trends evident in these articles reinforce the "traditional view that science is a rational, cumulative, corporate enterprise...realized through linguistic, rhetorical, and social choices." Furthermore, he asserts that these articles also reveal how a strong theory shapes the scientific activity and orders the social relations, creating "a kind of bureaucratization of the scientific community" (p. 183). In Chapter 9, "Codifying the Social Scientific Style: The *APA Publication Manual* as a Behaviorist Rhetoric," Bazerman examines the *APA Style Manual* and experimental reports in psychology to show how the official APA style, as it has evolved over seven decades, has influenced psychology by embodying "behaviorist assumptions about authors, readers, the subjects investigated, and knowledge itself" (p. 259).

Although it focuses on only two articles that appeared within a few months of each other in 1978, the final essay in this section, "Stories and Styles in Two Molecular Biology Review Articles" (1991), by Greg Myers, fits quite well with other studies that deal with trends over time. Myers' point is that these two review articles on the subject of RNA splicing demonstrate the genre's purpose: "a review shapes the literature of a field into a story in order to enlist the support of readers to continue that story" (p. 45). Thus, reviews are documents that occupy a special and powerful place in science, the social construction of science, and the rhetoric of science. As Myers asserts, they deserve our attention as much as the experimental articles they bring together and shape into a story.

STUDIES OF TWENTIETH-CENTURY WRITERS

One might expect that the high profile of science in our century, and the tendency within our culture to idolize individual scientists, would stimulate a number of studies that focus on the writing of great (or well-known) modern scientists. However, relatively few essays of this type have been published. Predictably, Albert Einstein, the most influential scientist of our century and himself a student of human communication, has attracted considerable attention; however, despite his frequent mention in studies that survey technical writing developments, Einstein's works have been the central focus of only three essays. In "The Relativity of Communication: Albert Einstein as Technical Writer" (1980), Michael J. Baresich examines three versions of Einstein's special theory of relativity to show how he varied tone, personal address, levels of diction, definitions, and examples to meet the special needs of three different audiences: technical, semitechnical, and nontechnical. Dennis E. Minor's "Albert Einstein on Writing" (1984) presents Einstein's early concern that his writing be clear and his reliance on a model for technical writing he found in Ernst Mach's *The Science of Mechanics* (1883). He found this model especially effective because it encouraged scientists to replicate the thinking process or, in Einstein's words, to duplicate "the struggle with their problems, [and] their trying everything to find a solution which came at last often by very indirect means...." (p. 15). Minor briefly examines one of Einstein's early papers to show how he used Mach's model. Finally, Richard D. Johnson-Sheehan in "Scientific Communication and Metaphors: An Analysis of Einstein's 1905 Special Relativity Paper," argues that what made Einstein's work in this paper exceptional among others on the same topic moving toward the same conclusion was Einstein's ability to use metaphor and, in particular, his creation of a "guiding metaphor" that lent itself to being adopted and modified by later scientists such as Planck and Minkowski.

Two other early 20th-century scientists, P. W. Bridgman and George Washington Carver, have attracted interest. In a brief note, "P.W. Bridgman on Style," John A. Muller and Linda Sladkey (1981) connect Bridgman's process of operational analysis—a tool that clarifies definitions by reducing concepts to a unique set of operations—to technical communication situations. In a more extended study, "George Washington Carver and the Art of Technical Communication" (1979), Steven Gresham examines Carver's contributions as a technical communicator by analyzing some of the technical bulletins he wrote during his tenure at the Agricultural Experiment Station at Tuskegee Institute. Gresham concludes that

Carver's accomplishment was to "create a verbal culture through which scientific and technical information could be transmitted to his audience of area farmers" (p. 224).

Beverly A. Sauer offers one of the more unusual and interesting studies of early 20th-century technical writing in her "Revisioning Sixteenth Century Solutions to Twentieth Century Problems in Herbert Hoover's Translation of Agricola's *De Re Metallica*" (1993). She argues that Hoover's interest in Argicola's "intellecutal achievements," the stated reason for Hoover's translation (1912), probably extended beyond the technical and into the economic and political assumptions that lay behind Agricola's arguments—assumptions that Hoover found useful during the Mine Strikes of 1922. Sauer uses this example to demonstrate that technical documents not only reflect the political idealogies of their writers, but also that those documents and assumptions can influence the use and shape of later technology.

The writings of three contemporary popularizers of science have been analyzed by scholars interested in the teaching and history of technical writing. Thomas M. Lessl's "Science and the Sacred Cosmos: The Ideological Rhetoric of Carl Sagan" (1985) is a rhetorical analysis of Sagan's *Cosmos* series. Lessl quickly separates Sagan's work generically from both scientific writing—the scientist-to-scientist writing that is usually studied in science-as-rhetoric literature—and science writing or journalism—a genre that is usually neutral in its approach. According to Lessl, Sagan's polemical, even evangelic, style "creates a mythic understanding of science which serves for television audiences the same needs that religious discourse has traditionally satisfied for churchgoers" (p. 175).

Jo Allen's brief analysis of Lewis Thomas's prose in "Dr. Lewis Thomas: Popularizer of Science" (1985) builds on a comparison of Thomas's use of qualifiers in two short papers—one for an audience of scholars, the other for an audience of laypersons—to show how he controls language in response to his audiences. In "Parallels in Scientific and Literary Discourse: Stephen Jay Gould and the Science of Form" (1986), Debra Journet analyzes two essays by Gould to demonstrate two important parallels between scientific and literary discourse: aesthetic appeal and the use of metaphor.

Watson and Crick and their writings on the structure of the DNA molecule have, especially in the past few years, attracted many researchers interested in the history of science writing and the rhetoric of science. In "Technical Writing and the Rhetoric of Science" (1978), S. Michael Halloran uses James Watson's *The Double Helix*—the account of how Watson and Crick discovered the molecular structure of DNA—and other professional writings by these two scientists to demonstrate that science is rhetorical. In his essay, Halloran reviews some of the important early work on the rhetoric of science and argues that technical writing is central to the liberal education of a technological society. Halloran returns to Watson and Crick again in "The Birth of Molecular Biology: An Essay in the Rhetorical Criticism of Scientific Discourse" (1984), to study their famous paper on DNA as a rhetorical case. In particular, he calls attention to the *ethos* that the Watson and Crick paper establishes and presents his analysis of how it operates. He ends this essay with a call for more studies of particular cases of scientific discourse to help build a more thorough understanding of the rhetoric of science. In "The Tale of DNA," Chapter 4 of *The Rhetoric of Science* (1990), Alan Gross compares Watson and Crick's initial paper in *Nature* and Watson's autobiographical memoir of the discovery. Despite their very different purposes, Gross finds a common ground in their subject matter and persuasive purpose: Both argue, from very different perspectives, that facts—of an experiment or of a person's life—occurred and that they generate "significant" knowledge. In Chapter 10 of his book, *A Rhetoric of Science: Inventing Scientific Discourse* (1989), Lawrence J. Prelli uses the *Nature* article on DNA in another rhetorical comparison. In this chapter, "Practicing Rhetorical Invention: Creating Scientifically Reasonable Claims," Prelli compares the success of the DNA article with the failure of the arguments presented by creationists that creationism should have equal time in science classes. Prelli's goals are (1) to demonstrate the practical usefulness of rhetorical theory in analyzing scientific discourse and (2) to support his thesis that "degrees of scientific reasonableness will vary according to (a) the availablity to a rhetor of appropriate (i.e., scientific) purposes, issues, and topics, and (b) the rhetorical wisdom of rhetors' choices from among available ideas" (p. 219). According to Prelli, the creationists failed in their legal battle (the case of *McLean* v. *Arkansas* in 1981–1982) because they could not "satisfy the minimal tests of reasonable scientific discussion" (p. 235). Watson and Crick were successful not only because of the elegance of their theory, but also because of their display of "consumate rhetorical skill"; they "chose wise and logical *topoi* as grounds from which to argue and render their claims reasonable for other scientists" (p. 249).

In Chapter 7 of *Shaping Written Knowledge* (1988), "Making Reference: Empirical Contexts, Choices, and Constraints in the Literary Creation of the Compton Effect," Charles Bazerman carefully studies Arthur Holly Compton's writing process, from his notes through to his final publications, to demonstrate how the influential scientist's work was influenced by the "communicative context," a context that was by the early 20th century already highly developed. In that context, Bazerman argues, Compton worked as an individual scientist, but one who has "structured opportunities, resources, and constraints out of which to construct claims and arguments that will move others within the same system to come to his view of experience." Thus, the scientist, as Bazerman shows in his analysis of Compton's work, "behaves normatively, creatively, and self-interestedly within a complex system" (p. 191). Finally, Debra Journet, in "Ecological Theories as Cultural Narratives: F. E. Clement's and H. A. Gleason's 'Stories' of Community Succession" (1991), examines the work of two American ecologists of the first half of the 20th century who debated questions of community succession (how ecologies change over time). She argues that when they develop theories about community succession these scientists are really constructing and testing narratives that depend, like literary narratives, on cultural assumptions and values.

Journet finds these two scientists, with their conflicting stories about ecological systems, valuable because they demonstrate the "importance of narrative as an interpretive and rhetorical strategy in scientific discourse" (p. 446).

STUDIES THAT FOCUS ON SPECIFIC EVENTS OR SITUATIONS

The studies in this section are interesting and useful; many are also potentially life-saving—if the lessons they present are heeded. All of these studies analyze recent historical events or situations to show how technical or professional language conveyed, failed to convey, distorted, or concealed important information.

The first three studies are evaluations of the language and strategies used to present information about nuclear energy, AIDS, and euthanasia. In "The Science Journalist and Early Popular Magazine Coverage of Nuclear Energy" (1981), Steven L. Del Sesto reviews articles on the peaceful uses of nuclear energy, published in the 1940s and 1950s in popular magazines. He shows that these articles were often overly optimistic, even sensational, and reflect not only the underdeveloped state of science journalism, but also the desire among scientists and publishers alike to show that nuclear power, so recently and so dramatically seen as the ultimate instrument of war, could be used for peaceful, nondestructive purposes. Carol Reeves, in "Establishing a Phenomenon: The Rhetoric of Early Medical Reports on AIDS" (1990), examines the first three reports on AIDS that appeared in *The New England Journal of Medicine* in 1981, and the rhetorical strategies that those scientists used as they attempted to convince the medical community of the deadly threat of AIDS. These early explanations of AIDS are, she argues, "illustrations of how belief, experience, and documented knowledge are linked in scientists' efforts to construct theories and then argue the validity of those theories" (p. 414), and are important documents for study as we try to understand the processes and patterns in medical rhetoric. In "Medicine, Rhetoric, and Euthanasia: A Case Study in the Workings of a Postmodern Discourse" (1993), Michael J. Hyde offers a critical reading of "It's Over, Debbie," and Dr. George Lundberg's decision to publish the controversial narrative in the *Journal of the American Medical Association* anonymously and without critical editorial comment, in the belief that the story presented in that way would do the "good" work of promoting public debate on the issue. Through his analysis of the story and the controversy that erupted around it and Dr. Lundberg, Hyde argues that the narrative's "ambiguity and undecidability" make it a "postmodern" text that worked very effectively to provoke an intense and complex debate about euthanasia, both within and outside of the medical establishment.

Two other essays approach the study of the rhetoric of science and technology by focusing on the rhetorical patterns writers follow in controversies. Randall L. Bytwerk, in "The SST Controversy: A Case Study of the Rhetoric of Technology" (1979), looks at the arguments on both sides and concludes that SST proponents made serious errors in analyzing their audiences. In "'Punctuated Equilibria': Rhetorical Dynamics of a Scientific Controversy" (1986), John Lyne and Henry F. Howe examine the rhetorical strategies of proponents of "punctuated equilibria" as they first attacked gradualistic assumptions, and then advanced their theory as a challenge to evolutionary biology. Lyne and Howe identify four "dimensions along which rhetorical dynamics" played a role in the controversy: the persona of the scientific writer, the changing audience, picturing strategies used in the discourse, and use of contrast.

In "The Social Drama of Recombinant DNA," Chapter 12 of *The Rhetoric of Science* (1990), Gross looks at a different kind of scientific controversy—that between science and society, not between scientists and scientists. His examination of the controversy over recombinant DNA research that played itself out in the 1970s and 1980s leads him to conclude that "the democratic process has a legitimate role in the determination of the general direction of scientific work" (p. 191), despite the potential for frustration, divisiveness, and the interruption of important research. That process is often one that tests the effectiveness of scientists, their patience, their ability to use language, and the rhetoric of science. (Prelli examines a similar conflict in his study of the creationists' efforts to add creationism to science texts in Arkansas. See the previous section.)

Several of this century's scientific and technical "disasters" have occasioned studies that examine how traditional rhetorical and language patterns obscured facts, events, and conclusions, thus contributing to the "disaster." The explosion of the space shuttle Challenger in January of 1986 generated a tremendous amount of national grief and finger pointing as to who was to blame. Several essays in business and technical writing journals see the disaster as a result of poor technical and corporate communication. In "Communication: The Missing Link in the Challenger Disaster" (1988), Vanessa Dean Arnold and John C. Malley study the Presidential Commission's report, isolate three major problems (faulty reporting of technical information; isolation of management; and selective listening), and propose several actions that should be taken in corporate environments to improve communication between technical and nontechnical personnel. Whereas Arnold and Malley approach the disaster from a managerial communication perspective, Roger C. Pace provides a more detailed analysis of the testimony collected by the Presidential Commission in his essay "Technical Communication, Group Differentiation, and the Decision to Launch the Space Shuttle Challenger" (1988). His study reveals failures of communication in four areas of differentiation (clarity, interrelatedness, centrality, and openness) and shows how these failures of individuals to make their interpretations of facts clear to the group (differentiation) led to major breakdowns in communication. In his analysis of the Challenger documents, "Communication Failures Contributing to the Challenger Accident: An Example for Technical Communicators" (1988), D. A. Winsor concludes that the miscommunication comes not only from "managers and engineers interpreting data from different perspectives" but also from "the difficulty of believing and then sending bad news, especially to superiors or outsiders" (p. 101). In her evaluation of the arguments used by engineers in the prelaunch discussions, Christine M. Miller (1993) found that the engineers' assumptions about science and technology affected their ability to frame effective arguments in that situation. In that debate, the engineers found that representing science and technology as "impartial was inconsistent with the amount of interpretive work" they had to do to support their position. Furthermore, the hierarchical structure of the agency made it very difficult for them to shift out of their roles as scientists so that they could effectively argue, based on their interpretation of what both they and the managers saw as inconclusive data.

Paul M. Dombrowski has published three essays on the communication issues that emerged from the Challenger disaster. In "The Lessons of the Challenger Disaster" (1991), Dombrowski looks at the event and the investigations from a social constructionist perspective. According to his analysis, NASA officials "construed information about O-ring charring in socially contingent ways" and pressured engineers to "work under similar assumptions." Furthermore, both investigations of the disaster emphasized "procedural concerns while largely neglecting personal judgment and responsibility, even though the evidence suggests a key role for personal and social judgment" (p. 211). In "Challenger Through the Eyes of Feyerabend" (1994), he supplements his earlier social constructionist evaluation by using Paul Feyerabend's ideas about the powerful role of "prior adherence to scientific theory [conceptualizations] in shaping subsequent perceptions of data" (p. 7) to help explain why the engineers' arguments were not effective. He goes on to argue that Feyerabend's philosophy can make an important contribution to science and technical writing by helping us break free from the "pre-eminence of the rigid, traditional scientistic and technicist views of the world" and consider a more pluralistic approach. In "Can Ethics Be Technologized? Lessons from Challenger, Philosophy, and Rhetoric" (1995), Dombrowski uses the Challenger disaster to show that ethics cannot be effectively technologized—that "collecting additional technical data cannot of itself prevent ethical lapses." Ethical choices must ultimately be "deliberated among people in an interminate way" (p. 146).

Thomas B. Farrell and G. Thomas Goodnight's study, "Accidental Rhetoric: The Root Metaphors of Three Mile Island" (1981), reconstructs in detail the discourse at Three Mile Island and then, based on an analysis of that discourse, finally coming to the rather bleak conclusion that "the limits of technical communicative discourse are severe, recurrent, and perhaps irreparable" (p. 271). Five years after the Farrell and Goodnight study, J. C. Mathes, in *Three Mile Island: The Management Communication Failure* (1986), gave his detailed evaluation of the communication problems at Babcock and Wilcox that delayed needed changes in instructions for the nuclear reactor. Mathes' work has, in turn, facilitated the work of subsequent researchers, for example, Carl G. Herndl, Barbara A. Fennell, and Carolyn R. Miller's "Understanding Failures in Organizational Discourse: The Accident at Three Mile Island and the Challenger Disaster" (1991). Herndl, Fennell, and Miller examine the documents surrounding these two events from a social constructionist point of view. Their work leads them to several conclusions: First, they distinguish between two kinds of communication failures (miscommunication and misunderstanding); second, they describe which research techniques are best at identifying discourse communities within groups of different sizes and natures; finally, they see their work illustrating the complexity of the term "discourse community."

M. Jimmie Killingsworth and Jacqueline S. Palmer, in their book-length study *Ecospeak: Rhetoric and Environmental Politics in America* (1992), and in a recent article entitled "The Discourse of 'Environmentalist Hysteria'" (1995), analyze "a bable of discourse communities" (ranging through the spectrum from purely scientific to intensely political) that is beginning to produce, they argue, "an emerging discourse that brings a plurality of knowledge cogently and coherently to bear on a problem afflicting the world public." Their cautious, yet hopeful view is based on the assumption that rhetoric is a means for building consensus, thus contributing to "the story of how human action reconciles conflicting demands in the search for the good life" (1992, pp. 20–21).

Finally, Martha Solomon, in "The Rhetoric of Dehumanization: An Analysis of Medical Reports of the Tuskegee Syphilis Project" (1985), deals with the role of language in allowing a "disaster" in medical ethics to go unnoticed, despite 13 published reports on the project between 1936 and 1973. The Tuskegee Syphilis Project involved a study of untreated syphilis in adult male African Americans in Macon County, Alabama, despite the existence of treatment, even in 1932 when the project began. Solomon's analysis of the published reports on the project reveals that the subjects were depicted as scene and agency and, thus, were dehumanized in the eyes of the physicians both writing and reading the scientific reports. The issues these physicians saw—or allowed themselves to see—were scientific, not ethical, and were determined by the conventional rhetoric of scientific reporting, which obscured questions about whether human beings were being treated ethically. Solomon's essay is important and disturbing; her work here, as well as the work of other scholars who have studied breakdowns in technical and scientific communication, demonstrate the importance of rhetoric, the value of rhetorical studies of professional communication, and the need to teach courses that raise questions about the relationship between language and ethics.

This same point, although directed more specifically at nontechnical, institutional communication, is made by Francis W. Weeks in "An Editorial: The Watergate Hearings and Business Communication" (1973). Even though he was writing while the hearings were still in progress, Weeks had seen and heard enough about the communication breakdowns within the Nixon Administration to point out, in his brief editorial, not only problems with confusing language and confusing organization in documents, but also the ways written documents reflected the insulated, dehumanized administrative style and the attitude of the administration. Surprisingly, Weeks' editorial is the only essay published in our journals which connects the Watergate scandal with professional writing principles.

CONCLUSION

Despite the incredible growth in the study of the history of business, technical, and scientific writing, large gaps in our knowledge still exist. That conclusion should come as no surprise. If nothing else, this bibliographical essay should have demonstrated the tremendous range within which students of the history of business and technical writing must operate. That range already includes all historical *time*; the full variety of *cultures* and *languages* inherent in Western civilization; a broad sample of the diverse *topics* that human beings solving problems in business, science, and technology must address; and the many different *disciplines and approaches* that can and should be used to help us understand the technical, scientific, and business documents we study. We have come a long way from modest beginnings 15 to 20 years ago; we still have much to study and learn. Let me make a few observations and suggestions about what we might keep in mind as we pursue our study of history.

First, we need to be increasingly aware of the need to place our studies in context. Particularly in studies of individual writers and works, we need to connect their work to other technical, scientific, or business writing done in the same period, as well as to all the writing (and thinking) of the period. We also need to consider the place of these authors and works in the changing patterns of writing over time. How writers and works fit and don't fit can teach us much about why writing is effective or ineffective. Furthermore, we need, for all periods, a better understanding of the interactions between the writing on technical, scientific, and business done in different languages and cultures.

Second, we should learn about and incorporate into our bag of analytical tools the insights and approaches used in a variety of disciplines and special areas of study, including literature, literary criticism, philosophy, history, the history of science, the rhetoric of science, linguistics, composition and rhetoric, classical studies, and speech communication. Scholars in these areas are working on the writers and documents we also must study and are articulating theories and approaches that parallel and inform ours. We must avoid the temptation to look only at our journals and to listen only at our professional meetings.

Third, as Tebeaux and Killingsworth (1992) suggest, we need more studies that seek to define the characteristics of, and changes in, the business, technical, and scientific writing done within particular historical periods. At this point, we seem to know most about 16th- and 17th-century British writing. The business, technical, and scientific writings of other periods need similar attention.

Fourth, we need more book-length studies that attempt to trace the history of business or technical writing over a single historical period or, even more helpful, although difficult, over longer periods of time. Thus far, we have only four books that deal with the history of business or technical writing (those by Adams [1993], Kynell [1996], Tebeaux [1997], and Yates [1989]), in addition to those books devoted to the rhetoric of science (most notably for our purposes, those by Bazerman

[1988], Gross [1990], and Prelli [1989]). These more extensive studies will come as our knowledge of authors, contexts, and periods grows over the next few years; furthermore, the very recent appearance of the books by Adams, Kynell, and Tebeaux suggests that more may be on the way. These and other studies that synthesize our knowledge will, of course, provide a framework in which further studies of authors, works, and periods can be more fruitful as our understanding of history grows, is tested, and is revised. That kind of activity is indicative of a discipline that is reaching maturity; what makes our work at this point both exciting and daunting is that we are still moving toward that goal.

Fifth, we will need, again as Tebeaux and Killingsworth (1992) suggest, to make clearer distinctions between technical and scientific writing so that our work can not only be more manageable, but also be better focused and, therefore, more productive. I have stressed here the need to make connections, to put authors and works in context. That task is of paramount importance not only between technical and scientific, but also between any kind of writing and all the writing done within a culture during a given period in its history. I have included here studies of business, technical, and scientific writing for several reasons, including the continuities that exist between them within the historical periods, as well as the fact that many of us teach and study all three kinds of writing. As the body of literature on the history of applied writing becomes larger, we will need to narrow our topical focus. That prospect is inevitable, but I hope that we will not, in our growth and success, lose the sense and goal of connectedness.

Sixth, although it may not be immediately apparent, studies in the history of technical and scientific writing have become more and more prolific, whereas those in the history of business writing have dropped off dramatically since a surge in the 1980s. I can only offer some tentative guesses about the reasons for this decline. First, I think we must consider the paradoxical difficulties inherent in doing research on institutional and business writing: the number of documents one must examine to gain a sense of shifting patterns is immense; however, finding and gaining permission to examine internal documents (documents which many companies might be reluctant to make available for various legal and business reasons) is fraught with problems. A preliminary study that identifies the locations and contents of available archives would provide a valuable starting point. Until scholars follow Yates' lead and find such archives, studies in the history of business writing will have to focus on business documents produced for public consumption. However, we have seen very few of these studies recently. A second reason for this decline of work on the history of business writing may be that researchers find themselves in environments that privilege quantitive research over qualitative (historical) research.

Finally, we must never forget that our work—even our studies of the history of business and technical writing—arises from, and should return to, the classroom. Although it is tempting to lose ourselves in the past, we must keep our grounding and sense of purpose with our students and our teaching. What we learn from our study of the history of business and technical writing can have tremendous impact in the classroom. Our work, if it is done well, should help us show our students that their study of writing, and the writing they will do in their professional careers, is the continuation of a rich, complex, and very old cultural tradition. It is a tradition we should all seek to understand better so that our contributions to it—as students, professionals, and teachers—will be informed, thoughtful, and (hopefully) more effective. Perhaps most importantly, study should help us see that no matter how technical or mundane the subject matter, all writing is still a matter of one human being trying to communicate with another.

REFERENCES

Adams, K. H. (1993). *A history of professional writing instruction in American colleges: Years of acceptance, growth, and doubt.* Dallas, TX: Southern Methodist University Press.

Adolph, R. (1971). *The rise of modern prose style.* Cambridge, MA: MIT Press.

Alexander, L. L. (1988). Cicero's arrangement in scientific writing. *Issues in Writing, 2*(1), 72–91.

Alford, E. M. (1988, April). Thucydides and the plague in Athens. *Written Communication, 5*(2), 131–153.

Allen, B., Qin, J. & Lancaster, F. W. (1994). Persuasive communities: A longitudinal analysis of references in the philosophical transactions of the Royal Society, 1665–1990. *Social Studies of Science, 24,* 279–310.

Allen, J. (1984). The literature of science and the technical writing curriculum. *Proceedings of the 31st International Technical Cummunication Conference,* RET 70–72.

Allen, J. (1985). Dr. Lewis Thomas: Populaizer of science. *Technical Communication, 32*(4), 72.

Allen, J. (1986). Agricola's Preface to *De Re Metallica. Technical Communication, 33*(2), 90–92.

Allen, J. (1991). Thematic repetition as rhetorical technique. *Journal of Technical Writing and Communication, 21*(1), 29–40.

Allen, J. (1992). Commentary: A response to J. T. H. Connor and Jennifer J. Connor's analysis. *Journal of Technical Writing and Communication, 22*(2), 203–209.

Anderson, F. D. (1982, Spring). [Review of the book *Galileo and the art of reasoning: Rhetorical foundations of logic and scientific method*]. *Philosophy and Rhetoric, 15*(2), 134–138.

Anderson, W. (1989, Winter). Scientific nomenclature and revolutionary rhetoric. *Rhetorica, 7*(1), 45–53.

Andrews, D. C., & Andrews, W. D. (1981). Nineteenth century American house pattern books: A rhetorical analysis. *Proceedings of the 28th International Technical Communication Conference,* E1–E4.

Arnold, V. D. (1989, September). A twenty-five year perspective on the pedagogy of business communication. *The ABC Bulletin, 52*(3), 3–6.

Arnold, V. D., & Malley, J. C. (1988, December). Communication: The missing link in the Challenger disaster. *The Bulletin of the Association of Business Communication, 51*(4), 12–14.

Baker, C. (1983). Francis Bacon and the technology of style. *The Technical Writing Teacher, 10,* 118–123.

Baresich, M. J. (1980). The relativity of communication: Albert Einstein as technical writer. *Journal of Technical Writing and Communication, 10*(2), 125–132.

Basquin, E. A. (1981). The first technical writer in English: Geoffrey Chaucer. *Technical Communication, 28*(3), 22–24.

Batschelet, M. W. (1988). Plain style and scientific style: The influence of the Puritan plain style sermon on early science writers. *Journal of Technical Writing and Communication, 18*(4), 287–295.

Batschelet, M. W. (1990). *Early American scientific and technical literature: An annotated bibliography of books, pamphlets, and broadsides.* Metuchen, NJ: Scarecrow Press.

Bazerman, C. (1988). *Shaping written knowledge: The genre and activity of the experimental article in science.* Madison, WI: University of Wisconsin Press.

Bazerman, C. (1991). How natural philosophers can cooperate: The literary technology of coordinated investigation in Joseph Priestley's *History and Present State of Electricity* (1767). In C. Bazerman & J. Paradis (Eds.), *Textual dynamics of the professions: Historical and contemporary studies of writing in professional communities* (pp. 13–44). Madison, WI: University of Wisconsin Press.

Bennett, J. (1971). Science and the plain style. In J. R. Bennett (Ed.), *Prose style: A historical approach through studies* (pp. 281–297). San Francisco: Chandler.

Bernheim, M. (1981). Benjamin Franklin: Communicator unlimited. *Journal of Business Communication, 18*(1), 35–43.

Briggs, J. C. (1989). *Francis Bacon and the rhetoric of nature.* Cambridge, MA: Harvard University Press.

Brockmann, R. J. (1982). A reformed writer in 1676. *Technical Communication, 29*(2), 48.

Brockmann, R. J. (1983). Bibliography of articles on the history of technical communication. *Journal of Technical Writing and Communication, 13*(2), 155–165.

Brockmann, R. J. (1988). Does Clio have a place in technical writing? Considering patents in a history of technical communication. *Journal of Technical Writing and Communication, 18*(4), 297–304.

Brogan, J. A. (1965, June). Lessons from Benjamin Franklin, America's first great technical writer. *IEEE Transactions on Engineering Writing and Speech,* 3–7.

Broman, T. H. (1991). J. C. Reil and the 'journalization' of physiology. In P. Dear (Ed.), *The literary structure of scientific argument* (pp. 13–42). Philadelphia: University of Pennsylvania Press.

Broughton, B. B. (1985). 'No man is allowed to spell ill': Modern communication advice from an eighteenth century expert. *Journal of Technical Writing and Communication, 15*(2), 157–161.

Broughton, B. B. (1989). The art of falconry: A surprising manual of rhetoric. *Journal of Technical Writing and Communication, 19*(4), 371–379.

Bush, D. W., Jr. (1978). The diction of Sir Humphrey Davy. *Technical Communication, 25*(1), 32–33.

Bytwerk, R. L. (1979, Summer). The SST controversy: A case study of the rhetoric of technology. *Central States Speech Journal, 30*(2), 187–198.

Campbell, J. A. (1970). Darwin and *The Origin of the Species*: The rhetorical ancestry of an idea. *Speech Monographs, 37,* 1–14.

Campbell, J. A. (1975, December). The polemical Mr. Darwin. *The Quarterly Journal of Speech, 61*(4), 375–390.

Campbell, J. A. (1986, November). Scientific revolution and the grammar of culture: The case of Darwin's *Origin. The Quarterly Journal of Speech, 72*(4), 351–376.

Campbell, J. A. (1989, Winter). The invisible rhetorician: Charles Darwin's 'third party' rhetoric. *Rhetorica, 7*(1), 55–85.

Campbell, J. A. (1993, Winter/Spring). The comic frame and the rhetoric of science, epistomology, and ethics in Darwin's *Origins. Rhetoric Society Quarterly, 24*(1/2), 27–50.

Campbell, J. A. (1993, Summer). Reply to Gaonkar and Fuller. *Southern Communication Journal, 58*(4), 312–318.

Cogan, M. (1981, Fall). Rhetoric and action in Francis Bacon. *Philosophy and Rhetoric, 14*(4), 212–233.

Connaughton, M. E. (1981). Technical writing in America: A historical perspective. In J. C. Mathes & T. E. Pinelli (Eds.), *Technical writings past, present, and future* (pp. 31–42). Hampton, VA: National Aeronautics and Space Administration.

Connor, J. J. (1988). Poetry at work: Historical examples of technical communication in verse. *Journal of Technical Writing and Communication, 18*(1), 11–21.

Connor, J. J. (1991, March). History and the study of technical communication in Canada and the United States. *IEEE Transactions on Professional Communication, 34*(1), 3–6.

Connor, J. J. (1993). Medical text and historical context: Research issues and methods in history and technical communication. *Journal of Technical Writing and Communication, 23*(3), 211–232.

Connor, J. J. (1994). Self-help medical literature in 19th-century Canada and the rhetorical convention of plain language. *Journal of Technical Writing and Communication, 24*(3), 265–283.

Connor, J. J., & Connor, J. T. H. (1988, March). Semantic bypassing in technical communication: The historical case of *Antiseptics. IEEE Transactions on Professional Communication, 31*(1), 13–17.

Connor, J. T. H., & Connor, J. J. (1992). Commentary on rhetorical analysis of William Harvey's *De Motu Cordus* (1628). *Journal of Technical Writing and Communication 22*(2), 195–201.

Connors, R. J. (1982). The rise of technical writing instruction in America. *Journal of Technical Writing and Communication, 12*(4), 329–351.

Corbett, E. P. J. (1989). What classical rhetoric has to offer the teacher and the student of business and technical writing. In M. Koger (Ed.), *Writing in the business professions* (pp. 65–72). Urbana, IL: National Council of Teachers of English and Association for Business Communication.

Croll, M. W. (1966). *Attic and Baroque prose style: The anti-Ciceronian movement.* Princeton, NJ: Princeton University Press.

Croll, M. W., & Crane, R. S. (1971). Reviews of R. F. Jones "Science and English Prose Style in the Third Quarter of the Seventeenth Century." In S. E. Fish (Ed.), *Seventeenth century prose* (pp. 90–93). New York: Oxford University Press.

Daniel, C. A. (1982, Spring). Sherwin Cody: Business communication pioneer. *Journal of Business Communication, 19*(2), 3–12.

Davie, D. (1963). *The language of science and the language of literature, 1700–1740.* London: Sheed, Ward.

Dear, P. (1991). *The literary structure of scientific argument*. Philadelphia: University of Pennsylvania Press.
Dear, P. (1991). Narratives, anecdotes, and experiments: Turning experience into science in the seventeenth century. In P. Dear (Ed.), *The literary structure of scientific argument* (pp. 135–163). Philadelphia: University of Pennsylvania Press.
Del Sesto, S. L. (1981). The science journalist and early popular magazine coverage of nuclear energy. *Journal of Technical Writing and Communication, 11*(4), 315–327.
Denton, L. W. (1985). The etiquette of American business correspondence. In G. H. Douglas & H. W. Hildebrandt (Eds.), *Studies in the history of business writing* (pp. 87–95). Urbana, IL: Association for Business Communication.
Dickson, D. R. (1985). Humanistic influences on the art of the familiar epistle in the Renaissance. In G. H. Douglas & H. W. Hildebrandt (Eds.), *Studies in the history of business writing* (pp. 11–21). Urbana, IL: Association for Business Communication.
Dombrowski, P. M. (1991, December). The lessons of the Challenger investigations. *IEEE Transactions on Professional Communication, 43*(4), 211–216.
Dombrowski, P. M. (1994). Challenger through the eyes of Feyerabend. *Journal of Technical Writing and Communication, 24*(1), 7–18.
Dombrowski, P. M. (1995, September). Can ethics be technologized? Lessons from Challenger, philosophy, and rhetoric. *IEEE Transactions on Professional Communication, 38*(3), 146–150.
Douglas, G. H. (1985). Business writing in America in the nineteenth century. In G. H. Douglas & H. W. Hildebrandt (Eds.), *Studies in the history of business writing* (pp. 125–133). Urbana, IL: Association for Business Communication.
Eisner, S. (1985). Chaucer as a technical writer. *The Chaucer Review, 19*(3), 179–204.
Fahnestock, J. (1996, Fall). Series reasoning in scientific argument: *Incrementum* and *gradatio* and the case of Darwin. *Rhetoric Society Quarterly, 26*(4), 13–40.
Farrell, T. B., & Goodnight, G. T. (1981, December). Accidental rhetoric: The root metaphors of Three Mile Island. *Communication Monographs, 48*(3), 271–300.
Fenno, C. R. (1986). Aristotle and the ways we work today part II: Classical rhetoric and the electronic office. *Proceedings of the 33rd Annual International Technical Communication Conference, 33*, 243–247.
Finocchiaro, M. A. (1977, Spring). Logic and rhetoric in Lavoisier's sealed note: Toward a rhetoric of science. *Philosophy and Rhetoric, 10*(2), 111–122.
Finocchiaro, M. A. (1980). *Galileo and the art of reasoning: Rhetorical foundations of logic and scientific method*. Dordrecht, The Netherlands: D. Reidel.
Finocchiaro, M. A. (1996). [Review of the book *Novelties in the heavens: Rhetoric and science in the Copernican controversy*]. *Philosophy and Science, 29*(2), 206–209.
Freeman, W. A. (1961, October). Geoffrey Chaucer, technical writer. *Soceity of Technical Writers and Publishers Review, 8*(4), 14–15.
Fuller, S. (1993, Summer). 'Rhetoric of science': A doubly vexed expression. *Southern Communication Journal, 58*(4), 306–311.
Gaonkar, D. P. (1993, Summer). The idea of rhetoric in the rhetoric of science. *Southern Communication Journal, 58*(4), 258–295.
Gates, R. L. (1990, Winter). Understanding writing as an art: Classical rhetoric and the corporate context. *The Technical Writing Teacher, 17*(1), 50–60.

Gould, S. J., & Lowontin, R. C. (1979). The spandrels of San Marco and panglossian paradigm: A critique of the adaptionist programme. *Proceedings of the Royal Society of London, Series B, 205,* 581–598

Grego, R. C. (1987). Science, late nineteenth-century rhetoric, and the beginnings of technical writing instruction in America. *Journal of Technical Writing and Communication, 17*(1), 63–78.

Gresham, S. (1979). George Washington Carver and the art of technical communication. *Journal of Technical Writing and Communication, 9*(3), 217–225.

Gresham, S. L. (1977). Benjamin Franklin's contributions to the development of technical communication. *Journal of Technical Writing and Communication, 7*(1), 5–13.

Gresham, S. L. (1978). When technical communicators face the past. *Technical Communication, 25*(3), 8–11.

Gresham, S. L. (1981). From Aristotle to Einstein: Scientific literature and the teaching of technical writing. In D. Stevenson (Ed.), *Courses, components, and exercises in technical communication* (pp. 87–93). Urbana, IL: National Council of Teachers of English.

Gross, A. (1993, Spring). Experiment as text: The limits of literary analysis. *Rhetoric Review, 11*(2), 290–300.

Gross, A. (1993, Spring). [Review of the book *Novelties in the heavens: Rhetoric and science in the Copernican controversy*]. *Rhetorica, 11*(2), 205–207.

Gross, A. (1993, Summer). What if we're not producing knowledge? Critical reflections on the rhetorical criticism of science. *Southern Communication Journal, 58*(4), 301–305.

Gross, A. (1995, Annual ed.). [Review of the book *The discourses of science*]. *Rhetoric Society Quarterly, 25,* 252–254.

Gross, A. G. (1988). Discourse on method: The rhetorical analysis of scientific texts. *PRE/TEXT, 9*(3–4), 170–185.

Gross, A. G. (1989). The rhetorical invention of scientific invention: The emergence and transformation of a social norm. In H. W. Simons (Ed.), *Rhetoric in the human sciences* (pp. 89–107). London: Sage.

Gross, A. G. (1990). *The rhetoric of science*. Cambridge, MA: Harvard University Press.

Hager, P. J., & Nelson, R. J. (1993, June). Chaucer's *A Treatise on the Astrolabe*: A 600-year-old model for humanizing technical documents. *IEEE Transactions on Professional Communication, 36*(2), 87–94.

Hagge, J. (1989). Ties that bind: Ancient epistolography and modern business communication. *Journal of Advanced Composition, 9,* 26–44.

Hagge, J. (1989, Winter). The spurious paternity of business communication principles. *Journal of Business Communication, 26*(1), 33–55.

Hagge, J. (1990). The first technical writer in English: A challenge to the hegemony of Chaucer. *Journal of Technical Writing and Communication, 20*(3), 269–289.

Halloran, S. M. (1984, September). The birth of molecular biology: An essay in the rhetorical criticism of scientific discourse. *Rhetoric Review, 3*(1), 70–83.

Halloran, S. M. (1978). Technical writing and the rhetoric of science. *Technical Communication, 25*(4), 7–13.

Halloran, S. M. & Bradford, A. N. (1984). Figures of speech in the rhetoric of science and technology. In R. J. Connors, L. S. Ede, and & A. A. Lundsford (Eds.), *Essays on classical rhetoric and modern discourse* (pp. 179–192). Carbondale and Edwardsville, IL: Southern Illinois University Press.

Halloran, S. M., & Whitburn, M. D. (1982). Ciceronian rhetoric and the rise of science: The plain style reconsidered. In J. J. Murphy (Ed.), *The rhetorical tradition and modern writing* (pp. 58–72). New York: MLA.

Hargis, C. C., Jr. (1962, January). America's first great technical writer. *STWP Review, 9*, 12–13.

Harmon, J. E. (1986). Perturbations in the scientific literature. *Journal of Technical Writing and Communication, 16*(4), 311–317.

Harmon, J. E. (1987). The literature of enlightment: Technical periodicals and proceedings in the 17th and 18th centuries. *Journal of Technical Writing and Communication, 17*(4), 397–405.

Harmon, J. E. (1989). Development of the modern technical article. *Technical Communication, 36*(1), 33–38.

Harmon, J. E. (1989, September). The structure of scientific and engineering papers: A historical perspective. *IEEE Transactions on Professional Communication, 32*(3), 132–138.

Harrington, E. W. (1948). *Rhetoric and the scientific movement: A study of invention.* Boulder, CO: University of Colorado Press.

Harrison, J. L. (1975). Bacon's view of rhetoric, poetry, and the imagination. *Huntington Library Quarterly, 20*, 107–125.

Henderson, J. R. (1996, Winter). [Review of the book *Medieval and Renaissance letter treatises and form letters: A census of manuscripts found in part of Western Europe, Japan, and the United States of America*]. *Rhetorica, 14*(1), 103–104.

Herndl, C. G., Fennell, B. A., & Miller, C. R. (1991).Understanding failures in organizational discourse: The accident at Three Mile Island and the shuttle *Challenger* disaster. In C. Bazerman & J. Paradis (Eds.), *Textual dynamics of the professions: Historical and contemporary studies of writing in professional communities* (pp. 279–305). Madison, WI: University of Wisconsin Press.

Hildebrandt, H. W. (1984). Aristotlian views of the 20th century: Presidential address November 22, 1983. *Journal of Business Communication, 21*(2), 45–53.

Hildebrandt, H. W. (1985). A 16th century work on communication: Precursor of modern business communication. In G. H. Douglas & H. W. Hildebrandt (Eds.), *Studies in the history of business writing* (pp. 53–67). Urbana, IL: Association for Business Communication.

Hildebrandt, H. W. (1988, Summer). Some influences of Greek and Roman rhetoric on early letter writing. *Journal of Business Communication, 25*(3), 7–27.

Hildebrandt, H. W., & Varner, I. (1985). The communication theory of Johann Carl May: Its influence on business communication in Germany. In G. H. Douglas & H. W. Hildebrandt (Eds.), *Studies in the history of business writing* (pp. 97–123). Urbana, IL: Association for Business Communication.

Holmes, F. L. (1991). Argument and narrative in scientific writing. In P. Dear (Ed.), *The literary structure of scientific argument* (pp. 164–180). Philadelphia: University of Pennsylvania Press.

Hull, D. L. (1987). *Business and technical communication: A bibliography 1975–1985.* Metuchen, NJ, and London: Scarecrow Press.

Hunt, B. J. (1991). Rigorous discipline: Oliver Heaviside versus the mathematicians. In P. Dear (Ed.), *The literary structure of scientific argument* (pp. 72–96). Philadelphia: University of Pennsylvania Press.

Hyde, M. J. (1993, May). Medicine, rhetoric, and euthanasia: A case study in the workings of a postmodern discourse. *Quarterly Journal of Speech, 79*(2), 201–224.

Johnson-Sheehan, R. D. (1995). Scientific communication and metaphors: An analysis of Einstein's 1905 special relativity paper. *Journal of Technical Writing and Communication 25*(1), 71–83.

Jones, D. R. (1985, Fall). A rhetorical approach for teaching the literature of scientific and technical writing. *The Technical Writing Teacher, 12*(2), 115–125.

Jones, J. (1991, July). [Review of the book *The rhetoric of science*]. *Journal of Business and Technical Communication, 5*(3), 330–333.

Jones, R. F. (1930). Science and English prose style in the third quarter of the seventeenth century. *PMLA, 45*, 977–1009.

Jones, R. F. (1951). Science and criticism. In *The seventeenth century: Studies in the history of English thought and literature from Bacon to Pope* (pp. 41–71). Stanford, CA: Stanford University Press.

Jones, R. F. (1971). Science and language in England of the mid-seventeenth century. In S. E. Fish (Ed.), *Seventeenth century prose* (pp. 94–111). New York: Oxford University Press.

Journet, D. (1984). Rhetoric and sociobiology. *Journal of Technical Writing and Communication, 14*, 339–50.

Journet, D. (1986). Parallels in scientific and literary discourse: Stephen Jay Gould and the science of form. *Journal of Technical Writing and Communication, 16*(4), 299–310.

Journet, D. (1991). Ecological theories as cultural narratives: F. E. Clements's and H. A. Gleason's "stories" of community succession. *Written Communication, 8*(4), 446–472.

Journet, D. (1991, July). [Review of the book *Early American scientific and technical literature: An annotated bibliography of books, pamphlets, and broadsides*]. *Journal of Business and Technical Communication, 5*(3), 321–322.

Kallendorf, C., & Kallendorf, C. (1985, Winter). The figure of speech, *ethos*, and Aristotle: Notes toward a rhetoric of business communication. *Journal of Business Communication, 22*(1), 35–50.

Kaufer, D. S. (1989, Winter). [Review of the book *Shaping written knowledge: The genre and activity of the experimental article in science*]. *Rhetoric Society Quarterly, 19*(1), 69–70.

Keith, W. (1993, Summer). Rhetorical criticism and the rhetoric of science: Introduction. *Southern Communication Journal, 58*(4), 255–257.

Kelley, S., & Kelley, P. M. (1980). Sir Charles Lyell: Geologist and technical communicator. *Technical Communication, 27*(2), 40.

Kilgore, D. (1981). *Moby Dick*: A whale of a handbook for technical writing teachers. *Journal of Technical Writing and Communication, 11*(3), 209–216.

Killingsworth, M. J. (1991, July). [Review of the book *A rhetoric of science: Inventing scientific discourse*]. *Journal of Business and Technical Communication, 5*(3), 322–325.

Killingsworth, M. J., & Palmer, J. S. (1992). *Ecospeak: Rhetoric and environmental politics in America.* Carbondale, IL: Southern Illinois University Press.

Killingsworth, M. J., & Palmer, J. S. (1995, February). The discourse of 'environmentalist hysteria.' *Quarterly Journal of Speech, 58*(1), 1–19.

Kniskern, W. (1986, June). Samuel Johnson: Technical writer. *IEEE Transactions on Professional Communication, 29*(2), 3–6.
Kynell, T. (1994). Considering our pedagogical past through textbooks: A conversation with John M. Lannon. *Journal of Technical Writing and Communication, 24*(1), 49–55.
Kynell, T. (1995). English as an engineering tool: Samuel Chandler Earle and the Tufts experiment. *Journal of Technical Writing and Communication, 25*(1), 85–92.
Kynell, T. C. (1996). *Writing in a milieu of utility: The move to technical communication in American engineering programs 1850–1950.* Norwood, NJ: Ablex.
Lawson, C. A. (1954). Joseph Priestley and the process of cultural evolution. *Science Education, 38*, 267–76.
Lay, M. (1980). A classical example of a procedure. *Technical Communication, 27*(4), 40.
Lay, M. (1995, May). Rhetorical analysis of scientific texts: Three major contributions. *College Composition and Communication, 46*(2), 292–302.
Leff, M. (1993, Summer). The idea of rhetoric as interpretive practice: A humanist's response to Gaonkar. *Southern Communication Journal, 58*(4), 296–300.
Lessl, T. M. (1985, May). Science and the sacred cosmos: The ideological rhetoric of Carl Sagan. *Quarterly Journal of Speech, 71*, 175–187.
Lipson, C. S. (1982). Descriptions and instructions in Medieval times: Lessons to be learnt from Geoffrey Chaucer's scientific instruction manual. *Journal of Technical Writing and Communication, 12*(3), 243–256.
Lipson, C. S. (1985). Francis Bacon and plain scientific prose: A reexamination. *Journal of Technical Writing and Communication, 15*(2), 143–155.
Lipson, C. S. (1990). Ancient Egyptian medical texts: A rhetorical analysis of two of the oldest papyri. *Journal of Technical Writing and Communication, 20*(4), 391–409.
Locker, K. O. (1985). The earliest correspondence of the British East India Company (1600–19). In G. H. Douglas & H. W. Hildebrandt (Eds.), *Studies in the history of business writing* (pp. 69–86). Urbana, IL: Association for Business Communication.
Locker, K. O. (1985). 'Sir, this will never do': Model dunning letters, 1592–1873. In G. H. Douglas & H. W. Hildebrandt (Eds.), *Studies in the history of business writing* (pp. 179–200). Urbana, IL: Association for Business Communication.
Locker, K. O., Miller S. L., Richardson M., Tebeaux, E., & Yates, J. (1996, June). Studying the history of business communications. *Business Communication Quarterly, 59*(2), 109–127.
Loges, M. (1994). *The Treatise of Fishing with an Angle*: A study of a fifteenth-century technical manual. *Journal of Technical Writing and Communication, 24*(1), 37–48.
Loges, M. (1995). A classical case of poor communication: P. G. T. Beauregard's battle orders and report of the first battle of Bull Run. *Journal of Technical Writing and Communication, 25*(3), 261–273.
Losano, W. A. (1979). The technical writer as naturalist: Some lessons from the classics. *Journal of Technical Writing and Communication, 9*(3), 227–237.
Lyne, J., & Howe, H. F. (1986, May). 'Punctuated equilibria': Rhetorical dynamics of a scientific controversy. *The Quarterly Journal of Speech, 72*(2), 132–147.
Markel, M. H. (1979). The rhetorical principles of Sir Thomas Browne. *Technical Communication, 26*(1), 8–9.
Masse, R. E., & Kelley, P. M. (1977). Teaching the tradition of technical and scientific writing. In T. M. Sawyer (Ed.), *Technical and professional communication* (pp. 79–87). Ann Arbor, MI: Professional Communications Press.

Mathes, J. C. (1986). *Three Mile Island: The management communication failure.* Ann Arbor, MI: College of Engineering, University of Michigan.

McNair, J. R. (1991). Ancient memory arts and modern graphics. *The Journal of Technical Writing and Communication, 21*(3), 259–269.

McRae, M. W. (Ed.). (1993). *The literature of science: Perspectives on popular scientific writing.* Athens, GA: The University of Georgia Press.

Mendelson, M. (1987, Spring). Business prose and the nature of the plain style. *Journal of Business Communication, 24*(2), 3–18.

Mendelson, M. (1989). The rhetorical case: Its Roman precedent and the current debate. *The Journal of Technical Writing and Communication, 19*(3), 203–226.

Miller, C. M. (1993). Framing arguments in a technical controversy: Assumptions about science and technology in the decision to launch the space shuttle Challenger. *Journal of Technical Writing and Communication, 23*(2), 99–114.

Miller, C. R. (1989, Winter). [Review of the book *Shaping written knowledge: The genre and activity of the experimental article in science*]. *Rhetorica, 7*(1), 101–114.

Miller, W. J. (1961, December). What can the technical writer of the past teach the technical writer of today. *IRE Transactions on Engineering Writing and Speech,* 69–76.

Minor, D. E. (1984). Albert Einstein on writing. *Journal of Technical Writing and Communication, 14*(1), 113–118.

Moran, M. (1985). The history of technical and scientific writing. In M. G. Moran & D. Journet (Eds.), *Research in technical communication: A bibliographical sourcebook* (pp. 25–38). Westport, CT, and London: Greenwood Press.

Moran, M. (1990). John White: Renaissance England's first important ethnographic illustrator. *Journal of Technical Writing and Communciation, 20*(4), 343–356.

Moran, M. (1995). Frank Aydelotte: AT&T's first writing consultant, 1917–1918. *Journal of Technical Writing and Communication, 25*(3), 231–241.

Moran, M. G. (1984). Joseph Priestley, William Duncan and analytical arrangement in 18th-century discourse. *Journal of Technical Writing and Communication, 14*(3), 207–215.

Moran, M. G. (1986). Gilbert White and the personal style. *Technical Communication, 33*(1), 56.

Moss, J. D. (1989, Winter). The interplay of science and rhetoric in seventeenth century Italy. *Rhetorica, 7*(1), 23–43.

Moss, J. D. (1993). *Novelties in the heavens: Rhetoric and science in the Copernican controversy.* Chicago and London: University of Chicago Press.

Muller, J. A., & Sladkey, L. (1981). P. W. Bridgman on style. *Technical Communication, 28*(1), 40.

Myers, G. (1991). Stories and styles in two molecular biology review articles. In C. Bazerman & J. Paradis (Eds.), *Textual dynamics of the professions: Historical and contemporary studies of writing in professional communities* (pp. 45–75). Madison, WI: University of Wisconsin Press.

Neeley, K. A. (1992, December). Woman as mediatrix: Women as writers on science and technology in the eighteenth and nineteenth centuries. *IEEE Transactions on Professional Communication, 35*(4), 208–216.

Nelson, R. J. (1988). Leonardo da Vinci as technical writer. *Technical Communication, 35*(1), 80.

Nelson, R. J. (1990, September). Ben Jonson's *Timber*: A compilation of verities. *IEEE Transactions on Professional Communication 33*(3), 145-148.
Novozhilov, Y. V., & Richardson, J. G. (1976). Fifty years after the death of Flammarion, the science popularizer. *Journal of Technical Writing and Communication, 6*(2), 89-96.
Nyhart, L. K. (1991). Writing zoologically: The *Zeitschrift fur wissenschaftliche Zoologie* and the zoological community in late nineteenth-century Germany. In P. Dear (Ed.), *The literary structure of scientific argument* (pp. 43-71). Philadelphia: University of Pennsylvania Press.
Olds, B. M. (1984). An Anglo-Saxon technical writer. *Technical Communication, 31*(3), 64.
Olsen, G. R. (1991, March). Eideteker: The professional vommunicator in the new visual culture. *IEEE Transactions on Professional Communication, 34*(1), 13-19.
Ovitt, G., Jr. (1981). A late Medieval technical directive: Chaucer's *Treatise on the Astrolabe. Proceedings of the 28th Annual International Technical Communication Conference*, E78-E81.
Ovitt, G., Jr. (1987). History, technical style, and Chaucer's *Treatise on the Astrolabe*. In M. Amsler (Ed.), *Creativity and the imagination: Case dtudies from the Classical Age to the twentieth century* (pp. 34-58). Newark, DE: University of Delaware Press.
Pace, R. C. (1988). Technical communication, group differentiation, and the decision to launch the space shuttle Challenger. *Journal of Technical Writing and Communication, 18*(3), 207-220.
Paradis, J. (1983). Bacon, Linnaeus, and Lavoisier: Early language reform in the sciences. In P. V. Anderson, R. J. Brockmann, and C. R. Miller (Eds.), *New essays in technical and scientific communication: Research, theory, and practice* (pp. 200-224). Farmingdale, NY: Baywood.
Paradis, J. G. (1981). The Royal Society, Herny Oldenburg, and some origins of the modern technical paper. *Proceedings of the 28th Annual International Technical Communication Conference*, E82-E86.
Patterson, J. S. (1984). Darwin and style. *Technical Communication, 31*(4), 64.
Pauly, J. (1985). The historical and cultural significance of direct-mail fund-raising letters. In G. H. Douglas & H. W. Hildebrandt (Ed.), *Studies in the history of business writing* (pp. 165-178). Urbana, IL: Association for Business Communication.
Pera, M. (1994). *The discourses of science* (C. Botsford, Trans.). Chicago: University of Chicago Press.
Pera, M. (1995, Annual ed.). [Response to Alan Gross's review of *The discourses of science*]. *Rhetoric Society Quarterly, 25,* 254-257.
Pera, M., & Shea, W. R. (Eds.). (1991). *Persuading science: The art of scientific rhetoric.* Canton, MA: Science History Publications.
Perelman, L. (1991). The medieval art of letter writing: Rhetoric as institutional expression. In C. Bazerman & J. Paradis (Eds.), *Textual dynamics of the professions: Historical and contemporary studies of writing in professional communities* (pp. 97-119). Madison, WI: University of Wisconsin Press.
Perelman, Ch. (1982, Spring). [Review of the book *Galileo and the art of reasoning: Rhetorical foundations of logic and scientific method*]. *Philosophy and Rhetoric, 15*(2), 134-138.
Polak, E. J. (1994). *Medieval and Renaissance letter treatises and form letters: A census of manuscripts found in part of Western Europe, Japan, and the United States of America.* Leiden, The Netherlands: E. J. Brill.

Prelli, L. (1993, Summer). Rhetorical perspective and the limits of critique. *Southern Communication Journal, 58*(4), 319–327.

Prelli, L. J. (1989). *A rhetoric of science: Inventing scientific discourse.* Columbia, SC: University of South Carolina Press.

Rathjens, D. (1987). The problem of synonymy: Bacon's third idol expanded. *Journal of Technical Writing and Communication, 17*(4), 373–384.

Reeves, C. (1990, July). Establishing a phenomenon: The rhetoric of early medical reports on AIDS. *Written Communication, 7*(3), 393–416.

Richardson, M. (1980, Spring). The earliest business letters in English: An overview. *Journal of Business Communication, 17*, 19–31.

Richardson, M. (1984). The *dictamen* and its influence on fifteenth-century English prose. *Rhetorica, 2*, 207–226.

Richardson, M. (1985). Business writing and the spread of literacy in late Medieval England. In G. H. Douglas & H. W. Hildebrandt (Eds.), *Studies in the history of business writing* (pp. 1–9). Urbana, IL: Association for Business Communication.

Richardson, M. (1985). The first century of English business writing, 1417–1525. In G. H. Douglas & H. W. Hildebrandt (Eds.), *Studies in the history of business writing* (pp. 23–44). Urbana, IL: Association for Business Communication.

Richardson, M. (1985). Methology for researching early business writing in English. In G. H. Douglas & H. W. Hildebrandt (Eds.), *Studies in the history of business writing* (pp. 45–51). Urbana, IL: Association for Business Communication.

Rivers, W. E. (1979, Fall). Lord Chesterfield on the craft of business writing: The relationship of writing and reading. *Journal of Business Communication, 17*, 3–12.

Rivers, W. E. (1980, September). The place of business writing in English departments: A justification. *ADE Bulletin, 65*, 27–31.

Rivers, W. E. (1985). 'Elegant simplicity': Lord Chesterfield's ideal for business communication. In G. H. Douglas & H. W. Hildebrandt (Eds.), *Studies in the history of business writing* (pp. 135–143). Urbana, IL: Association for Business Communication.

Rivers, W. E. (1991) Eighteenth-century antecedents for Chesterfield's concept of style: A reconsideration of some business writing traditions. In B. Sims (Ed.), *Studies in technical communication: Proceedings of the 1991 CCCC and NCTE meetings* (pp. 1–15). Denton, TX: University of North Texas Press.

Roberts, L. (1991). Setting the table: The disciplinary development of eighteenth-century chemistry as read through the changing structure of its tables. In P. Dear (Ed.), *The literary structure of scientific argument* (pp. 99–132). Philadelphia: University of Pennsylvania Press.

Rosner, L. (1991). Eighteenth-century medical education and the didactic model of experiment. In P. Dear (Ed.), *The literary structure of scientific argument* (pp. 182–193). Philadelphia: University of Pennsylvania Press.

Rosner, M. (1983, Fall). Style and audience in technical writing. *The Technical Writing Teacher, 11*(1), 38–45.

Rude, C. D. (1992, June). [Review of the book *The rhetoric of scientific inquiry*]. *IEEE Transactions on Professional Communication, 35*(2), 88–90.

Russell, D. R. (1991). *Writing in the academic disciplines, 1870–1990.* Carbondale, IL: Southern Illinois University Press.

Russell, D. R. (1994). [Review of the book *A history of professional writing instruction in American colleges: Years of acceptance, growth, and doubt.*] *Journal of Technical Writing and Communication, 24*(4), 488–491.

Rutter, R. (1991). History, rhetoric, and humanism: Toward a more comprehensive definition of technical writing. *The Journal of Technical Writing and Communication, 21*(2), 133–153.

Sauer, B. A. (1993). Revisioning sixteenth century solutions to twentieth century problems in Herbert Hoover's translation of Agricola's *De Re Meticallica*. *Journal of Technical Writing and Commuication, 23*(3), 269–286.

Schmelzer, R. W. (1977). The first textbook on technical writing. *Journal of Technical Writing and Communication, 7*(1), 51–54.

Scott, G. L. (1982). The scientific poetry of Erasmus Darwin. *Technical Communication, 27*(3), 16–20.

Secor, M. (1993, Winter). [Review of the book *Novelties in the heavens: Rhetoric and science in the Copernican controversy*]. *Rhetoric Society Quarterly, 23*(1), 65–68.

Selzer, J. (Ed.). (1993). *Understanding scientific prose*. Madison, WI: University of Wisconsin Press.

Shapin, S. (1984). Pump and circumstance: Robert Boyle's literary technology. *Social Studies of Science, 14*, 481–520.

Shirk, H. N. (1988). Technical writing's roots in computer science: The evolution from technician to technical writer. *Journal of Technical Writing and Communication, 18*(4), 305–323.

Shulman, J. J. (1960, January). The anonymous technical writer in history. *STWP Review, 7*, 22–26.

Shulman, J. J. (1960, July). Technical writers who became famous as scientists. *STWP Review, 7*, 17–21.

Shulman, J. J. (1963a). Cotton Mather, America's first great technical writer. *STWP Review, 9*, 12–13.

Shulman, J. J. (1963b, April). Cotton Mather, America's first great technical writer. *STWP Review, 10*, 22–26.

Skelton, T. (1986). Aristotle and the ways we work today: Classical rhetoric in technical writing textbooks. *Proceedings of the 33rd Annual International Technical Communication Conference*, 238–242.

Smith, A. B., Jr. (1969, Spring). Historical development of concern for business English instruction. *Journal of Business Communication, 6*(3), 33–44.

Smith, B. (1984). Audience analysis in the 17th century. *Technical Communication, 31*(1), 48.

Solomon, M. (1985). The rhetoric of dehumnaization: An analysis of medical reports on the Tuskegee Syphilis Project. *The Western Journal of Speech Communication, 49*, 233–247.

Stephens, J. (1975). *Francis Bacon and the style of science*. Chicago: University of Chicago Press.

Stephens, J. (1975). Rhetorical problems in Renaissance science. *Philosophy and Rhetoric, 8*(4), 213–229.

Stephens, J. (1983). Style as therapy in Renaissance science. In P. V. Anderson, R. J. Brockmann, & C. R. Miller (Eds.), *New essays in technical and scientific communication: Research, theory, and practice* (pp. 187–199). Farmingdale, NY: Baywood.

Tebeaux, E. (1981). Franklin's *Autobiography*—important lessons in tone, syntax, and persona. *Journal of Technical Writing and Communication, 11*(4), 341–349.

Tebeaux, E. (1990). Books of secrets—Authors and their perception of audience in procedure writing of the English Renaissance. *Issues in Writing, 3*(1), 41–67.

Tebeaux, E. (1991). The evolution of technical description in Renaissance English technical writing, 1475–1640: From orality to textuality. *Issues in Writing, 4*(1), 59–107.

Tebeaux, E. (1991, July). Visual language: The development of format and page design in English Renaissance technical writing. *Journal of Business and Technical Communication, 5*(3), 246–274.

Tebeaux, E. (1991, October). Ramus, visual rhetoric, and the emergence of page design in medical writing of the English Renaissance: Tracking the evolution of readable documents. *Written Communication, 8*(4), 411–445.

Tebeaux, E. (1992, January). Renaissance epistolography and the origins of business correspondence, 1569–1640: Implications for modern pedagogy. *Journal of Business and Technical Communication, 6*(1), 75–98.

Tebeaux, E. (1993, April). Technical writing for women of the English Renaissance: Technology, literacy, and the emergence of a genre. *Written Communication, 10*(2), 164–199.

Tebeaux, E. (1997). *The emergence of a tradition: Technical writing in the English Renaissance, 1475–1640.* Amityville, NY: Baywood.

Tebeaux, E., & Killingsworth, M. J. (1992, Spring). Expanding and redirecting historical research in technical writing: In search of our past. *Technical Communication Quarterly, 1*(2), 5–32.

Tebeaux, E., & Lay, M. (1992, December). Images of women in technical books from the English Renaissance. *IEEE Transactions on Professional Communication, 35*(1), 196–207.

Walen, T. (1985). A history of specifications: Technical writing in perspective. *Journal of Technical Writing and Communication, 15*(3), 235–245.

Wallace, K. R. (1943). *Francis Bacon on communication and rhetoric.* Chapel Hill, NC: University of North Carolina Press.

Wallace, K. R. (1963). Imagination and Francis Bacon's view of rhetoric. In R. E. Nebergall (Ed.), *Dimensions of rhetorical scholarship* (pp. 65–81). Norman, OK: University of Oklahoma Press.

Wallace, W. A. (1989). Aristotelian science and rhetoric in transition: The Middle Ages and the Renaissance. *Rhetorica, 7*(1), 7–21.

Walzer, A. E. (1987, February). Logic and rhetoric in Malthus's *Essay on the Principle of Population*, 1798. *The Quarterly Journal of Speech, 73*(1), 1–17.

Warnick, B. (1983, Fall). A rhetorical analysis of episteme shift: Darwin's *Origin of the Species*. *The Southern Speech Communication Journal, 49*, 26–42.

Weeks, F. W. (1973). An editorial: The Watergate hearings and business communication. *Journal of Business Communication, 10*(4), 3–6.

Wells, S. (1996, February). Women write science: The case of Hannah Longshore. *College English, 58*(2), 176–191.

Wenzel, J. W. (1974, October). Rhetoric and anti-rhetoric in early American scientific societies. *Quarterly Journal of Speech, 60*(3), 328–336.

Whalen, T. (1985). A history of specifications: Technical writing in perspective. *Journal of Technical Writing and Communication, 15*(3), 235–245.

Whalen, T. (1986). Gaius Julius Ceasar, technical writer. *Technical Communication, 33*(3), 200.

Whitburn, M. D. (1976). Personality in scientific and technical writing. *Journal of Technical Writing and Communication, 6,* 299–306.
Whitburn, M. D. (1977). The past and the future of scientific and technical writing. *Journal of Technical Writing and Communication, 7*(2), 143–149.
Whitburn, M. D. (1978). The plain style in technical writing. *Journal of Technical Writing and Communication, 8,* 349–358.
Whittaker, D. A. (1979). Priestley's personal style. *Technical Communication, 26*(3), 28.
Winsor, D. A. (1988, September). Communication failures contributing to the Challenger accident: An example for technical communicators. *IEEE Transactions on Professional Communication, 31*(3), 101–107.
Wolff, L. M. (1979, Winter). A brief history of the art of dictamen: Medieval origins of business letter writing. *Journal of Business Communication, 16*(2), 3–11.
Verrept, S. A. (1985, Fall). Review essay: *Colloquia et Dictionariolum Septem Linguarum...Antverpiae* 1616: A starting point for systematic research on cross-cultural (business) communication. *Journal of Business Communication, 22*(4), 17–23.
Yates, J. (1982, Summer). From press book and pigeonhole to vertical filing: Revolution in storage and access systems for correspondence. *Journal of Business Communication, 19*(3), 5–26.
Yates, J. (1989). *Control through communication: The rise of system in American management.* Baltimore and London: Johns Hopkins University Press.
Zappen, J. P. (1975). Francis Bacon and the rhetoric of science. *College Composition and Communication, 26,* 244–247.
Zappen, J. P. (1977). Francis Bacon and the topics. *Proceedings of the 24th International Technical Communication Conference,* 309–312.
Zappen, J. P. (1985). Historical perspectives on the philosophy and the rhetoric of science: Sources for a pluralistic rhetoric. *PRE/TEXT, 6*(1–2), 10–29.
Zappen, J. P. (1987, Fall). Historical studies in the rhetoric of science and technology. *The Technical Writing Teacher, 14*(3), 285–298.
Zappen, J. P. (1991). Scientific rhetoric in the nineteenth and early twentieth centuries: Herbert Spencer, Thomas H. Huxley, and John Dewey. In C. Bazerman & J. Paradis (Eds.), *Textual dynamics of the professions: Historical and contemporary studies of writing in professional communities* (pp. 145–167). Madison, WI: University of Wisconsin Press.
Zappen, J. P. (1991, June). [Review of the book *A rhetoric of science: Inventing scientific discourse*]. *IEEE Transactions on Professional Communication, 34*(2), 123–124.

Author Index

A

Ackerman, J., 42, *46*
Adams, K. H., 1, *15*, 207, *221*, 257, 292, 293, *294*
Adolph, R., 271, *294*
Agg, T. R., 182, *193*
Alexander, L. L., 261, *294*
Alford, E. M., 262, *294*
Allen, B., 279, *294*
Allen, J., 256, 263, *294*
Allen, J., 14, 91, *101*, 234, 239, *243*, *244*, 269, 286, *294*
Allen, N., 210, *221*
American Psychiatric Association, 144, *151*
Anderson, F. D., 269, *294*
Anderson, F. H., 54, 60, *60*
Anderson, W., 274, *294*
Andrews, D. C., 283, *294*
Andrews, W. D., 283, *294*
Anglin, J. P., 119, *119*
Arms, V. M., 212, *221*
Arner, R. D., 86, *88*
Arnold, V. D., 260, 289, *294*
Astell, M., 117, *120*
Aune, J. A., 132, *151*
Aydelotte, F., 4, *15*, 178, *193*

B

Bacon, F., 7, 11, 12, *15*, 22, 28, *45*, 49–60, *60*, *61*, 125, 266, 267, 268, 270, 271, 276, 279
Baker, C., 268, *295*
Baresich, M. J., 285, *295*
Barton, B. F., 153, *169*
Barton, M. S., 153, *169*
Basquin, E. A., 263, *295*
Bathe, D., 64, 67, 72, 75, 78, 82, 83, 85, 86, 87, *88*
Bathe, G., 64, 67, 72, 75, 78, 82, 83, 85, 86, 87, *88*
Batschelet, M. W., 251, 281, *295*
Battalio, J., 212, *221*

Bazerman, C., 1, 9, 12, *15*, 22, 23, 34, 41, 42, 45, *45*, 212, *221*, 253, 263, 272, 273, 274, 284, 287, *295*
Bennett, J., 270, *295*
Berkenkotter, C., 42, *46*, 212, *221*
Berlin, J., 208, 209, 210, 211, *221*
Bernhardt, S. A., 212, *221*
Bernheim, M., 282, *295*
Bertin, J., 154, 166, *169*
Birk, W. O., 181, 182, *193*
Black, M., 133, *151*
Bleuler, E., 144, *151*
Blyler, N. R., 206, 215, *221*, 226
Boas, M., 51, *61*
Boerhaave, H., 139, 140, *151*
Boiarsky, C., 241, *243*
Bolter, J. D., 207, 211, *221*
Bosley, D. S., 237, 238, 241, *243*
Bourne, W., 156, *169*
Bowman, J. P., 210, *224*
Bradford, A. N., 6, *16*, 52, 55, *61*, 273, *298*
Bredvold, L. I., 60, *61*
Brenner, T. B., 236, *244*
Briggs, J. C., 268, *295*
Bright, T., 137, *151*
Britton, W. E., 188, *193*, 232, *243*
Brockmann, R. J., 14, 93, 97, 100, *101*, 201, 216, *221*, 257, *296*
Brogan, J. A., 281, *295*
Broman, T. H., 277, *295*
Broughton, B. B., 263, 275, *295*
Brown, C., 219, *225*
Brown, L. A., 158, *169*
Brown, S., 182, *195*
Bruffee, K. A., 235, 236, *243*
Burleson, B. R., 59, *61*
Burnett, R. E., 237, *243*
Burton, R., 137, *151*

Burtt, E. A., 50, *61*
Bush, D. W., Jr., 277, *295*
Bytwerk, R. L., 288, *295*

C

Caldwell, D. L., 190, *194*
Callon, M., 22, *46*
Camden, C. C., 119, *120*
Campbell, J. A., 256, 276, *295*
Campbell, M., 119, *120*
Cardinale, S., 108, *122*
Carey, M., 65, *88*
Carroll, J., 218, *221*
Carson, K. D., 241, *243*
Carson, P. P., 241, *243*
Cassian, J., 135, 136, *151*
Cellier, E., 115, 116, 117, 119, *120*
Chatburn, G. R., 179, *193*
Christian, B., 239, *243*
Clark, A., 119, *120*
Cogan, M., 60, *61*, 268, *296*
Collins, H., 22, *46*
Collins, W. E., 190, *193*
Connaughton, M. E., 173, *193*, 282, *296*
Connor, J. J., 251, 252, 269, 270, 278, 283, 284, *296*
Connor, J. T. H., 269, 270, 283, *296*
Connors, R. J., 14, 93, 97, 100, *101*, 201, 216, *221*, 257, *296*
Cooper, M. M., 211, 214, 218, *222*
Corbett, E. P. J., 87, *88*, 89, 260, *296*
Crane, R. S., 271, *296*
Crawford, P., 108, 110, *120*
Creek, H. L., 93, *101*, 174, 181, *193*
Cressy, D., 119, *120*
Croll, M. W., 50, 51, 60, *61*, 271, *296*
Crosland, M., 27, 28, *46*
Crowley, S., 132, 149, *151*
Cruse, E., 85, *88*
Culler, J., 134, *151*
Cumming, W, P., 155, 159, 165, 169, *169*
Cunningham, W., 161, *169*

D

Daniel, C. A., 259, *296*
Dansereau, D. F., *245*
Dautermann, J. 205, *226*
David, K., 240, *245*
Davie, D., 273, *296*
Davis, R. M., 189, 191, *193*
Dawson, T., 109, 112, *120*

Dear, P., 253, 269, *297*
Debs, M. B., 237, *243*
December, J., 203, *222*
Dell, S. A., 239, *243*
De Man, P., 149, *151*
Del Sesto, S. L., 288, *297*
Denton, L. W., 283, *297*
Derks, S., 87, *88*
Derrida, J., 133, 147, 149, 150, *151*
Desaguliers, J. T., 28, *46*
Dickson, D. R., 264, *297*
Diethelm, O., 144, 145, *151*
Digges, L., 163, *169*
Doak, J., 216, *225*
Dobrin, D. N., 209, 214, *222*, 228, *243*
Dobson, T., 67, 68, 69, 70, 71, 72, 81, 83, 86, *88*
Doheny-Farina, S., *224*
Dombrowski, P. M., 207, 215, *222*, 228, 229, 232, 235, *243*, 290, *297*
Dorff, D. L., 209, *222*
Douglas, G. H., 283, *297*
Driskill, L., 209, *222*
Dudley, J. W., 190, *193*
Dudley-Evans, T., 40, *46*
Duffy, T. M., 218, *222*
Duin, A. H., 205, 209, 212, 215, *222*, 224
Dury, J., 58, *61*
Dzujna, C. C., 214, *222*

E

Eagleton, T., 150, *151*
Earle, S. C., 176, 178, *193*
Ede, L., 234, 237, *243*
Ehninger, D., 86, *88*
Eisner, S., 264, *297*
Elyott, T., 107, 119, *120*
Evans, O., 12, 15, 63–88, *88*
Ezell, M. M., 110, *120*

F

Fahnestock, J., 277, *297*
Faigley, L., 206, 209, 214, *222*, 235, *244*
Farrell, T. B., 290, *297*
Fatout, P., 184, *193*
Feenberg, A., 216, 217, 219, 220, *222*
Fennell, B. A., 8, *16*, 290, *299*
Fenno, C. R., 261, *297*
Ferguson, E. S., 64, 70, *89*
Ficino, M., 137, *138*
Finnocchiaro, M. A., 268, 269, 274, *297*
Fitting, R. U., 180, *193*

AUTHOR INDEX   311

Fitzherbert, J., 106, *120*
Flatley, M. E., 210, *224*
Flower, L., 212, 214, *222*
Foster, W., 182, *193*
Foucalt, M., 149, *151*
Fountain, A. M., 182, 185, *193*, *195*
Franklin, B., 24–25, 28, 35–36, 40, *46*, 67, 69, *89*, 281, 282
Freedman, M., 187, 192, *194*
Freeman, W. A., 3, *16*, 263, *297*
Freud, S., 145, 146, 148, *151*
Fruchtman, J., 27, *46*
Fujimura, J. H., 45, *46*
Fuller, S., 256, *297*
Fulton, J. F., 26, *46*
Funkhouser, H. G., 26, *46*

G

Galen, 134, 135, 137, *151*
Gaonkar, D. P., 255, 256, *297*
Gartenberg, P., 107, *120*
Garvey, W. D., 22, *46*
Gates, R. L., 260, *297*
Gaum, C., 181, *194*
Geertz, C., 238, *244*
Gerson, S., 212, *215*
Gerson, S. J., 215, *222*
Gerson, S. J., 215, *222*
Giddens, A., 214, *222*
Gilbert, W., 24, *46*
Gilbertson, M. K., 154, 166, *170*
Glanvill, J., 271, *273*
Goodnight, G. T., 290, *297*
Gould, J. R., 184, 186, *194*, *195*
Gould, S. J., 254, *298*
Gotswami, D., 63, *89*
Gralath, D., 28, *46*
Graves, H., 181, *194*
Greene, B. G., 241, *244*
Greenhood, D., 159, 162, *169*
Grego, R. C., 258, *298*
Gresham, S., 2–3, *16*, 285, *298*
Gresham, S. L., 256, 281, *298*
Grey, E., 109, 110, 113, 119, *120*
Gribbons, J., 212, *222*
Gross, A. G., 87, *89*, 253, 255, 261, 267, 269, 272, 277, 279, 280, 287, 289, 293, *298*
Grymeston, E., 107, 111, *120*
Guillemeau, J., 113, *120*
Gurak, L. J., 203, *225*

H

Haas, C., 203, *222*
Hager, P. J., 263, *298*
Hagge, J., 3, 4, 5, *16*, 229, 233, *244*, 259, 260, 264, *298*
Hall, A. V., 181, *194*
Halloran, S. M., 6, *16*, 52, 55, *61*, 261, 271, 273, 286, *298*, *299*
Hand, H. E., 189, *194*
Hansen, C. J., 205, 215, *222*
Harbarger, S. A., 4, 12, *16*, 92–100, *101*, 179, 180, 183, 192, *194*
Harding, S., 53, 59, *61*
Hargis, C. C., Jr., 281, *299*
Harmon, J. E., 272, 278, 279, *299*
Harrington, E. W., 260, *299*
Harriot, T., 13, *16*, 155–167, 169, *169*
Harrison, J. L., 268, *299*
Hawisher, G. E., 205, *223*
Haydn, H., 50, *61*
Hayhoe, G. F., 202, *223*
Hayes, J. R., 212, 214, *222*
Hays, R., 188, *194*
Heibert, I. A., 27, *46*
Heilborn, J. L., 43, *46*
Heinroth, J. C., 142, 143, *151*
Henderson, J. R., 252, *299*
Herndl, C. G., 8, *16*, 290, *299*
Herrington, A., 63, *89*
Hildebrandt, H. W., 260, 261, 264, 275, *299*
Hildreth, W. H., 97, *101*
Hippocrates, 134, 135, *152*
Hoecker, J. J., 44, *46*
Hoerter, G. E., 236, *244*
Hoffman, F., 139, 140, *152*
Hoffman, L., 181, *194*
Hofstede, G. H., 238, *244*
Hogrefe, P., 110, *120*
Holbrook, S. E., 92, 93, *101*
Holmes, F. L., 271, *299*
Hopton, A., 161, 162, *170*
Horton, W. K., 216, 217, *223*
Houston, R. A., 119, *120*
Howe, H. F., 289, *301*
Howell, W. S., 6, *16*, 60, *61*, 86, *89*
Huckin, T., 42, *46*
Huckin, T. N., 209, 212, 214, *221*, *223*
Hull, D. L., 251, *299*
Hull, S., 106, *120*
Hulton, P., 156, 165, *170*
Hunt, B. J., 278, *299*

Hunter, L., 109, 110, 119, *120*
Hunter, L. C., 72, 85, *89*
Hyde, M. J., 288, *300*
Hythecker, V. I., *245*

I

Irwin, C., 241, *243*

J

Jackson, S. W., 135, 137, 140, 141, 150–151, *152*
Jardine, L., 55, 60, *61*
Johnson, M., 133, *152*
Johnson, R., 198, 210, *223*
Johnson-Eilola, J., 11, 14, 207, 211, 212, 213, 214, 215, 216, 218, 219, *223*, *225*
Johnson-Sheehan, R. D., 285, *300*
Johnston, J., 241, *244*
Jones, B. G., 237, *244*
Jones, D. R., 2, *16*, 256, *300*
Jones, J., 253, *300*
Jones, R. F., 51, *61*, 124, 125, *130*, 271, *300*
Jones, T., 66, *67*
Jordan, S., 190, *194*
Journet, D., 251, 286, 287, 288, *200*

K

Kallendorf, C., 261, *300*
Kallendorf, C., 261, *300*
Kanter, R. M., 214, *223*
Karis, W. M., 237, *244*
Kaufer, D. S., 253, *300*
Keith, W., 255, *300*
Keller, E. F., 53, 59, *61*
Kelley, P. M., 256, 277, *300*, *301*
Kelley, S. A., 277, *300*
Kendall, R., 190, *194*
Kent, T., 234, *244*
Kilgore, D., 282, *300*
Killingsworth, M. J., 154, 166, *170*, 219, 221, *223*, 237, *244*, 253, 266, 291, 292, 293, *300*, *306*
Kinosita, K., 241, *244*
Klaver, P., 192, 193, *194*
Klein, W., 212, *225*
Klibansky, R., 138, *152*
Kline, S. L., 59, *61*
Kniskern, W., 273, *301*
Knoblauch, C. H., 204, 205, *223*
Knorr-Cetina, K. D., 21, *46*
Kobe, K. A., 186, *194*
Korman, H., 216, *224*
Kostelnick, C., 153, *170*

Kraepelin, E., 144, *152*
Krafft-Ebing, R., von, 143, *152*
Kramnick, I., 27, *47*
Kreth, M., 229, 233, *244*
Kuhn, T., 227, 235, *244*
Kuhn, T. S., 51, *62*
Kupperman, K. O., 155, *170*
Kynell, T. C., 1, 11, 12, *16*, 91, 101, *102*, 201, 207, 209, 214, 215, *223*, *244*, 258, 259, 292, 293, *301*

L

Laboucheix, H., 27, 44, *47*
Lakoff, G., 133, *152*
Lambiotte, J. G., *245*
Lancaster, F. W., 279, *294*
Landow, G. P., 207, *223*
Lannon, J. M., 192, *194*, 214, *223*, 259
Laqueur, T., 119, *121*
Larson, C. O., *245*
Latour, B., 21, 22, *47*
Lavoisier, A., 7, *16*, 274, 275
Law, J., 22, *46*
Lawson, C., 41, *47*
Lawson, C. A., 274, *301*
Lawson, W., 106, *121*
Lay, M., 212, 214, *223*, *224*, 253, 254, 266, 277, *301*, *306*
Lay, M. M., 108, 113, *122*, 237, 239, 240, *244*, *245*
LeFevre, K. B., 235, *244*
Leff, M., 256, *301*
Leigh, D., 107, 111, *121*
Leitch, V. B., 133, *152*
Lessl, T. M., 286, *301*
Lipson, C. S., 3, *16*, 52, 55, 60, *62*, 238, *244*, 262, 263, 268, *301*
Little, S. B., *224*
Locker, K. O., 7, 8, *16*, 252, 264, 275, 281, *301*
Loeb, H. M., 216, *226*
Loges, M., 263, 282, *301*
Losano, W. A., 278, *301*
Lowontin, R. C., 254, *298*
Lucar, C., 162, *170*
Lunsford, A., 234, 237, *243*
Lyley, J., 124, *130*
Lyne, J., 289, *301*

M

MacDonald, M., 138, *152*
MacNamara, D. J., 186, *194*

AUTHOR INDEX

Magruder, W. T., 93, *102*
Malley, J. C., 289, *294*
Mandel, S., 190, *194*
Maria, Q. H., 109, 119, *121*
Markel, M., 214, *224*
Markel, M. H., 270, *301*
Markham, G., 106, 107, 109, 110, 111, 112, 119, *121*
Martin, B., 70, *89*
Masek, R., 119, *121*
Masse, R. E., 256, *301*
Mathes, J. C., 192, 193, *194*, 290, *302*
Mattill, J. I., 192, *194*
McEvoy, J. G., 27, *47*
McGavin, D., 212, 219, *225*
McKee, J. H., 174, *193*
McMullen, N., 119, *121*
McNair, J. R., 279, *302*
McRae, M. W., 253, 254, *302*
Mead, R., 139, 140, 141, *152*
Mehlenbacher, B., 218, *222*
Mendelson, M., 261, 262, *302*
Merton, R. K., 22, *47*
Meyer, A., 144, *152*
Miller, C., 228, 232, 233, *244*
Miller, C. M., 290, 302
Miller, C. R., 8, 9, *16*, 207, 209, 212, 216, *224*, 229, *244*, 253, 290, *299*, *302*
Miller, D. J., 216, *225*
Miller, S. L., 252, *300*
Miller, T. P., 212, *224*
Miller, W. J., 2, *16*, 256, 260, *302*
Mills, G. H., 173, 187, 188, 190, 192, *194*
Minor, D. E., 285, *302*
Mirel, B., 218, *224*
Moore, P., 212, *224*, 229, 232, 233, 235, *244*
Moran, M. G., 6, 11, 13, *16*, 26, 41, *47*, 105, *121*, *170*, 250, 251, 256, 259, 264, 265, 274, 275, *302*
Moss, J. D., 269, *302*
Muller, J. A., 285, *302*
Murrell, J., 109, 112, *121*
Musson, A. E., 88, *89*
Myers, G., 22, 45, *47*, 284, 285, *302*

N

Najjar, H., 40, *47*
Napier, R., 138, 139, 141, *152*
Neeley, K. A., 280, *302*
Negroponte, N., 217, *224*
Nelson, J. R., 93, *102*, 177, 179, 180, 181, *194*
Nelson, R. J., 263, 270, *298*, *302*
Newton, I., 45, *47*, 272
Niles, H., 87, *89*
Norris, C., 133, *152*
Novozhilov, Y. V., 277, *303*
Nyhart, L. K., 278, *303*

O

O'Brien, H. R., 175, *194*
O'Donnell, A., 236, *245*
Odell, L., 63, *89*, 235, *245*
Olds, B. M., 263, *303*
Olsen, G. R., 279, *303*
Ong, W. J., 119, *121*
Ornatowski, C. M., 207, *224*, 239, *245*
Ornstein, M., 60, *62*
Ovitt, G., Jr., 3, *16*, 263, 264, *303*

P

Pace, R. C., 289, *303*
Pacey, A., 53, 59, *62*
Palmer, J. E., 218, *222*
Palmer, J. S., 291, *300*
Panofsky, E., 138, *152*
Papert, S., 212, *226*
Paradis, J., 7, *17*, 217, *224*, 263, 271, 274, *303*
Park, C. W., 94, *102*, 179, 181, *194*
Parker, J. W., 181, *194*
Parsons, G. M., 238, *245*
Partridge, J., 109, 110, 112, *121*
Patterson, J. S., 277, *303*
Pauly, J., 283, *303*
Pearsall, T. E., 201, *224*
Penrose, J. M., 210, *224*
Pera, M., 255, *303*
Perelman, Ch., 268, *303*
Perelman, L., 263, *303*
Pierce, J. R., 186, *195*
Pinch, T., 22, *46*, 212, *224*
Platt, H., 109, 110, *121*
Polak, E. J., 252, *303*
Pollard, A. W., 106, *121*
Postman, N., 206, *224*
Power, K., 189, *195*
Prelli, L. J., 253, 287, 289, 293, *304*
Price, J., 216, *224*
Priestley, J., 12, *17*, 21, 24–45, *47*, 274
Prior, M., 119, *121*
Purver, M., 51, 54, *62*

Q

Qin, J., 279, *294*
Quick, D., 63, *89*
Quinn, D. B., 155, *170*

R

Racker, J., 187, *195*
Rafoth, B. A., 235, *245*
Rathbone, R. R., 186, *195*
Rathborne, A., 163, 164, *170*
Rathjen, D., 268, *304*
Rawlins, C., 238, *245*
Read, H., 154, 166, *170*
Reeves, C., 288, *304*
Redgrave, G. R., 106, *121*
Redish, J., 229, 233, *244*
Rehling, L., 239, 240, *245*
Reynolds, T. S., 5, *17*
Richardson, J. G., 277, *303*
Richardson, M., 252, 262, *301*, *303*
Richeson, A. W., 156–157, 159, 161, 162, *170*
Rickard, T. A., 4, *17*, 95, *102*, 176, 179, 180, *195*
Rider, M. L., 185, 186, *195*
Rip, A., 22, *46*
Rivers, W. E., 1, 3, 15, *17*, 105, *121*, 153, *170*, *304*
Roberts, L., 274, *304*
Robinson, E., 88, *89*
Rocklin, T., 236, *245*
Roesslin, E., 113, *121*
Rorty, R., 147, *152*, 227, 235, *245*
Rosner, L., 275, *304*
Rosner, M., 259, *304*
Rubin, D. L., 235, *245*
Rude, C. D., *224*, 253, *304*
Rudwick, M., 22, *47*
Rush, B., 142, *152*
Russell, D. R., 1, *17*, 91, 93, *102*, 257, *304*
Rutter, R., 2, *17*, 256, 257, *305*

S

Sauer, B. A., 212, *224*, 286, *305*
Saxl, F., 138, *152*
Schaffer, S., 38, *47*
Schlesinger, E. K., 191, *195*
Schmelzer, R. W., 259, *305*
Schofield, R. E., 25, 28, *47*
Schofield, R. S., 119, *121*
Schrage, M., 207, *224*
Scott, G. L., 275, *305*
Scott, J. C., 240, *245*

Secor, M. J., 269, *305*
Seeley, B. E., 5, *17*
Selber, S. A., 11, 14, 202, 203, 204, 205, 211, 212, 213, 214, 215, 216, 219, *223*, 224
Selfe, C. L., 11, 14, 205, 207, 211, 212, 219, 222, *223*, *224*, 225
Selfe, R. J., 212, *225*
Sellers, C., 85, *89*
Selzer, J., 209, 214, *224*, 253, 254, *305*
Shannon, C. E., 236, *245*
Shapin, S., 272, *305*
Sharp, J., 113, 114, 115, *122*
Shea, W. R., 255, *303*
Sheridan, P. J., 190, *195*
Sherman, T. A., 187, 188, *195*
Shirk, H. N., 11, 13, 207, 211, *225*, 284, *305*
Shirley, J. W., 156, 158, *170*
Shriver, K. A., 154, 166, *170*
Shulman, J. J., 260, 278, 281, *305*
Silker, C. M., 203, *225*
Simon, J., 119, *122*
Skelton, T., 261, *305*
Slack, J. D., 216, *225*
Sladkey, L., 285, *302*
Smeaton, J., 70, *89*
Smith, A. B., Jr., 259, *305*
Smith, B., 268, *305*
Smith, C., 210, *225*
Smith, H. L., 108, *122*
Smith, R. W., 188, *195*
Snyder, I., 211, *225*
Solomon, M., 291, *305*
Sontag, S., 136, 138, 150, *152*
Southard, S. G., 238, *245*
Souther, J. W., 187, *195*, 201, *225*
Spilka, R., 212, *226*
Sprat, T., 125, *130*
Staudenmaier, J. M., 206, *226*
Stephens, J., 52, 55, 56, 60, 62, 267, 268, *305*
Sternberg, R. J., 237, *246*
Stevenson, D. W., 192, 193, *194*
Stick, D., 155, *170*
Stone, L., 119, *122*
Stoner, R. B., 231, *245*
Stotsky, S., 212, *226*
Stoughton, B., 180, *195*
Stoughton, M. R., 180, *195*
Sullivan, D., 205, 212, *226*, 228, 238, *245*
Sullivan, P., 215, *226*
Swales, J., 40, *47*, 212, *226*
Swarts, H., 212, 214, *222*

Sweigert, R., Jr., 187, *195*
Sypherd, W. O., 182, 183, *195*

T

Talbot, A. H., 109, 110, 119, *122*
Taylor, C. A., 212, *226*
Taylor, E. G. R., 155, 156, 163, *170*
Tebeaux, E., 1, 10, 11, 13, 14, *17*, 91, *102*, 106, 108, *122*, 154, 155, *170*, 214, *226*, 239, 240, 244, *245*, 252, 264, 265, 266, 271, 279, 281, 292, 293, *301*, *305*, *306*
Teklinski, B., 209, *226*
Telleen, J. M., 175, *195*
Tellenbach, H., 134, *152*
Terpstra, V., 240, *245*
Thomas, J. D., 187, *195*
Thomas, K., 119, *122*
Thralls, C., 206, 212, *226*
Tillinghast, M., 117, *122*
Trelease, S. F., 95, *102*, 180, *195*
Tredgold, T., 66, 81, *89*
Turkle, S., 212, *226*
Turnbull, A. D., 79, *89*
Tusser, T., 106, *122*

U

Ulman, J. N., Jr., 186, *195*
U.S. Department of Health and Human Services, 146, 147, 148, *152*

V

Van den Daele, W., 53, *62*
Varner, I., 275, *299*
Varner, I. I., 241, *245*
Vaughn, J., 239, *245*
Veiga, N. E., 239, *245*
Verept, S. A., 270, *307*
Vickers, B., 60, *62*

W

Waddell, C., 212, *226*
Wahlstrom, B. J., *224*
Wallace, K. R., 54, 55–56, 60, *62*, 268, *306*
Wallace, W. A., 279, *306*
Walter, J. A., 173, 186, 187, 188, 190, 191, 192, *194*
Walzer, A. E., 277, *306*

Ward, J., 68, 86, *89*
Waring, E., 70, *89*
Warnick, B., 276, *306*
Warren, T. L., 201, *224*
Weaver, W., 236, *245*
Webster, C., 58, 60, *62*
Weeks, F. W., 291, *306*
Weinberger, J., 53, 59, *62*
Weisman, H. W., 123, *130*
Wellborn, G. P., 186, *195*
Wells, S., 282, *306*
Wenzel, J. W., 283, *306*
Whalen, T., 262, *306*
Whitburn, M. D., 13, *130*, 261, 271, 273, *299*, *307*
Whitney, C., 53, 59, *62*
Whittaker, D. A., 274, *307*
Whittemore, N. T., 107, *120*
Willey, B., 50, *62*
Williams, T. R., 154, *170*
Williams, W. M., 237, *245*
Williamson, G., 51, 55, 60, *62*
Willis, T., 139, 141, *152*
Wilson, J. H., 187, *195*
Wing, D., 119, *122*
Winner, L., 52, *62*, 206, *226*
Winsor, D., 239, *246*
Winsor, D. A., 212, *226*, 289, *307*
Witte, S. P., 209, 214, *222*
Wolff, L. M., 262, 263, *307*
Woolever, K. R., 216, *226*
Woolgar, S., 21, *47*
Woolley, H., 111, 113, 114, 117, *122*
Wright, C. D., 87, *89*
Wright, L., 106, 119, *122*
Wrightson, K., 119, *122*

Y

Yates, J., 252, 283, 292, *301*, *307*
Yule, E. S., 95, *102*, 180, *195*

Z

Zakon, R. H., 197, *226*
Zappen, J. P., 11, 12, 238, *246*, 250, 252, 253, 267, 280, *307*
Ziman, J., 22, *47*

Subject Index

A
"–Acedia" sin im-agery of Middle Ages, 135–136
Argumentative *fields*, 9
Audience
 importance of analysis, 3, 77, 126, 234, 285
 importance of engagement, 74, 80–81
 importance of inclusiveness, 242
 needs and English Renaissance writers, 106–107, 113
 and style of early midwifery books, 113, 114–115
 and "usability" of Hariot-White-deBry map, 155, 165–168
 writer-reader relationship, 187, 236

B
Bacon, Francis
 "corporate approach" to science, 51
 democratic science/plain style, 49, 52–55, 57, 58, 59
 institutionalized science/complex style, 49, 51–52, 56
 linguistic principles, 7, 55–58
 positivistic science/plain style, 49, 56–57
 and scientific cooperation, 22
 "terminological reforms," 7
Born writer myth, 126
British Royal Society
 and cooperative model, 22
 and plain style standard, 50–51, 125, 271

C
Carver, George Washington, 285–286
Chaucer, priority debate in English technical writing, 3–4, 263–264
Civil Rights Movement, attitude questioning and promotion of equality, 230–235, 238
Competition
 in scientific texts, 21–22
 vs. Priestley's model, 43
Complementarity principle, 166
Computer usage
 and democratic communication, 207
 and least transformative activities, 198
Computers and technical communication, 14
 critical perspective, 199–200, 216–220
 pedagogical perspective, 199–200, 201–207, 220
 rhetorical perspective, 199–200, 207–212, 220
 spatial dynamic perspective, 199–200, 213–216, 220
Corporate communication histories, 7–8
Corporate culture, as a sociological community with the potential for change, 237–239
Council for Programs in Technical and Scientific Communication (CPTSC), 201–202
Craft knowledge. *See* Empirical experience
Critics of science, feminists, 53, 59
Current-traditional rhetoric, 208, 209
Curricular histories, 4–5
 and "two-culture" split, 175, 184

D
Darwin, Charles, studies, 276
Deconstructing text methodology, 131–133, 150
Decontextualization model, 218
Derrida, Jacques, and "written-ness" notion, 147
Discourse
 modes, 2
 See also Failures in organizational discourse
Dudley, Juanita Williams, and character of technical writing classes, 190

E
Earle, Samuel Chandler, and early technical writing theory, 176–179

317

SUBJECT INDEX

Empirical experience
  and early education in chemical technology/medicinal preparations, 109
  to Joseph Priestley, 38–39
Engineering studies
  evolution, 174–176
  and vocation vs profession status, 92, 96–97, 99, 258
English for Engineers, 95
Evans, Oliver
  failure of accusatory/arrogant voice, 80–81
  fitting style and audience, 77–78
  "projector" image, 64
  rhetorical decisions/consequences, 63, 79, 83–85
  success of humble persona image, 69, 76
Expanding and contracting model of information design/use, 213–216
Experimental articles in science, conventionalized formulations, 9–10

F

Failures in organizational discourse
  Shuttle Challenger explosion, 8–9, 212, 289–290
  Three Mile Island, 8–9, 290–291
Feenberg, A., critical theory of technology, 204, 216–217, 219–220
Franklin, Benjamin
  and advantage of appearing hesitant, 69
  and Joseph Priestley, 25, 28
  and knowledge of classical rhetoric, 67
French Royal Academy, and cooperative model, 22

G

Gender
  base of approaches to computer use, 212
  by vs. about women, 113–115
  and collaborative teams, 237, 239
  and earliest domestic medicine books, 108–113
  early proposal writing by women, 115–117
  and educational opportunities of Puritan reformers, 58
  historical importance of women in technical communication, 11, 12, 13, 91, 105, 239–240
  and information limits, 231
  and service-related teaching, 12
  See also War and changes in communication
General Electric, 185

General Motors, 185
Genre studies, 9–10
Government communication histories, 7–8

H

Habermas's theory of communication, 53, 59, 219
Hammond Reports, 183–184
Hand, Harry E., study of technical writing errors, 189
Harbarger, Sada A.
  as an enthusiastic teacher, 94
  experiment in model of cooperation, 97
  importance of teaching rhetoric, 98–99
  as a pioneer, 95–98
  textbook organization by forms, 180
Hariot-White-deBry 1590 map, 165–168, 169
  accuracy, 160
  historical background, 154–157
  information about surveying technology, 157–165
  multiple purposes, 167–168
Historical perspectives
  and emergent principles to Priestley, 33–36
  episodic vs. time line, 15, 227
  "Generals-and-kings" view, 3
  and natural philosophy to Priestley, 25–29
How-to book genre, 10
Humoral theory imagery, 134–135, 144
Hypertext, 198, 211, 215, 216

I

Instrumental view of technology, 216
Internet
  and globalization, 240–241
  time line, 197

J

Johnson, Samuel, as a technical writer, 273

L

Linnaean descriptive biology, and influence of Bacon, 7

M

Makin, Bathsua, and modern approach to document design, 117
Marketable technical communication, 65, 95
  and rhetorical arrangement, 12
Mechanical imagery in the 18th century, 139–141

SUBJECT INDEX 319

Melancholia imagery in the Renaissance, 136–138
  and humanism values attribution of greater creativity, 138
Metaphor
  and Albert Einstein's technical writing, 285
  benefit, 149
  historical use, 13, 149
  and our conceptual system, 133, 140, 146
  and presuppositions, 133
  use in biomedical communication, 16, 131, 150
  use in science, 53, 59
  See also Watson and Crick
Morill Act, 92, 174

N

Newton, Isaac
  influence on style, 272
  and Priestley's criticism of hidden thinking practices, 41

P

"Plain style," 13, 123, 268, 270
  and communication of personality, 127
  as a modern ideal, 125–127
  and Puritan reformers, 50
  and scientific revolution of 17th century, 124–125
  vs. ornate style, 124
Political "problem of technology," 53
Priestly, Joseph
  and 18th-century electricity, 24–27
  communication of enthusiasm, 39
  and coordination with other forms of knowledge, 37–38
  first use of bar graph for historical duration visual, 26
  historicizing theory, 34–36
  "ladders," 41
  and modern footnote practices, 28, 40
  and review of the literature genre, 25
  and the shared experience of scientific research, 12, 24, 36–38, 42–43, 274
  specific accounts/general claims, 29–32
  and theory of analytic arrangement, 6, 30–33, 40–42
Puritan reformers
  and Bacon's rhetoric, 50, 53, 58
  and technical writing, 281

R

Research methods in technical communication history, 2–10, 249–250, 252
  ancient and classical literature, 260–262
  meta-analysis, 14, 198
  primary sources, 13, 14, 106
  studying a microcosm of social change, 118–119
  surveying instruments and techniques in cartographic study, 157–165
  tracing writing by/for women in the English Renaissance, 105–108
Rhetorical strategies
  conceptual approaches, 208–212
  histories, 5–7
  interdisciplinary in business and technical communication, 252–256
  vs. instrumental discourse, 233
Rickard, T. A., nonforms text, 180
Rolling codification in modern scientific articles, 42
Rush, Benjamin, utilitarian view, 142

S

"Saving the phenomenon" experience, 32
Scientific cooperation model, 12, 22, 23–24
  and emergence of societies and journals, 23
Scottish Common Sense Realism, 209
Sign system. *See* Visuals in technical communication
Social constructionism theory, 235–236
Society for the Promotion of Engineering Education (SPEE), 5, 12, 94, 97, 99–100, 175, 179, 181, 183–184
Society for Technical Communications, 185, 188
Society of Technical Writers. *See* Society for Technical Communications
Specific events, case studies, 8–9
Subjective approaches to rhetorical theory, 210–211

T

Teaching English for engineers, 180
  low status in English faculty, 92, 93, 178–179, 181, 258
  service-related, 12, 93, 188, 189
  stressing English as a tool, 96
  vs. the "culture obsession," 96–97, 98, 174, 177
Teaching technical communication
  and the Depression, 181–183

formation of a discipline, 179–180, 190–192
reasons for computer use, 204–207
team-teaching, 186
uses of historical models, 2–3, 256–257
and a "woman's place," 92
See also Teaching English for engineers

Technical communication
ancient roots, 173
collaboration, 236–237
and experiments in stylist tools for nonspecialist audiences, 127–129
humanistic resurgence, 228, 232
importance of White-Hariot-deBry map, 155
and inference of historical literacy levels, 106
key American movements, 14–15
key European movements, 13–14
as a network of social practices and needs, 228, 229
proposal writing as a new form, 190
social perspective, 206, 242
"solid tradition," 2
value of rhetorical devices, 129–130
and writing as a specialist industry, 185–187, 191
See also Computers and technical communications; Visuals in technical communication

Technical communicators, 3–4
critical choices between rhetorical alternatives, 63
need to acknowledge humanistic concerns, 233–234
redefinition of "workplace" and women, 91
roles in computer age, 207

Telleen, J. Martin, pioneering technical writing teacher, 175

Textual devices for scientific communication, 23, 24

Transactional theories and rhetorical framework, 211–212

Transmission model of technology, 216–217

U

Utilitarian values, 214, 220

V

Visuals in technical communication
history, 153
and integrated sign system, 154
"product view," 153
vs. verbal functions, 166

W

War and changes in communication
Civil War and technology status, 174, 214
English Civil War and women's writing, 110
importance of cartography to military, 156
Sputnik and war of technology, 188
World War II and a new technological imperative, 183–184

Watson and Crick, using metaphors to build models, 6, 286–287

Westinghouse, 185

www.ingramcontent.com/pod-product-compliance
Lightning Source LLC
Chambersburg PA
CBHW072121290426
44111CB00012B/1738